Monogamy

Mating Strategies and Partnerships in Birds, Humans and Other Mammals

Why do males of some species live with a single mate when they are capable of fertilizing more than one female's eggs? Why do some females pair only with one male, and not with several partners? Why do birds usually live in pairs and feed chicks together whilst mammals often live in larger groups with females rearing their young without male help? These questions form the central theme of this book. Social monogamy is a complex, multi-faceted phenomenon that does not always correspond with reproductive monogamy, so a paired male may not necessarily be raising his own offspring. Exploring the variables influencing and maintaining the fascinating diversity of social, sexual and reproductive monogamous partnerships in birds, mammals and humans, this book provides clues to the biological roots of monogamy for students and researchers in behavioural ecology, evolutionary anthropology, primatology, zoology and ornithology.

ULRICH H. REICHARD is a research scientist at the Max-Planck-Institute for Evolutionary Anthropology in Leipzig, Germany. His research interests include the evolution of mating systems and reproductive strategies of socially monogamous animals. For over 13 years he has studied the behaviour of wild white-handed gibbons (*Hylobates lar*) in Thailand, the only apes where the core unit is a socially monogamous group.

CHRISTOPHE BOESCH is a Scientific Director of the Max-Planck-Institute for Evolutionary Anthropology in Leipzig, Germany. His long-term research into the behaviour of chimpanzees from the Taï National Park in the Côte d'Ivoire has earned him numerous academic accolades. He is also co-editor of *Behavioural Diversity in Chimpanzees and Bonobos* with Gottfried Hohman and Linda Marchant (2002, ISBN 0 521 80354 3 (hb) & 0 521 00613 9 (pb)).

MONOGAMY

Mating Strategies and Partnerships in Birds, Humans and Other Mammals

Edited by

Ulrich H. Reichard

Max-Planck-Institut für evolutionäre Anthropologie

Christophe Boesch

Max-Planck-Institut für evolutionäre Anthropologie

CAMBRIDGE
UNIVERSITY PRESS

PUBLISHED BY THE PRESS SYNDICATE OF THE UNIVERSITY OF CAMBRIDGE
The Pitt Building, Trumpington Street, Cambridge, United Kingdom

CAMBRIDGE UNIVERSITY PRESS
The Edinburgh Building, Cambridge CB2 2RU, UK
40 West 20th Street, New York, NY 10011-4211, USA
477 Williamstown Road, Port Melbourne, VIC 3207, Australia
Ruiz de Alarcón 13, 28014 Madrid, Spain
Dock House, The Waterfront, Cape Town 8001, South Africa

http://www.cambridge.org

First published 2003

Printed in the United Kingdom at the University Press, Cambridge

Typefaces Ehrhardt MT 9.5/12 pt and Quadraat   *System* LaTeX 2_ε [TB]

A catalogue record for this book is available from the British Library

Library of Congress Cataloguing in Publication data

Monogamy: mating strategies and partnerships in birds, humans and other mammals / edited by
Ulrich Reichard & Christophe Boesch.
p. cm.
Includes bibliographical references.
ISBN 0 521 81973 3 – ISBN 0 521 52577 2 (paperback)
1. Sexual behavior in animals. 2. Sex customs. 3. Psychology, Comparative. 4. Mammals–Behavior.
5. Birds–Behavior. I. Reichard, Ulrich, 1964– II. Boesch, Christophe.
QL761.M65 2003 591.56′2–dc21 2002041553

ISBN 0 521 81973 3 hardback
ISBN 0 521 52577 2 paperback

Contents

Contributors

CHRISTOPHE BOESCH
Department of Primatology
Max-Planck-Institute for Evolutionary Anthropology
Deutscher Platz 6
04103 Leipzig
Germany
boesch@eva.mpg.de

PETER N. M. BROTHERTON
Biodiversity Programme Unit
English Nature
Northminster House
Peterborough PE1 1UA
UK
peter.brotherton@english-nature.org.uk

JOANNA FIETZ
Department of Experimental Ecology
University of Ulm
Albert-Einstein-Allee 11
D-89081 Ulm
Germany
Joanna.Fietz@t-online.de

ANNE W. GOLDIZEN
Department of Zoology and Entomology
School of Life Sciences
University of Queensland
Brisbane, QLD 4072
Australia
agoldizen@zen.uq.edu.au

ECKHARD W. HEYMANN
Department of Ethology & Ecology
German Primate Center
Kellnerweg 4
37077 Göttingen
Germany
eheyman@gwdg.de

PETER M. KAPPELER
Department of Ethology & Ecology
German Primate Center
Kellnerweg 4
37077 Göttingen
Germany
pkappel@gwdg.de

ROLAND W. KAYS
Curator of Mammals
New York State Museum
CEC 3140
Albany, NY 12230
USA
rkays@mail.nysed.gov

RYOSUKE KISHIMOTO
Nagano Nature Conservation Research Institute
2054-120 Kitago
Nagano 381-0075
Japan
kishi@nacri.pref.nagano.jp

PETR E. KOMERS
MSES
207 Edgebrook Close NW
Calgary, Alberta T3A 4W5
Canada
pkomers@mses.ca

BOBBI S. LOW
School of Natural Resources and Environment
University of Michigan
Ann Arbor, MI 48109-1115
USA
bobbilow@umich.edu

ANDERS P. MØLLER
Laboratoire de Parasitologie Evolutive
CNRS UMR 7103
Université Pierre et Marie Curie
Bâtiment A, 7ème étage
7 quai St. Bernard, Case 237
75252 Paris Cédex 5
France
amoller@snv.jussieu.fr

ULRICH H. REICHARD
Department of Primatology
Max-Planck-Institute for Evolutionary Anthropology
Deutscher Platz 6
04103 Leipzig
Germany
reichard@eva.mpg.de

DAVID O. RIBBLE
Department of Biology
Trinity University
715 Stadium Drive
San Antonio, TX 78212
USA
dribble@trinity.edu

CAREL P. VAN SCHAIK
Department of Biological Anthropology and Anatomy
Duke University
Box 90383
Durham, NC 27708-0383
USA
vschaik@acpub.duke.edu

SIMONE SOMMER
Department of Ecology & Conservation
University of Hamburg
Martin-Luther-King-Platz 3
20146 Hamburg
Germany
Simone.Sommer@zoologie.uni-hamburg.de

BEVERLY I. STRASSMANN
Department of Anthropology
1020 LSA Bldg
University of Michigan
Ann Arbor, MI 48109-1382
USA
bis@umich.edu

LIXING SUN
Department of Biological Sciences
Central Washington University
Ellensburg, WA 98926-7537
USA
Lixing@cwu.edu

RICHARD H. WAGNER
Konrad Lorenz Institute
Austrian Academy of Sciences
Savoyenstrasse 1a
1160 Vienna
Austria
R.Wagner@klivv.oeaw.ac.at

Acknowledgements

We very much thank all the contributors to this book. The great efforts they took to structure their chapters according to the underlying theme, follow our instructions on terminology and definitions, as well as comply with other requests, all of this collegial teamwork allowed a collection of individual papers to develop into a coherent compendium about social monogamy.

Tracey Sanderson has been supportive of this project from the beginning, always providing fast, critical, straightforward and convincing comments on important matters, just as editors would wish to receive from their publisher. Paula Ross proofread the manuscript with great care and humour, and on more than one occasion made helpful suggestions on matters of substance. We also thank Anke Behlert, Karen Chambers, Birgitt Glöckel, Cornelia Koring, Anja Kösler and Claudia Nebel for editing and modifying figures, tables and reference lists. And we thank Daniel Stahl who provided his customary statistical expertise. Finally, we gratefully acknowledge the Max-Planck-Society for their continuous support of long-term anthropological research.

Introduction

CHAPTER 1
Monogamy: past and present

Ulrich H. Reichard

MONOGAMY APPROACHED

Scientists have been interested in social monogamy in animals for a long time (Wickler & Seibt, 1983; Dewsbury, 1988). By the end of the Victorian era, animal monogamy was already intuitively fascinating perhaps because it seemed to mirror most closely the living ideal of Western, industrialized societies: human monogamy as reflected in an early description of social monogamy as 'animal marriage' (Wundt, 1894). Part of social monogamy's attractiveness probably stemmed from both the assumed and postulated mutual faithfulness of the individuals in these partnerships. Unfortunately, *social* monogamy, which primarily describes a demographic and close sociospatial relationship between one male and one female, has often been mixed both with exclusive mating as well as joint reproduction and biparental care of offspring. This historical conflation has resulted in misunderstandings and disagreement about the biological meaning of the 'monogamy phenomenon' as well as conflicting expectations and interpretations among scientists about selective forces and behavioural traits associated with it (e.g. Wickler & Seibt, 1983; Murray, 1984; Mock, 1985; Dewsbury, 1988; Mock & Fujioka, 1990; Gubernick, 1994).

Variation in duration and levels of monogamous partnerships

The definition of *social* monogamy is that a female and a male each have only a single partner of the opposite sex at a time (Black, 1996; Gowaty, 1996*a*). The temporal component is important for understanding socially monogamous partnerships because only few sexually reproducing organisms truly establish a unique, holistic, monogamous lifetime partnership in which partnership duration equals the partners' remaining lifespan (i.e., Southgate *et al.*, 1998). The overwhelming majority of socially monogamous animals and humans, however, practise serial or sequential social monogamy, i.e., if a first partner dies, or is otherwise lost, another partner will be accepted (Wickler & Seibt, 1983).

Socially monogamous partnerships are highly variable in duration along a continuous time axis. They may be long-term and last until one partner dies (pair-living mammals: Hendrichs, 1975; Getz *et al.*, 1987; Sommer, chapter 7; Sun, chapter 9; birds: some geese, ducks, and swans: Black *et al.*, 1996; Williams & McKinney, 1996). Or they may be rather short-term, the result of active partnership termination and pairing with a new partner while the old partner is still alive. Active partnership termination and pairing with a new partner occurs in mammal species, including humans, with short- and long-term socially monogamous partnerships (e.g., Blurton Jones *et al.*, 2000; Marlowe, 2000; Sommer & Reichard, 2000), and also in birds (Desrochers & Magrath, 1996; Catry *et al.*, 1997; Hatchwell *et al.*, 2000). Relatively short, socially monogamous partnerships characterize some bird species in which all pair relationships break up at the end of each breeding season and pairing with a new partner the following year is not uncommon (McKinney, 1986; Ens *et al.*, 1996; Ligon, 1999). At the short-term end of the duration continuum, partnerships are found that fulfil only the minimal temporal requirement for social monogamy. Conventionally, this scenario has been that of a male and female staying together for at least a single breeding season (Wittenberger & Tilson, 1980; Birkhead & Møller, 1996), including a considerable post-fertilization period during which a recognizable pair relationship is maintained.

Besides temporal variation of pair relationships usually subsumed under the umbrella category 'monogamous', surprising behavioural and genetic mating system variations across levels of monogamous relationships have been revealed over the past two decades (Mock, 1985; Black, 1996; Hughes, 1998). Contrary to the early assumption of sexual exclusivity between socially paired individuals, it is now evident that sperm competition can also play a prominent role in partnerships of socially monogamous individuals, who may interact sexually and reproduce with multiple partners (Birkhead & Møller, 1998; Petrie *et al.*, 1998).

The advent of new molecular techniques in the mid-1980s (Jeffreys *et al.*, 1985) gave rise to a deepened understanding of the structure of monogamous mating systems (cf. Hughes, 1998). For the first time, it became possible to distinguish between social and genetic mating systems based on the assignment of genetic parentage. Genetic parentage analyses fundamentally changed our perception about the biological meaning(s) of social monogamy (Birkhead & Møller, 1996; Møller, chapter 2). Trivers (1972) had foreseen what is now recognized as a common male reproductive strategy when he predicted that pair-living males would cooperate with a female partner to raise offspring and maintain social monogamy, but copulate and reproduce with additional females. Soon after Trivers's (1972) claim, the first observational evidence confirmed the flexible mating patterns of pair-living individuals. Copulations outside social pair relationships of otherwise pair-living individuals were recorded and became quickly known as extra-pair copulations or EPCs (e.g., Bray *et al.*, 1975; Beecher & Beecher, 1979; Gladstone, 1979; McKinney *et al.*, 1984; Birkhead *et al.*, 1987; Møller, 1988*b*). But not only pair-living males search for EPC opportunities to increase their reproductive success. Far from being passive recipients of male extra-pair advances, there is growing evidence that socially monogamous females often play an active role in structuring sexual and reproductive relationships (Hrdy, 1986; Gowaty, 1996*b*). Pair-living females seek sexual contact with social and extra-pair partners (Kempenaers *et al.*, 1992; Sheldon, 1994; Smiseth & Amundsen, 1995; Kempenaers, 1997; Otter *et al.*, 1998; Berteaux *et al.*, 1999; Hasselquist & Sherman, 2001), often control the sexual activities (Møller, 1988*b*; Wagner, 1992 chapter 6; Ahnesjö *et al.*, 1993) and probably the paternity (Birkhead & Møller, 1993; Gowaty, 1996*a*; Pizzari & Birkhead, 2000; Shellman-Reeve & Reeve, 2000) of their offspring.

It was, however, not until the new molecular techniques were developed and applied that the occurrence of extra-pair fertilizations (EPFs) and the evolutionary significance of flexible mating behaviours of pair-living males and females were fully acknowledged. Alternative mating strategies of socially monogamous birds are now considered to be a widespread, regular phenomenon, forming an integral part of socially monogamous male and female reproductive strategies (reviews in Birkhead & Møller, 1992, 1995, 1996; Møller, 1998, 2000; Hasselquist & Sherman, 2001). Besides birds, monogamous social pairing combined with flexible reproduction has been observed in diverse animals, including socially monogamous lizards (Bull *et al.*, 1998) and cooperative and/or pair-living mammals (Keane *et al.*, 1994; Sillero-Zubiri *et al.*, 1996; Goossens *et al.*, 1998; Spencer *et al.*, 1998; Fietz *et al.*, 2000).

A brief word on terminology

Behavioural and genetic studies of socially monogamous birds (cf. Birkhead & Møller, 1992) and mammals (e.g., Reichard, 1995; Fietz *et al.*, 2000) have convincingly demonstrated that in a majority of contexts an uncritical use of the simple term 'monogamous' to describe and understand relationships of pair-living species is inadequate (Gowaty, 1996*a*). It ignores the flexibility, variability, and complexity of different levels of relationships between socially monogamous animals. The language that is needed is one that openly and precisely reveals our knowledge about male–female relationships at the social, sexual, and genetic levels in order to be able to identify the evolutionary consequences of variations in social association. This is particularly important where the social, sexual and reproductive systems of a pair-living species do not coincide (cf. Hughes, 1998).

This book uses a consistent terminology to reflect the complexity and current knowledge of, and the intended focus on, monogamous male–female relationships. ***Social monogamy*** refers to a male and female's social living arrangement (e.g., shared use of a territory, behaviour indicative of a social pair, and/or proximity between a male and female) without inferring any sexual interactions or reproductive patterns. In humans, social monogamy equals monogamous marriage. ***Sexual monogamy*** is defined as an exclusive sexual relationship between a female and a male based on observations of sexual interactions. The term ***genetic monogamy*** is used when DNA analyses can confirm that a female–male pair reproduce exclusively with each other. A combination of terms indicates examples where levels of relationships coincide, e.g., sociosexual and sociogenetic monogamy describe corresponding social and sexual, and social and genetic monogamous relationships, respectively. Lastly, the term *monogamous social system* is used synonymously with social monogamy, and the term *monogamous mating* *system* is synonymous with known monogamous sexual and genetic relationships.

DEVELOPING A THEORETICAL FRAMEWORK: PATHWAYS TO SOCIAL MONOGAMY

Far from being a unitary phenomenon with a simple evolutionary explanation, social monogamy is multifaceted, having evolved along different pathways in different animal lineages. Three components influence the occurrence of social monogamy: the amount of paternal care, the access mode to resources, and partner choice. A clear grasp of the importance of each of these factors, and how they interact and influence each other, provides the necessary foundation for understanding the different routes to social monogamy. Such a framework is outlined in Figure 1.1. The three bold arrows in the figure point back and forth between the three main components of social monogamy's evolution (paternal care, resource access, mate choice), indicating that these factors influence each other. The smaller arrows link the main steps leading to social monogamy. Because interactions between factors promoting the evolution of social monogamy are complex, it is not useful to connect with small arrows all the possible influences on the steps or specific conditions leading to social monogamy. This complexity seems to be particularly true for paternal care, which can influence the evolution of social monogamy at different stages and to various degrees (see The importance of paternal care, below). Therefore, an absence of connecting arrows does *not* equal an absence of a link between factors or conditions. For all pathways outlined in Figure 1.1 it was assumed that social monogamy would be the resulting, visible outcome of the compromise between the sexes' interests in maximizing their own reproduction (cf. Davies, 1991, 1992; Chapman & Partridge, 1996).

The importance of paternal care

It has repeatedly been suggested that the need for biparental care led to socially monogamous pairs (Lack, 1968; Ligon, 1999; Burley & Johnson, 2002). When a female cannot rear young successfully without the direct help (e.g., feeding, carrying, warming) of a male, social monogamy is likely to become the reproductive strategy that best maximizes male and female fitness (Kleiman, 1977; Wittenberger & Tilson, 1980; Kleiman & Malcolm, 1981; Malcolm, 1985; Birkhead & Møller, 1996). This hypothesis attracted much attention, probably because it is in socially monogamous species that biparental care is more common and obvious, particularly in birds (Gowaty, 1996*c*), although it also occurs in some small mammals (Gubernick, 1984; Gubernick & Teferi, 2000). Biparental care has also been suggested as one important factor explaining the prevalence of human social monogamy (Marlowe, 2000). The hypothesis originally postulated that male care could not be shared between offspring/broods of several females without decreasing either female or both adults' reproductive success below what could be achieved under a monogamous parenting system. Where male care can be shared, socioreproductive polygyny may develop. When male contributions to infant survival are critical, biparental care and social monogamy may become the straightforward parenting and social systems.

Additionally, even where paternal care is not critical for offspring survival and females can raise some offspring alone, paternal care may influence female mate choice at a later stage along the pathways leading to social monogamy. Under conditions where males occupy separate territories and naturally vary in quality, e.g., parenting ability, a male's parenting potential may become important for female mate choice. Females choosing parenting males as mates may increase paternal care and increasingly constrain males to social monogamy (Burley & Johnson, 2002). Aspects of male care may also be important in situations when monogamously paired females are confronted with the arrival of another female. If social monogamy maximizes female reproductive success, paired females may base their decision to enforce social monogamy aggressively, on the importance of and hence potential loss of paternal care when male assistance has to be shared with a second female (Slagsvold & Lifjeld, 1994). Such examples illustrate how intimately resource access strategies (males occupy separate ranges), paternal care, and mate choice (paired females benefit from social monogamy and prevent polygynous parenting) will often be linked along pathways to social monogamy.

Resource access and mate choice

Males do not provide a measurable amount of paternal care in all socially monogamous animal lineages (Clutton-Brock, 1991). In fact, direct male care is absent in the majority of socially monogamous mammals

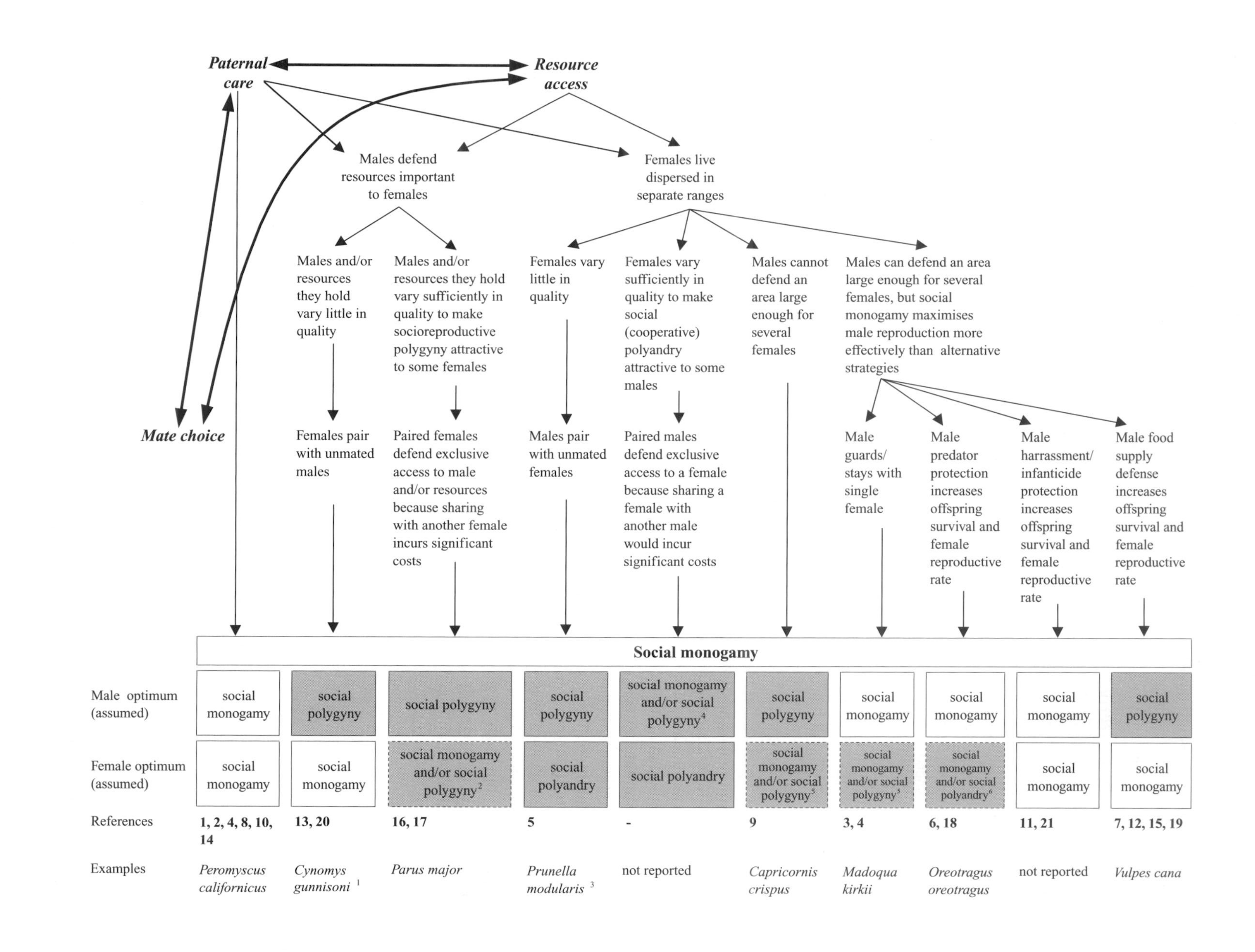

Paternal care
Resource access
Mate choice
Males defend resources important to females
Females live dispersed in separate ranges
Males and/or resources they hold vary little in quality
Males and/or resources they hold vary sufficiently in quality to make socioreproductive polygyny attractive to some females
Females vary little in quality
Females vary sufficiently in quality to make social (cooperative) polyandry attractive to some males
Males cannot defend an area large enough for several females
Males can defend an area large enough for several females, but social monogamy maximises male reproduction more effectively than alternative strategies
Females pair with unmated males
Paired females defend exclusive access to male and/or resources because sharing with another female incurs significant costs
Males pair with unmated females
Paired males defend exclusive access to a female because sharing a female with another male would incur significant costs
Male guards/ stays with single female
Male predator protection increases offspring survival and female reproductive rate
Male harrassment/ infanticide protection increases offspring survival and female reproductive rate
Male food supply defense increases offspring survival and female reproductive rate
Social monogamy
Male optimum (assumed)
Female optimum (assumed)
References
Examples
social monogamy
social polygyny
social polygyny
social polygyny
social monogamy and/or social polygyny[4]
social polygyny
social monogamy
social monogamy
social monogamy
social polygyny
social monogamy
social monogamy
social monogamy and/or social polygyny[2]
social polyandry
social polyandry
social monogamy and/or social polygyny[5]
social monogamy and/or social polygyny[5]
social monogamy and/or social polyandry[6]
social monogamy
social monogamy
1, 2, 4, 8, 10, 14
13, 20
16, 17
5
-
9
3, 4
6, 18
11, 21
7, 12, 15, 19
Peromyscus californicus
Cynomys gunnisoni[1]
Parus major
Prunella modularis[3]
not reported
Capricornis crispus
Madoqua kirkii
Oreotragus oreotragus
not reported
Vulpes cana

(Komers & Brotherton, 1997). When Komers and Brotherton (1997) tested the relationship between paternal care, female dispersion, and social monogamy across mammals they did not find the expected three-way interaction between the three character states, nor a two-way interaction, either between paternal care and female grouping or paternal care and mating system, and, finally, no single effect of paternal care. Further analyses revealed that social monogamy in mammals probably evolved significantly more often in the *absence* of paternal care than in its presence and that the associated evolution between the two character states of paternal care and social monogamy presumably occurred by chance when social monogamy was already present. The authors concluded that paternal care was a poor predictor of social monogamy, a finding already independently suggested for primates by Wright (1990), Tardif (1994), and Dunbar (1995). With decreasing importance of paternal care, the evolution of social monogamy may increasingly depend on how resources necessary for males and females to reproduce are distributed and which strategies individuals follow to maximize their reproduction.

Generally, differential investment in gametes, i.e., anisogamy (Parker *et al.*, 1972) is thought to set the stage for the evolution of sex-specific reproductive strategies (but see Snowdon, 1997). It is males who invest little in gametes and produce mobile sperm, whereas females make substantial energetic contributions to each gamete and produce large, nutritious eggs. Because of the inequality in gamete production, the difference in the potential reproductive rates of males and females is significant (Clutton-Brock & Parker, 1992): a male can usually fertilize many more eggs than a single female can develop to ovulation. Consequently, for most females access to resources important for successful, regular egg development limits their reproductive success (Trivers, 1972). Hence, females are expected to follow strategies that best maximize their resource access, which often equals their access to food (Janson, 1992). Ecological factors have been suggested as the major influence on the distribution of females (Clutton-Brock & Harvey, 1977; Emlen & Oring, 1977; Clutton-Brock, 1989). When resources occur in relatively small patches, uniformly distributed in time and space, and predation pressure is low, females may be able to afford to live non-gregariously to minimize feeding competition (Alexander, 1974; Wrangham, 1980; van Schaik, 1983).

In contrast to females, male reproduction is less dependent on access to food resources but is limited primarily by access to fertile females (Bateman, 1948; Trivers, 1972; Clutton-Brock, 1989), hence male spatial distribution often follows female spatio-temporal distribution (Andelman, 1986; Altmann, 1990; Mitani *et al.*, 1996). At first glance, the evolution of social monogamy may then be understood as a function of females living in exclusive ranges and males mapping their lives onto these females. However, although females living separately from each other represents an important precondition for the evolution of social monogamy in mammals (Komers & Brotherton, 1997; van Schaik & Kappeler, chapter 4), it is not a sufficient explanation because males

Figure 1.1. Pathways to social monogamy.

Key to references: **1** Arcese, 1989; **2** Bart & Tornes, 1989; **3** Brotherton & Rhodes, 1996; **4** Clutton-Brock, 1991; **5** Davies, 1992; **6** Dunbar & Dunbar, 1980; **7** Geffen & Macdonald, 1992; **8** Gubernick & Teferi, 2000; **9** Kishimoto & Kawamichi, 1996; **10** Kleiman, 1977; **11** McKinney, 1986; **12** Moehlman, 1989; **13** Pribil & Picman, 1996; **14** Ribble, chapter 5; **15** Rutberg, 1983; **16** Slagsvold, 1983; **17** Slagsvold & Lifjeld, 1994; **18** Sommer, 2000; **19** Sun, chapter 9; **20** Travis *et al.*, 1995, 1996; **21** van Schaik & Dunbar, 1990.

[1] In Gunnison's prairie dogs (*Cynomys gunnisoni*) social monogamy was associated with an even resource distribution, whereas social polygyny and polygynandry prevailed with increasing environmental patchiness (Travis *et al.*, 1995, 1996).

[2] Female optimum varies with status: paired females' optimum is social monogamy, whereas yet unpaired females' optimum is social polygyny.

[3] Dunnocks (*Prunella modularis*), have a flexible mating system and only some pairs of the study population were socially monogamous (Davies, 1992).

[4] Male optimum varies with status: paired males' optimum is social monogamy, whereas yet unpaired males' optimum is social polyandry.

[5] Female optimum is variable and depends on intrinsic female quality and/or environmental condition.

[6] Female optimum varies with status: paired females' optimum is social monogamy, whereas yet unpaired females' optimum is social polyandry.

could just as well adopt spatio-reproductive strategies other than social monogamy (Sandell & Liberg, 1992). Males could aim to realize socioreproductive polygyny by defending an area large enough to encompass several separate female ranges (e.g., Ferron & Oullet, 1989; Green *et al.*, 1998; Berteaux *et al.*, 1999; Fisher & Owens, 2000). Alternatively, they could try to defend two separate ranges and become polyterritorial (Temrin & Arak, 1989), although polyterritoriality may be a risky strategy in terms of lost matings with the primary female (Alatalo *et al.*, 1987). Finally, males could rove around to visit females living in different areas successively as they become fertile, as found in primates like the orangutan, *Pongo pygmaeus* (van Schaik & van Hooff, 1996) or the aye-aye, *Daubentonia madagascariensis* (Sterling & Richard, 1995).

Pathways to social monogamy: when females live apart from each other

Given the different options males have when females live in separate ranges, social monogamy is expected to evolve under only one of the following three conditions: first, when males are physically or otherwise unable to realize any of the above-mentioned polygynous strategies (Parker, 1974; Emlen & Oring, 1977; Barlow, 1988); second, when the chance to find another breeding partner is low so that staying with and guarding a current partner becomes a male's best option (cf. van Rhijn, 1991; Brotherton & Rhodes, 1996); third, when staying with one female provides reproductive benefits to males not otherwise accessible under alternative mating strategies.

The first condition for social monogamy is straightforward and depends largely on intrinsic male capacities. The second condition addresses difficulties for males to find another breeding partner, which may result from different situations. It may arise when a single brood or young is produced during a short breeding season and nearly all adults start reproduction at about the same time, leaving negligible polygyny potential for males (Ligon, 1999). Finding another partner may also become difficult where breeding space is highly limited (Freed, 1987), for example, in circumstances imposed by ecological constraints, and in populations with a male-biased sex ratio. Male-biased adult sex ratios may result, for example, in birds, from female-biased adult mortality resulting from migration or predation during incubation (McKinney, 1986; Møller, 1994). Alternatively, though, a male-biased sex ratio may lead to an increase of socially polyandrous units due to changing cost:benefit equations when male–male competition for breeding positions increases, as noticed by Davies (1992) in a dunnock (*Prunella modularis*) population after harsh winters.

Under the third condition, males are theoretically able to achieve socioreproductive polygyny but they live socially monogamously (e.g., Brotherton & Manser, 1997; Reichard, chapter 13), suggesting that social monogamy conveys higher fitness returns to males than, e.g., roving and searching for additional females (Parker, 1974; Wickler & Seibt, 1981). Reproductive benefits to males from staying with a single female may exceed reproductive success from alternative strategies when the continuous male presence with a female and her offspring significantly increases offspring survival and/or female reproductive rate (Clutton-Brock, 1989, 1991; van Schaik & Dunbar, 1990).

Territorial behaviour of mammalian males has been interpreted as increasing offspring survival and female reproductive rate sufficiently to make social monogamy beneficial to males (Rutberg, 1983). However, although access to an exclusive food supply benefits a female and her offspring and may influence a female's partner choice, a prime function of male territoriality is probably to attract reproductive partners (Baker, 1983; Gosling, 1986; Carranza *et al.*, 1990), and only secondarily to provide food resources. Territorial defence by socially monogamous males may therefore have become important for the maintenance of social monogamy, but this defence perhaps evolved secondarily after social monogamy was already in place for other reasons (cf. Brotherton & Rhodes, 1996).

Male protection from predation is another form of indirect male care that has been proposed as a positive influence on offspring survival and female reproductive rate (Dunbar & Dunbar, 1980). This pathway to social monogamy assumes that for non-gregarious females the permanent company of a male relieves them from predator vigilance, allows them optimized feeding, and leads to an increased female reproductive rate. Such reasoning is supported by the observation that, e.g., primate males are particularly effective in detecting and deterring predators (van Schaik & van Nordwijk, 1989; van Schaik & Hörstermann, 1994). However, empirical evidence for the predicted causal link between predation pressure and social monogamy is scarce. Moreover, the predation-hypothesis for the evolution of social monogamy makes a contradicting assumption. On one hand, predation pressure is assumed to be so strong that predator protection increases offspring survival and hence socially monogamous males' reproductive success beyond what could be achieved by alternative strategies. On the other hand, however, the belief is that in order for females to be able to live a non-gregarious lifestyle in the first place, predation pressure must be almost negligible. If predation pressure is important, social polygyny, perhaps with small female group size, seems more likely to have evolved than social monogamy (Isbell, 1994), and under conditions of moderate predation pressure it seems questionable to assume that male predator protection could sufficiently compensate for lost mating opportunities. The contradiction can only be resolved if social monogamy were assumed to have evolved at a very specific predation pressure magnitude where predation pressure is just low enough to allow females to live non-gregariously, but at the same time high enough to adequately compensate protecting males for lost matings. Alternatively, and similar to the resource defence function of male territoriality, perhaps a special male role in predator vigilance, detection, and deterrence evolved after social monogamy already existed because the male was already close to the female. Even if male predator protection did not cause social monogamy to evolve it may have become important thereafter because it may have further increased reproductive benefits derived from social monogamy.

Van Schaik and Kappeler (1993, 1997, chapter 4) suggested another pathway to social monogamy via a specific male role. They proposed that in infant-carrying primates the risk of infanticide by strange males was the selective advantage for males to stay with a female. The hypothesis predicts that when males protect females and infants from infanticide through their continuous presence, then they achieve higher fitness than males that leave a female after impregnation to search for another mate. The risk and occurrence of male infanticide is widespread among mammals (Hausfater & Hrdy, 1984; van Schaik, 2000*b*), and also occurs in some birds (Møller, 1988*a*; Veiga, 2000). Infanticide by immigrating males has been interpreted as part of a male reproductive strategy whereby males stop females from investing in an offspring of another male to speed up the female's return to receptivity (Hrdy, 1979). Male infanticide will be an adaptive strategy under specific conditions: (i) the probability must be zero or close to zero that the infanticidal male sired the infant(s); (ii) the mother returns to menstrual cycling earlier than if the infant(s) had lived; and (iii) the infanticidal male has an increased probability of siring the female's next infant(s) (Hrdy *et al.*, 1995; van Schaik, 2000*a*). Empirical support for a direct link between male infanticide and the evolution of social monogamy is still lacking for primates as well as rodents (Blumstein, 2000) and birds (Veiga, 2000). Nonetheless, the concept of male infanticide appears as valid for socially monogamous as for other one-male social mating systems in which infanticide has been documented (e.g., Hanuman langur, *Semnopithecus entellus*: Sommer, 1994). Since male changes do occur in some socially monogamous species (e.g., gibbons: Treesucon & Raemaekers, 1984; Brockelman *et al.*, 1998; Reichard, chapter 13), a potential for male infanticide does exist. Therefore, social monogamy may have evolved as an infanticide prevention strategy in some lineages (van Schaik & Kappeler, chapter 4).

Finally, when females live in separate territories and males join them the evolution of social monogamy – instead of social polyandry – may depend on intrinsic female qualities. This hypothesis resembles the reversed version of the polygyny-threshold hypothesis for females (see below) and suggests a reversed polyandry-threshold model for the evolution of social monogamy. It assumes that males would accept cooperative polyandry only if the loss from shared mating is compensated by the increased quality of offspring. Otherwise, males would opt for a socially monogamous strategy. Two pathways to social monogamy appear plausible. First, where variation in female quality is low, males pair with yet unpaired females to avoid potential costs of shared mating and reproduction. Second, even where the polyandry-threshold is exceeded, and females vary sufficiently in quality to make social polyandry a beneficial strategy for some males, already paired males may prevent other males from joining and forming socially polyandrous units in order to avoid the reproductive costs of shared matings. Studies of cooperatively polyandrous species indicate that indeed males often share matings (Heymann, 2000) and paternity (Hartley *et al.*, 1995), suggesting that cooperative polyandry may be costly to a dominant male, who may then, therefore, enforce social monogamy.

Pathways to social monogamy: when males defend a territory

So far, the focus has been on situations where resource access strategies of females would lead to females living apart from each other and males pairing with single females. This situation seems more common in mammals than in birds. However, another important starting point for pathways to social monogamy exists. Social monogamy may evolve when males defend a territory to attract a precious resource – female(s) (e.g., Alatalo *et al.*, 1986). And this male strategy seems more common in birds than in mammals.

When males live in separate territories, and females join them, social monogamy evolves if males, or the resources they hold, vary little in quality and an individual female gains most from pairing with an unmated male (Orians, 1969; Slagsvold & Drevon, 1999). This hypothesis reverses the logic of the polygyny-threshold hypothesis originally proposed to explain the evolution of social polygyny in birds in relation to different habitat qualities (Verner, 1964; Verner & Willson, 1966). The polygyny-threshold hypothesis predicted that if males, or the resources they defend, vary sufficiently in quality, some females would breed more successfully as the second female with a high quality male or in a high quality territory than as the sole female with a low quality male or in a poor quality territory (Orians, 1969; Slagsvold & Drevon, 1999). Consequently, when the polygyny-threshold is exceeded, social polygyny is expected, whereas below the threshold, females would pair monogamously. This hypothesis also assumes that females generally profit from social monogamy by avoiding the costs associated with sharing either paternal care or territorial resources.

The final pathway to social monogamy proposed here predicts that social monogamy evolves even when males and/or the resources they hold vary sufficiently in quality to allow some females successfully to become socially polygynous. Under such conditions, social monogamy evolves when adding a second female to a pair or a home range would considerably decrease the paired female's reproductive success due to the costs of sharing territorial resources and/or male care. Hence, a paired female may aggressively enforce social monogamy (Rutberg, 1983; Slagsvold & Lifjeld, 1994; Kokita, 2002).

Empirical support for pathways to social monogamy

Paternal care is important

The most convincing examples of the selective importance of biparental/male care come from species where loss of male care results in complete breeding failure. Such examples fit the classical approach in which paternal care cannot be shared, i.e., is 'essential', and only if this condition is met will the evolution of social monogamy be triggered (cf. Wittenberger & Tilson, 1980). However, as outlined above in the discussion of pathways to social monogamy, even when male care is not directly essential for offspring survival, paternal care may still exert an important influence on social monogamy's evolution.

In birds, males can perform most parental duties equally well as females, and biparental care is the norm (Clutton-Brock, 1991; Gowaty, 1996*c*). Evidence of one of the benefits of direct paternal care is provided by the observation that reduction in female reproductive success due to the absence of a male mate was positively correlated with a male's provisioning contribution (Møller, 2000): the more the male was absent, and therefore not providing food, the lower the female's reproductive success. There is little doubt that in species where biparental care occurs, removing the male has a negative effect on offspring survival (Mock & Fujioka, 1990; Ligon, 1999). An *essential* role of male helping behaviour is, however, not always evident (Clutton-Brock, 1991; Gowaty, 1996*a*). Perhaps a male's contribution to incubation qualifies as critical male assistance and hence a strong selective pressure for biparental care and social monogamy in some birds (Ligon, 1999). In the socially monogamous, biparental western sandpiper (*Calidris mauri*), removal of the female parent during incubation led to males abandoning the nest before the eggs hatched (Erckmann, 1983). Møller (2000) determined the importance of male parental care (feeding) across birds by measuring the reduction in female reproductive success in the absence of male care, and providing several examples in which male absence would lead to complete breeding failure. In addition to the importance of shared incubation, Bart and Tornes (1989) identified other stages of the nesting cycle where male loss would considerably decrease reproductive success, including laying, nestling, and post-nestling, and which were supported by additional studies (Wolf *et al.*, 1988; Rees *et al.*, 1996).

Because of gestation and lactation, female mammals are particularly well adapted to care for young without male help. This is reflected in the sparse distribution of paternal care in mammals, with only about 5% of species showing a measurable amount of direct male care (Clutton-Brock, 1991). Paternal behaviour occurs primarily in rodents, carnivores, and primates (Kleiman & Malcolm, 1981; Wright, 1990; Woodroffe & Vincent, 1994), and despite its general rarity seems essential in some species (Moehlman, 1989). In an experimental laboratory study of the Djungarian hamster (*Phodopus campbelli*), Wynne-Edwards (1987) showed that the presence of a pair-male significantly affected offspring survival. Females raised almost all young (95%) when males were present, whereas solitary females lost more than half of their offspring (53%). Even assistance from a littermate sister could only marginally compensate for the loss of male assistance (61% surviving young), verifying that male care was essential. In an experimental study under natural conditions, Gubernick and Teferi (2000) also confirmed an essential role for the male California mouse (*Peromyscus californicus*). Significantly fewer offspring emerged from the burrows where a female's male partner was captured compared to controls (81% emerging offspring father-present; 26% emerging offspring father-absent). This effect was not due to different female reproductive performance because those females whose males were captured subsequently raised a second litter with a new male and achieved significantly greater reproductive success than for the litter without male assistance. Ribble (chapter 5) also concluded that male warming of offspring was essential in *P. californicus*: a relatively smaller litter size and greater relative litter mass compared with other small rodents resulted in comparatively greater heat loss of offspring. Male presence appeared essential for offspring survival during the first days after birth when the female is away from the nest. Finally, in a mark-recapture study of the nocturnal, tree-hole nesting fat-tailed dwarf lemur (*Cheirogaleus medius*), females also failed to raise young without male help (Fietz, 1999*a*, *b*).

An interesting parallel emerges from a bird–mammal comparison. In birds, male presence appeared important when incubation needed to be shared during early infancy. And similarly, in the mammalian examples, shared attendance of a burrow or nest-hole directly after birth appeared to be the strongest selective force for paternal investment. Biparental care may ultimately be linked to those species in which offspring are sensitive to heat loss early in their development, so that while one parent forages, the other must remain with the offspring in order to mitigate the thermal disadvantages connected to small litter size (cf. van Rhijn, 1991; Ribble, chapter 5).

In general, biparental care has a positive effect on offspring survival (but see Royle *et al.*, 2002), although it is also highly variable within and between species as well as across mating systems. Therefore, it is unlikely to provide a universal explanation for the evolution of social monogamy (Mock, 1985; Bart & Tornes, 1989; Gowaty, 1996*a*; Møller, 2000). Because of its general positive effect on offspring survival, paternal care may also reflect the maintenance of social monogamy rather than its ultimate function (Gowaty, 1996*c*; Komers & Brotherton, 1997; Tullberg *et al.*, 2002).

Males defend a resource important to females

The reversed polygyny-threshold hypothesis

Despite some criticism (Davies, 1989; Ligon, 1999), examples exist for birds, which support the logic of the reversed polygyny-threshold hypothesis for the occurrence of social monogamy, because females vary their socioreproductive strategy in response to varying resource availability. When males or the resources they hold do not vary significantly in quality, females prefer to pair with an unmated male.

In the well-studied red-winged blackbird (*Agelaius phoenicus*) females generally preferred to settle in the territories of unpaired males because socioreproductive polygyny decreased female reproductive success (Pribil & Picman, 1996). However, if territory quality was manipulated experimentally, with some territories providing better nesting opportunities than others, female mate choice could be reversed and females preferred to settle with an already paired male (Pribil & Searcy, 2001), suggesting that indeed social monogamy was usually resource dependent. Similarly, in an extended version of the polygyny-threshold theory Slagsvold and Drevon (1999) directly manipulated male quality (phenotypic degree of bright coloration) and demonstrated that females would mate monogamously when male quality did *not* vary but changed their choice when male quality *did* vary.

The polgyny-threshold hypothesis was particularly suited to situations where males occupied their territories first, e.g., as found in some migratory birds. In these species, newly-arrived females may be able to fly around to inspect and assess different males and/or the territories those males are defending, freely basing their eventual choice of male and territory on male and/or territorial quality (e.g., Møller, 1994). These results suggest that females' choice of mating strategy (polygynous vs. monogamous) may be based either on the resources a male can provide or directly on a male's quality. This situation rarely emerges in mammals because females are often philopatric (Greenwood, 1980) and do not migrate between seasons, staying in the same social groups for long periods. Furthermore, even dispersing females often remain close to their natal area (e.g., Sterck & Korstjens, 2000), which limits their opportunities to visit and inspect a variety (or any) of groups in which they might take up future residence. Finally, dispersing females may encounter hostile conspecifics and may neither be able to evaluate several territories/groups prior to emigration nor to freely choose which group or male to join (e.g., Dietz & Baker, 1993).

The limitations on female decision making in choosing a particular male and/or territory make it difficult to test when a female would do better by choosing an unpaired male compared with an already paired male. In humans, however, Borgerhoff Mulder (1990) found some indirect evidence supporting the idea that a low resource holding potential results in socially monogamous marriage: among the Kipsigis of Kenya, wealth was a strong, although not the only, predictor of socioreproductive polygyny. Hence, when this logic is reversed, a low resource holding potential may explain why some Kipsigi men had to remain socially monogamous. Another example of resource-dependent development of social monogamy is provided by Gunnison's prairie dog (*Cyanomis gunnisoni*): social monogamy was associated with an even resource distribution, whereas social polygyny and polygynandry prevailed with increasing environmental patchiness (Travis *et al.*, 1995, 1996).

Overall, the polygyny-threshold model appears of limited value in studies of mammalian mating systems because socioecological conditions of mammals rarely meet the models' various assumptions (see Clutton-Brock, 1989; Borgerhoff Mulder, 1990). These limitations make it difficult to explore whether environmental and/or male quality may be important pathways leading to socioreproductive monogamy in mammals. In contrast, the reversed polygyny-threshold hypothesis may explain social monogamy in some birds (but see Davies, 1989).

When a paired female does not share resources

This hypothesis assumes that the polygyny threshold is exceeded because either environmental quality or male quality varies (or both vary) sufficiently to allow socioreproductive polygyny to become an optional reproductive strategy for some females. However, the presence of a second breeding female would be costly for an already paired female, who would thus aggressively enforce social monogamy (Arcese, 1989; Slagsvold *et al.*, 1992; Ahnesjö *et al.*, 1993). Female–female aggression is the suggested mechanism that maintains social monogamy. The suggested potential costs to a resident female bird from an additional female are, e.g., loss or decrease of parental care, loss of food or nest sites, egg destruction, egg dumping, displacement from the territory, and increased predation risk (Slagsvold, 1983; Slagsvold & Lifjeld, 1994).

Female–female aggression: birds

An example from the bird literature is Veiga's (1992) study of the predominantly monogamous house sparrow (*Passer domesticus*). Male reproductive success increased with the number of females acquired. Nonetheless, many males remained socially monogamous despite the abundance of unpaired females, suggesting that the polygyny threshold should have been exceeded for many females of the population. Even when reproductive options were manipulated by giving males more nest-boxes, experimental males could not attract additional females any more frequently than control males. The prevalence of social monogamy was explained by the observation that females already present on a male's territory prevented other females from settling, presumably to avoid costs associated with resource sharing (Veiga, 1992). In cases where two females did breed on one male's territory, Veiga (1992) proposed that the costs of expelling female competitors were higher than benefits derived from maintaining social monogamy. Another example is the song sparrow (*Melospiza melodia*), in which settled females would also try to prevent unpaired females from settling, and where unpaired females could only successfully join an established pair if the resident female was already incubating and could not defend her monogamous status (Arcese, 1989). In the European starling (*Sturnus vulgaris*), Sandell and Smith (1996) experimentally manipulated the distance between nest-boxes, thereby artificially increasing the chance for additional females to settle. Most males remained socially monogamous as long as nest-boxes were close to one another because primary females would prevent secondary females from settling. Only when nest-boxes were sufficiently far apart that costs of defending two boxes presumably exceeded the benefits of social monogamy could two females breed on one male home range. Finally, aggression between females, presumably to prevent male socioreproductive polygyny, may be intense even when males are polyterritorial (Slagsvold *et al.*, 1992).

Female–female aggression: mammals

Aggression between females has also been suggested as an important mechanism for maintaining social monogamy in pair living mammals. For example, gibbons (*Hylobates lar*), small apes of the tropical forests of Asia, show mainly intrasexual territorial aggression in preventing male social polygyny (Brockelman & Srikosamatara, 1984; Mitani, 1984). In golden lion tamarins (*Leontopithecus rosalia*), a small Neotropical primate in which sociosexual monogamous as well as sociosexual polyandrous units occur, resident breeding females always vigorously chased non-group females away (Baker & Dietz, 1996) and no female was recorded as having ever successfully joined a group that already contained a breeding female (Dietz & Baker, 1993). However, social polygyny via a familial route was found in a few cases where group habitats appeared rich enough to allow mature daughters to share their mothers' ranges, suggesting that a resident, paired female would only accept an increase in female group size if the cost of sharing resources would be compensated by an exceptionally rich territory.

The above examples of female–female aggression are suggestive of the hypothesis stated earlier, i.e., that adding a second female to a pair or home range significantly decreases the paired female's reproductive success as a result of sharing territorial resources and/or male care, and they do not contradict the evolution of social monogamy via this route. However, most studies do not provide the data and quantitative tests that would document the reproductive consequences of both socially monogamous and socially polygynous females, evidence necessary to evaluate the potential costs to a paired female when a second female settles in her range. Specific studies are needed to address the link between social monogamy and the interests of a paired female since female–female aggression could occur for reasons other than maintaining this form of monogamy (Wittenberger & Tilson, 1980).

Females live separate from each other

When females fail to attract several males

Social or cooperative polyandry is relatively rare in birds and mammals but regularly occurs in, for example, shorebirds, pukekos, saddle-back tamarins, and some other species (Goldizen, 1987; Ligon, 1999). Under conditions that would generally favour the evolution of social polyandry (cf. Erckmann, 1983), social monogamy may evolve when females vary little in quality because males may then prefer to pair with a yet unpaired female.

Examples where social monogamy may have developed because females failed to attract more than one male are rare. Perhaps this pathway was relevant for some dunnocks (*Prunella modularis*) of the Cambridge University Botanic Garden population studied by Davies (1992). Dunnocks have a variable socioreproductive mating system, with individuals being either socially monogamous, socially polyandrous, socially polygynous, or socially polygynandrous (Davies & Lundberg, 1984). Socially monogamous pairs comprised 25–63% of the population's social units during 11 years (Davies, 1992). Among the different socioreproductive units, socially polyandrous females achieved the greatest reproductive success and also defended the largest ranges (Davies, 1992). In contrast, females that occupied small areas were mostly socially monogamous. This observation is in line with the hypothesis that the strongest females of the highest quality could perhaps defend the largest ranges and also attract more males than lower quality females inhabiting smaller ranges. Males would thus only accept social polyandry if the potential loss of shared matings would be compensated by high female quality, as expressed in these females' occupancy of a large territory. Perhaps variation in female quality could also explain the occurrence of social monogamy in the Otika pukeko (*Porphyrio porphyrio*) study population (Jamieson, 1997). Pukekos usually live in socially polyandrous units, but at the Otika site, social monogamy was relatively common, perhaps in relation to the low quality of certain females.

The few examples mentioned here to point towards a causal relationship between individual female quality and the evolution of social monogamy lack the required quantitative measures of individual quality. Currently, therefore, it can only be stated that the observations do not contradict the expectations derived from the hypothesis. Examples for mammals are even more sparse than for birds. However, studies focusing on individual quality variation in female primates with a variable sociosexual mating system, such as the Callitrichines (Goldizen, 1987; Heymann, 2000) appear potentially rewarding.

When males fail to become socially polygynous

The basic condition for this pathway to social monogamy is that females live dispersed in separate ranges. When males fail to establish ranges large enough to encompass several female ranges, social monogamy may be the resulting compromise between the sexes.

An example of this pathway to social monogamy in birds is the willow ptarmigan (*Lagopus lagopus*), most of which live in pairs. Hannon and Martin (1992) found no difference in the reproductive success of socially monogamous compared with socially polygynous females, indicating that female reproduction was independent of the two alternative male strategies. In contrast, reproductive profit for males came most from social polygyny. However, because females were largely aggressive towards each other, which resulted in spatial separation (Hannon, 1983), only a small fraction of males (5–20% per year) could establish a territory large enough for two females (Hannon & Martin, 1992). Socioreproductive polygynous males achieved higher 'scores' on a relative scale indexing male quality based on physiological parameters than did socioreproductive monogamous males, suggesting that because those males were only able to defend ranges large enough for one female, physical limitations indeed restricted most males to socioreproductive monogamy (Hannon & Dobush, 1997). Another interesting example where male physical condition may also limit males to social monogamy is the tree swallow (*Tachycineta bicolor*), studied by Kempenaers and colleagues (2001). Males of the study population were divided into two main categories: (i) resident breeders who formed socioparental monogamous relationships with a female; and (ii) floating males who were considered non-breeders because they did not establish a recognizable partnership with any female. Surprisingly, the floating males achieved high reproductive success in the population through extra-pair copulations. In addition to considerable reproductive success, floater males also appeared to be in better physical shape (heavier) than breeding residents. It was suggested (Kempenaers *et al.*, 2001) that males in good physical condition could afford to float, avoid the burden of paternal duties, and possibly breed polygynously, whereas males in worse body condition would become socially monogamous, accepting the duty of paternal care and the additional cost of a potentially high frequency of extra-pair young in the broods they fed.

In mammals, physical strength seems to limit most Japanese serow males (*Capriconis crispus*) from becoming socially polygynous with two females. Males only achieved social polygyny when one of the female ranges with which the male's overlapped was considerably smaller than the average female home range (Kishimoto & Kawamichi, 1996; Kishimoto, chapter 10). Male home ranges were on average only about 10% larger than average female home ranges, thus allowing only males that happened to overlap with a female inhabiting an average size home range and a female with a particularly small range to become socially polygynous. Therefore, males generally appeared physically unable to defend sufficiently large areas to realize socioreproductive polygyny. Otherwise, at least the strongest males in the population would have defended two average-sized female ranges.

Considering that males can usually maximize their reproductive success with a socioreproductive polygynous strategy, it seems plausible that social monogamy may occasionally occur as a consequence of males being unable to gain or defend access to several females.

When guarding a single female becomes a male's best option

Maintaining breeding status may be a critical factor influencing partner fidelity and continuous male–female association. Where loss of a breeding position is very costly, individuals may employ strategies to control their partner's activities in order to forestall mate change, thus leading to social monogamy.

In Macaroni penguins (*Eudyptes chrysolophus*), breeding success following mate change was not associated with reproductive disadvantages, and as many newly formed pairs as reunited pairs raised one offspring to fledging (Williams, 1996). However, more than a third of the males that had divorced a female remained without a partner the following breeding season and did not breed again for from two to four years, although they retained their old nest site (Williams, 1996). The long period after divorce without a breeding partner, despite retention of the nest site, indicates how important it is for these penguins to stay with their current mates. Similarly, in Tropical house wrens (*Trogolodytes aedon*), a limited number of suitable breeding territories coupled with a low number of available partners probably resulted in social monogamy. Already breeding males were forced to mate-guard their female partners continuously in order to maintain their breeding positions (Freed, 1987). An important guarding role of a male has also been suggested for barnacle geese (*Branta leucopis*), where male presence significantly affected female and offspring survival (Black & Owen, 1989).

A mammalian example of the evolution of social monogamy via mate guarding is the prairie vole (*Microtus ochrogaster*). Females lived dispersed and males were assumed to optimize their reproductive success by staying with one female because of the high probability of again successfully mating with the female when she entered her postpartum oestrus (Getz *et al.*, 1987). Intensive mate guarding was also suggested as the key male strategy and the origin of socioreproductive monogamy in Africa's small Kirk's dik-dik antelope (*Madoqua kirkii*: Brotherton & Komers, chapter 3). Females experienced short fertility periods of only one day, which made male roving in search of fertile females a risky and probably inferior male strategy compared with social monogamy. Moreover, male–male competition for territory ownership was intense, which promoted year-round maintenance of existing pair relationships (Brotherton & Rhodes, 1996; Brotherton & Komers, chapter 3). Females probably also preferred the constant presence of only one male which could protect them from the harassment of other males (Brotherton *et al.*, 1997). Another mammalian example of the mate-guarding route to social monogamy is the mara (*Dolichotis patagonum*), a large rodent that relies on a swiftly depleted, slowly replenished, and patchily dispersed food supply during the wet season. Like dik-dik females, mara females experienced a short fertility period of only a few hours that lead to male mate guarding as a male's most successful reproductive strategy (Taber & Macdonald, 1992).

When males protect a female and offspring from predators

Protecting offspring and females from predator attacks, and remaining vigilant while the female forages, has been offered as an explanation for the evolution of social monogamy in klipspringer, *Oreotragus oreotragus* (Dunbar & Dunbar, 1980; Roberts & Dunbar, 2000). The predation rate is high in this species, and both sexes alternate in watching for predators. However, males were more vigilant and they also detected predators earlier, thus allowing females to forage more efficiently (Dunbar & Dunbar, 1980). A second example is the Malagasy giant jumping rat (*Hypogeomys antimena*), a large rodent endemic to Madagascar. During a period of annual peak predation, male but not female rats showed behaviours leading to an increase in predator contacts, indicating that males played an important role in trying to protect offspring from predators (Sommer, 2000). The influence of predation on the evolution of social monogamy in birds remained difficult to assess because no specific evidence was found in support of this route to social monogamy.

Few examples supported the predicted causal link between predation pressure and the evolution of social monogamy. Nonetheless, predation represents a strong selective force in the lives of animals (Alexander, 1974) and it is possible, for example, that predation protection evolved secondarily after socially monogamous pair relationships were already in place (Roberts & Dunbar, 2000; Brotherton & Komers, chapter 3). Further studies are needed that test whether male predation protection was a selective force for the evolution of social monogamy in some species or if it evolved as a consequence of pair living.

When males defend an exclusive food resource

Although this hypothesis is commonly included among those proposed for the evolution of socioreproductive monogamy (Wittenberger & Tilson, 1980; Rutberg, 1983), convincing empirical support is largely absent. The resource-defence hypothesis predicts that all intruders in a territory are evicted by the male, irrespective of whether the intruders are male or female. Without this level of intervention, a resident female would lose the benefit of her pair-mate's territorial defence. Territorial male defence was suggested as the origin for social monogamy in gibbons (Rutberg, 1983; Leighton, 1987); however, contrary to the prediction, territorial aggression was found to be mainly sex-specific (Mitani, 1984; Cowlishaw, 1992; Reichard & Sommer, 1997; Reichard, chapter 13).

This hypothesis is questionable because it assumes that socially monogamous males were unable to defend an area large enough to encompass two or more separate female ranges. Because otherwise, males would be expected to become socioreproductively polygynous. Socioreproductive polygyny would also benefit female interests since it further reduces male feeding competition. Furthermore, Komers and Brotherton (1997) showed that at least mammalian socially monogamous female home range sizes were smaller compared with home range sizes of individual socially polygynous females, which would also facilitate the evolution of socioreproductive polygyny instead of social monogamy.

The only possible examples where resource exploitation and perhaps male defence of a feeding territory may have played an important role in the evolution of social monogamy were found in Blanford's foxes (*Vulpes cana*) and the beaver (*Castor fiber*). Blanford's foxes are small, nocturnal canids that live in marginally overlapping territories. Their specialized, insectivorous diet is mainly found in dry creekbeds (Geffen & Macdonald, 1992), and this dependence on such a specific insect prey usually does not permit them to forage in groups of more than two adult individuals. Adding a breeding female to a group would require a considerable increase in territory size and would probably result in higher territorial defence costs for the male than benefits he would derive from socioreproductive polygyny (Geffen & Macdonald, 1992). Beavers consume a specialized diet of deciduous tree species, with comparatively low nutritional value and a slow regrowth rate. Providing enough food for offspring during winter months requires collecting plant material from a large territory. Maintaining a large-sized territory is an important male task, although females also engage in territorial behaviour, both of which probably limit beaver males to social monogamy (Sun, chapter 9).

The hypothesis that males defend a feeding territory for females and offspring, a strategy that increases male reproductive success beyond that of alternative strategies, due to an increased female reproductive rate, still awaits convincing empirical support. Although under specific conditions male territorial defence provides females and offspring with access to an exclusive resource, the causal link to social monogamy still must be quantitatively validated. As with some other explanations, it appears that male territorial defence in socially monogamous animals may have evolved secondarily.

When males protect females and offspring from harassment and infanticide

The hypothesis suggests that the threat of male harassment, and particularly the risk of infanticide in infant-carrying species, may have selected for the evolution of social monogamy (van Schaik & Kappeler, 1993, 1997, chapter 4).

Support for the hypothesis comes from observations of waterfowl, where paired males tried to protect and shield females from harassment by other males (McKinney *et al.*, 1984). Male harassment is potentially costly. In ducks, for example, forced copulation can lead to female death (McKinney & Stolen, 1982). For mammals, van Schaik and Dunbar (1990) and recently van Schaik and Kappeler (1993, 1997, chapter 4) suggested that year-round male–female association (including social monogamy) evolved in infant-carrying species in response to the risk of infanticide. There are examples of male infanticide in birds and mammals (Crook & Shields, 1985; Freed, 1986; Møller, 1994; van Schaik & Janson, 2000). However, a convincing example for the causal link between infanticide and social monogamy is still missing. On the other hand, that the suggested evolutionary step to uniform socially monogamous male–female pairs was caused by the threat of infanticide, cannot yet be excluded (van Schaik & Kappeler, chapter 4). So far, only indirect circumstantial evidence supports the hypothesis (see Reichard, chapter 13) and further tests are necessary to evaluate the importance of male infanticide risk for the evolution of monogamy.

SEXUAL CONFLICT AND SOCIAL MONOGAMY

Under social monogamy, it is more likely to be assumed that the sexes have identical interests, perhaps because pair partners usually maintain close spatial association and often perform spectacular, well-coordinated, pair-specific display behaviour before, during, and after pairing (e.g., duetting: Haimoff, 1986; tail-twining: Anzenberger, 1988; pair dancing: Vincent, 1995). Where male care is considered essential for females to achieve any reproductive success, male and female interests are probably closest and both sexes may benefit equally from socioreproductive monogamy (cf. Birkhead & Møller, 1996; Figure 1.1). Outside this scenario, however, adopting an idealized view of harmonious, monogamous pair relationships where partners share mutual socioreproductive interests is probably misleading (cf. Parker, 1984; Davies, 1989; Gowaty, 1996*a*) for at least two reasons. First, assuming shared interests that lead to social monogamy is just one possible evolutionary pathway to social monogamy, as shown above (see also Wittenberger & Tilson, 1980; Wickler & Seibt, 1981). Second, pair *formation* is not equivalent to an absence of sexual conflict, and the costs and benefits of mating may be asymmetric for males and females (Parker, 1979, 1984). Conflict of interests between pair partners may arise over the duration and intensity of sociosexual relationship(s), particularly the frequency of mating, as well as parental duties. Also, in socially monogamous relationships individuals are expected to take advantage of their social partner's contributions whenever possible (Trivers, 1972; Davies, 1989; Lessells, 1999; Johnstone & Keller, 2000; Royle *et al.*, 2002).

An especially informative example that demonstrates the dynamics of sexual conflict and its resolution is given by Davies's (1992) study of the dunnock (*Prunella modularis*). Dunnocks show a variable mating system in which social monogamy, social polygyny, social polygynandry and social polyandry occur (Davies & Lundberg, 1984). Davies (1989, 1992) could show that females achieve highest reproductive success with cooperative social polyandry and least with social polygyny, whereas the opposite held for males, which achieved highest success with social polygyny and lowest with cooperative polyandry. Depending on varying ecological conditions and power symmetries, individuals were able to achieve their preferred option, despite the conflicting preferences of others, resulting in various social compositions (Davies, 1989).

Potential sexual conflict also characterizes most pathways to social monogamy in Figure 1.1. Expected optima, representing the socioreproductive strategy with the theoretically greatest pay-off for male and female, were presented below each pathway to social monogamy and shaded grey boxes were put around optima to visualize where the compromise (social monogamy) potentially conflicted with preferred options. 'Optimal social strategies' were derived from sexual selection theory and adapted, where necessary, to particular conditions (e.g., assumed optimal strategies for paired vs. unpaired individuals varied). Sociosexual polygyny was principally assumed to be an optimal male strategy whereas sociosexual monogamy or sociosexual polyandry were the assumed optimal female strategies (cf. Parker, 1984).

Most pathways to the social-monogamy-compromise conflicted with the 'optimal strategies' of one or both sexes. Only where biparental care and infanticide protection were important factors did social monogamy coincide with the greatest payoff for both sexes (Figure 1.1). Hence, these appear as the clearest pathways along which social monogamy is expected to correspond to genetic monogamy given the sexes' shared interests. In contrast, when optimal reproductive interests of one or both partners do not overlap with social monogamy, further conflict between the sexes on the sexual and genetic level is expected. Different pathways contain the potential for different conflicts along the evolutionary routes to social monogamy, and may explain the variations between the social and the genetic system of socially monogamous species that were identified in the 1990s and early 2000s (cf. Petrie & Kempenaers, 1998; Petrie *et al.*, 1998; Hasselquist & Sherman, 2001).

CONCLUSIONS

Social monogamy does not evolve from a common, single origin but arises independently through different evolutionary pressures and along different pathways in different lineages. Social monogamy in birds and mammals, including humans, represents just one possible outcome of the compromise between the reproductive interests and strategies of the sexes under specific conditions. The three components that influence the occurrence of social monogamy are: the magnitude of paternal care, the mode of resource access, and mate choice. These components interact with each other and produce ten distinct pathways to social monogamy, most of which are supported by empirical studies. The evolution of social monogamy in different animal groups is the focus of the first part of this book (chapters 2–5).

Parts II and III focus on the reproductive strategies of males and females. With the exception of chapters 11 and 16, the authors focus on a particular species or taxon (chapter 15) to identify the specific conditions and evolutionary pressure(s) responsible for social monogamy and corresponding genetic mating systems, as well as how those systems are maintained.

The influence of paternal care on the evolution of social monogamy has always been an important focus (cf. Kleimann, 1977; Wittenberger & Tilson, 1980). Where paternal care becomes highly important it may directly lead to social monogamy. But, paternal care often plays a more subtle role along the routes to social monogamy and depends largely on the balance between the amount of care provided by males and how much is needed or wanted by females. Whenever direct or indirect paternal investment in offspring is important for female reproductive success it will influence whether or not females are willing to share a male with another female. Similarly, under certain conditions, males may avoid sharing a female, and in both cases, if either of two prime preconditions are met, social monogamy is expected to evolve. Females must live non-gregariously in separate ranges, or males must defend resources that are important to females, which are often territories. The conditions of females living apart from each other and males defending a territory represent basic male and female strategies to maximize access to the resources relevant for their respective reproduction and to lay the foundation of the pathways to social monogamy. Part of the dialogue presented here is an interesting difference that emerged between birds and mammals. Male defence of resources necessary for females appears to be a more frequent starting point for the evolution of bird social monogamy, whereas females living in separate ranges appears to be a more common starting point for the evolution of social monogamy in mammals. If this assumption holds, it may explain why more socially monogamous mammals are also genetically monogamous (Foltz, 1981; Ribble, 1991; Heller *et al.*, 1993; Brotherton *et al.*, 1997; Sommer & Tichy, 1999) than are socially monogamous birds, because a greater potential for the sexes to share a monogamous reproductive interest exists when females live apart than when males occupy territories as a way to attract a mating partner (see Figure 1.1). The hypothesis, that the discrepancy between social and genetic mating systems may be predicted by evolutionary pathways to social monogamy and the intensity of sexual conflict, will have to be scrutinized in further studies.

This volume's approach to social monogamy builds on and extends earlier concepts of the evolution of monogamy (Emlen & Oring, 1977; Wittenberger & Tilson, 1980) because it integrates pathways rarely considered in a comprehensive framework. In the classical approach of Emlen and Oring (1977), the evolution of social monogamy was reduced to two elements: (i) neither sex has the opportunity to monopolize additional members of the opposite sex, and (ii) the absence of an environmental potential for polygamy. It has since become clear that additional avenues to social monogamy do exist. As shown by Davies (1992), the outcome of social conflict may be just as important as ecological factors in shaping social monogamy and other social arrangements. Even when the polygyny-threshold is exceeded, social monogamy may evolve because payoff expectations for involved individuals vary, influencing their mate choice and reproductive decisions accordingly. The classical four or five explanations for the evolution of social monogamy (Wittenberger & Tilson, 1980; Clutton-Brock, 1989) were refined to incorporate specific conditions favouring the evolution of social monogamy, like the threat of infanticide (van Schaik & Dunbar, 1990) or the influence of individual quality variation.

Because models of the mating system's evolution still appear to be biased by a single-sex perspective, additional research is needed. Identifying ecological factors and tracing individual decisions that may lead to social monogamy is difficult enough in one sex. However, to understand fully what shapes social monogamy, male and female perspectives must be combined. Furthermore, the interplay between the sexes' interests, on the social, sexual, and reproductive levels, as well as resolutions of the conflicts between them, must be examined in more detail. These studies should test for what the optimal strategies for each sex would look like and what constraints may exist, and should then analyse how these aspects correspond to compromises between partners. Furthermore, it will be important to look at the interplay between factors like paternal care and individual quality, which may produce varying affects along the steps to social monogamy, for a better understanding of individual behavioural variation on the levels of social, sexual, and genetic partnerships. For example, in male care, at what cut-off point in relation to a female's needs does a female resist the arrival of a second female on her breeding ground, and how does this influence (additional?) male parental behaviour? Or, what are the necessary female quality characters that allow for social variation within a population? The answers to questions like these will be critical if the fine structure of social monogamy is to be clearly understood.

Since the 1980s, the gross settings in which social monogamy can evolve have been identified. However, information on the finely-tuned evaluations of individual decisions about whether or not to establish socially monogamous relationships, with whom, and for how long, are still relevant lines to pursue. An interesting area for continued research on social monogamy and its corresponding genetic systems is the variation in individual female quality (Hoi-Leitner *et al.*, 1999; Forstmeier *et al.*, 2001). Because the opportunity to increase reproductive success via the number of breeding partners is limited when individuals live predominantly in socially monogamous units, variation in individual quality may become an increasingly important factor (Anderson, 1994). The present model has begun to recognize and integrate individual quality variation on a theoretical level, and future research will evaluate how important this aspect is for the origin and maintenance of social monogamy.

Acknowledgements
This chapter has greatly profited from the stimulating comments of Christophe Boesch, Nick Davies, Tim Clutton-Brock, Joe Manson, Anders Pape Møller, and Linda Vigilant. I thank the Max-Planck-Society for supporting my work on monogamy.

References

Ahnesjö, I., Vincent, A., Alatalo, R. V., Halliday, T. & Sutherland, W. J. (1993). The role of females in influencing mating patterns. *Behavioural Ecology*, **4**, 187–9.

Alatalo, R. V., Gottlander, K. & Lundberg, A. (1987). Extra-pair copulations and mate guarding in the polyterritorial pied flycatcher, *Ficedula hypoleuca*. *Behaviour*, **101**, 139–55.

Alatalo, R. V., Lundberg, A. & Glynn, C. (1986). Female pied flycatchers choose territory quality and not male characteristics. *Nature*, **323**, 152–3.

Alexander, R. D. (1974). The evolution of social behaviour. *Annual Review of Ecology and Systematics*, **5**, 325–83.

Altmann, J. (1990). Primate males go where the females are. *Animal Behaviour*, **39**, 193–5.

Andelman, S. J. (1986). Ecological and social determinants of cercopithecine mating patterns. In *Ecological Aspects of Social Evolution: Birds and Mammals*, ed. D. Rubenstein & R. Wrangham, pp. 201–16. Princeton, New Jersey: Princeton University Press.

Anderson, M. (1994). *Sexual Selection*. Princeton: Princeton University Press.

Anzenberger, G. (1988). The pair bond in the titi monkey (*Callicebus moloch*): intrinsic versus extrinsic contributions of the pairmates. *Folia Primatologica*, **50**, 188–203.

Arcese, P. (1989). Intrasexual competition and the mating system in primarily monogamous birds: the case of the song sparrow. *Animal Behaviour*, **38**, 96–111.

Baker, A. J. & Dietz, J. M. (1996). Immigration in wild groups of golden lion tamarins (*Leontopithecus rosalia*). *American Journal of Primatology*, **38**, 47–56.

Baker, R. R. (1983). Insect territoriality. *Annual Review of Entomology*, **28**, 65–89.

Barlow, G. W. (1988). Monogamy in relation to resources. In *The Ecology of Social Behaviour*, ed. C. N. Slobodchikoff, pp. 55–79. London: Academic Press.

Bart, J. & Tornes, A. (1989). Importance of monogamous birds in determining reproductive success. *Behavioral Ecology and Sociobiology*, **24**, 109–16.

Bateman, A. J. (1948). Intra-sexual selection in *Drosophila*. *Heredity*, **2**, 349–68.

Beecher, M. D. & Beecher, I. M. (1979). Sociobiology of bank swallows: reproductive strategy of the male. *Science*, **205**, 1282–5.

Berteaux, D., Bêty, J., Regifo, E. & Bergeron, J.-M. (1999). Multiple paternity in meadow voles (*Microtus pennsylvanicus*): investigating the role of the female. *Behavioral Ecology and Sociobiology*, **45**, 283–91.

Birkhead, T. R. & Møller, A. P. (1992). *Sperm Competition in Birds: Evolutionary Causes and Consequences*. London: Academic Press.

(1993). Female control of paternity. *Trends in Ecology and Evolution*, **8**, 100–4.

(1995). Extra-pair copulations and extra-pair paternity in birds. *Animal Behaviour*, **49**, 843–8.

(1996). Monogamy and sperm competition in birds. In *Partnerships in Birds: The Study of Monogamy*, ed. J. M. Black, pp. 323–43. Oxford: Oxford University Press.

(1998). *Sperm Competition and Sexual Selection*. London: Academic Press.

Birkhead, T. R., Atkin, L. & Møller, A. P. (1987). Copulation behaviour of birds. *Behaviour*, **101**, 101–38.

Black, J. M. (1996). Pair bonds and partnerships. In *Partnerships in Birds: The Study of Monogamy*, ed. J. M. Black, pp. 3–20. Oxford: Oxford University Press.

Black, J. M. & Owen, M. (1989). Parent–offspring relationships in wintering barnacle geese. *Animal Behaviour*, **37**, 187–98.

Black, J. M., Choudhury, S. & Owen, M. (1996). Do Barnacle Geese benefit from lifelong monogamy? In *Partnerships in Birds: The Study of Monogamy*, ed. J. M. Black, pp. 91–117. Oxford: Oxford University Press.

Blumstein, D. T. (2000). The evolution of infanticide in rodents: a comparative analysis. In *Infanticide by Males and its Implications*, ed. C. P. van Schaik & C. H. Janson, pp. 178–97. Cambridge: Cambridge University Press.

Blurton Jones, N. G., Marlowe, F. W., Hawkes, K. & O'Connel, J. F. (2000). Paternal investment and hunter-gatherer divorce rates. In *Adaptation and Human Behavior*, ed. L. Cronk, N. Chagnon & W. Irons, pp. 69–90. New York: Aldine de Gruyter.

Borgerhoff Mulder, M. (1990). Kipsigis women's preferences for wealthy men: evidence for female choice in mammals? *Behavioral Ecology and Sociobiology*, **27**, 255–64.

Bray, O. E., Kennely, J. K. & Guareno, J. K. (1975). Fertility of eggs produced on territories of vasectomized red-winged blackbirds. *Wilson Bulletin*, **87**, 187–95.

Brockelman, W. Y. & Srikosamatara, S. (1984). Maintenance and evolution of social structure in gibbons. In *The Lesser Apes: Evolutionary and Behavioural Biology*, ed. H. Preuschoft, D. J. Chivers, W. Y. Brockelman & N. Creel, pp. 298–323. Edinburgh: Edinburgh University Press.

Brockelman, W. Y., Reichard, U., Treesucon, U. & Raemaekers, J. J. (1998). Dispersal, pair formation and social structure in gibbons (*Hylobates lar*). *Behavioral Ecology and Sociobiology*, **42**, 329–39.

Brotherton, P. N. M. & Manser, M. B. (1997). Female dispersion and the evolution of monogamy in the dik-dik. *Animal Behaviour*, **54**, 1413–24.

Brotherton, P. N. M. & Rhodes, A. (1996). Monogamy without biparental care in a dwarf antelope. *Proceedings of the Royal Society of London, Series B*, **263**, 23–9.

Brotherton, P. N. M., Pemberton, J. M., Komers, P. E. & Malarky, G. (1997). Genetic and behavioural evidence of monogamy in a mammal, Kirk's dik-dik (*Madoqua kirkii*). *Proceedings of the Royal Society of London, Series B*, **264**, 675–81.

Bull, C. M., Cooper, S. J. B. & Baghurst, B. C. (1998). Social monogamy and extra-pair fertilization in an Australian lizard, *Tiliqua rugosa*. *Behavioral Ecology and Sociobiology*, **44**, 63–72.

Burley, N. T. & Johnson, K. (2002). The evolution of avian parental care. *Philosophical Transactions of the Royal Society of London, Series B*, **357**, 241–50.

Carranza, J., Alvarez, F. & Redondo, T. (1990). Territoriality as a mating strategy in red deer. *Animal Behaviour*, **40**, 79–88.

Catry, P., Ratcliff, N. & Furness, R. W. (1997). Partnerships and mechanisms of divorce in the great skua. *Animal Behaviour*, **54**, 1475–82.

Chapman, T. & Partridge, L. (1996). Sexual conflict as fuel for evolution. *Nature*, **381**, 189–90.

Clutton-Brock, T. H. (1989). Mammalian mating systems. *Proceedings of the Royal Society of London, Series B*, **236**, 339–72.

(1991). *The Evolution of Parental Care*. Princeton: Princeton University Press.

Clutton-Brock, T. H. & Harvey, P. H. (1977). Primate ecology and social organization. *Journal of Zoology, London*, **183**, 1–39.

Clutton-Brock, T. H. & Parker, G. A. (1992). Potential reproductive rates and the operation of sexual selection. *Quarterly Reviews in Biology*, **67**, 437–56.

Cowlishaw, G. (1992). Song function in Gibbons. *Behaviour*, **121**, 131–53.

Crook, J. R. & Shields, W. M. (1985). Sexually selected infanticide by adult male barn swallows. *Animal Behaviour*, **33**, 754–61.

Davies, N. B. (1989). Sexual conflict and the polygamy threshold. *Animal Behaviour*, **38**, 226–34.

(1991). Mating systems. In *Behavioural Ecology: An Evolutionary Approach*, ed. J. R. Krebs & N. B. Davies, pp. 263–94. Oxford: Blackwell Science Publications.

(1992). *Dunnock Behaviour and Social Evolution*. Oxford: Oxford University Press.

Davies, N. B. & Lundberg, A. (1984). Food distribution and a variable mating system in the dunnock, *Prunella modularis*. *Journal of Animal Ecology*, **53**, 895–912.

Desrochers, A. & Magrath, R. D. (1996). Divorce in the European Blackbird: seeking greener pastures? In *Partnerships in Birds: The Study of Monogamy*, ed. J. M. Black, pp. 344–401. Oxford: Oxford University Press.

Dewsbury, D. A. (1988). The comparative psychology of monogamy. In *Nebraska Symposium on Motivation 1987: Comparative Perspectives in Modern Psychology*, Volume 35, ed. R. A. Dienstbier & D. W. Leger, pp. 1–50. Lincoln: University of Nebraska Press.

Dietz, J. M. & Baker, A. J. (1993). Polygyny and female reproductive success in golden lion tamarins, *Leontopithecus rosalia*. *Animal Behaviour*, **46**, 1067–78.

Dunbar, R. I. M. (1995). The mating system of callitrichid primates: I. Conditions for the coevolution of pair bonding and twinning. *Animal Behaviour*, **50**, 1057–70.

Dunbar, R. I. M. & Dunbar, E. P. (1980). The pairbond in klipspringer. *Animal Behaviour*, **28**, 219–29.

Emlen, S. T. & Oring, L. W. (1977). Ecology, sexual selection and the evolution of mating systems. *Science*, **197**, 215–23.

Ens, B., Choudhury, S. & Black, J. M. (1996). Mate fidelity and divorce in monogamous birds. In *Partnerships in Birds: The Study of Monogamy*, ed. J. M. Black, pp. 344–401. Oxford: Oxford University Press.

Erckmann, W. J. (1983). The evolution of polyandry in shorebirds: an evaluation of hypotheses. In *Social Behavior of*

Female Vertebrates, ed. S. K. Wasser, pp. 113–68. New York: Academic Press.

Ferron, J. & Ouellet, J. P. (1989). Temporal and intrasexual variations in the use of space with regards to social organization in the woodchuck (*Marmota monax*). *Canadian Journal of Zoology*, **67**, 1642–9.

Fietz, J. (1999*a*). Monogamy as a rule rather than exception in nocturnal lemurs: the case of the fat-tailed dwarf lemur, *Cheirogaleus medius*. *Ethology*, **105**, 259–72.

(1999*b*). Demography and floating males in a population of *Cheirogaleus medius*. In *New Directions in Lemur Studies*, ed. B. Rakotosamimanana, H. Rasaminanana & J. U. Ganzhorn, pp. 159–72. New York: Kluwer Academic/Plenum.

Fietz, J., Zischler, H., Schwiegk, C., Tomiuk, J., Danzman, K. H. & Ganzhorn, J. U. (2000). High rates of extra-pair young in the pair-living fat-tailed dwarf lemur, *Cheirogaleus medius*. *Behavioral Ecology and Sociobiology*, **49**, 8–17.

Fisher, D. O. & Owens, I. P. F. (2000). Female home range size and the evolution of social organization in macropod marsupials. *Journal of Animal Ecology*, **69**, 1083–98.

Foltz, D. W. (1981). Genetic evidence for long-term monogamy in a small rodent, *Peromyscus polionotus*. *American Naturalist*, **117**, 665–75.

Forstmeier, W., Leisler, B. & Kempenaers, B. (2001). Bill morphology reflects female independence from male parental help. *Proceedings of the Royal Society of London, Series B*, **268**, 1583–8.

Freed, L. A. (1986). Territory takeover and sexually selected infanticide in tropical house wrens. *Behavioral Ecology and Sociobiology*, **19**, 197–206.

(1987). The long-term pair bond of tropical house wrens: advantage or constraint? *American Naturalist*, **130**, 507–25.

Geffen, E. & Macdonald, D. W. (1992). Small size and monogamy: spatial organization of Blanford's foxes, *Vulpes cana*. *Animal Behaviour*, **44**, 1123–30.

Getz, L. L., Hofmann, J. E. & Carter, C. S. (1987). Mating systems and fluctuations of the prairie vole, *Microtus ochrogaster*. *American Journal of Zoology*, **27**, 909–20.

Gladstone, D. E. (1979). Promiscuity in monogamous colonial birds. *American Naturalist*, **114**, 545–57.

Goldizen, A. W. (1987). Facultative polyandry and the role of infant-carrying in wild saddle-back tamarins (*Saguinus fuscicollis*). *Behavioral Ecology and Sociobiology*, **20**, 99–109.

Goossens, B., Graziani, L., Waits, E. F., Magnolon, S., Coulon, J., Bel, M.-C., Taberlet, P. & Allainé, D. (1998). Extra-pair paterntiy in the monogamous Alpine marmot revealed by nuclear DNA microsatellite analysis. *Behavioral Ecology and Sociobiology*, **43**, 281–8.

Gosling, L. M. (1986). The evolution of mating strategies in male antelopes. In *Ecological Aspects of Social Evolution: Birds and Mammals*, ed. D. Rubenstein & R. Wrangham, pp. 244–81. Princeton: Princeton University Press.

Gowaty, P. A. (1996*a*). Battles of the sexes and origins of monogamy. In *Partnerships in Birds: The Study of Monogamy*, ed. J. M. Black, pp. 21–52. Oxford: Oxford University Press.

(1996*b*). Multiple mating by females selects for males that stay: another hypothesis for social monogamy in passerine birds. *Animal Behaviour*, **51**, 482–4.

(1996*c*). Field studies of parental care in birds: new data focus questions on variation in females. In *Advances in the Study of Behavior*, Volume 25, ed. C. T. Snowdon & J. S. Rosenblatt, pp. 476–531. New York: Academic Press.

Green, K., Mitchell, A. T. & Tennant, P. (1998). Home range and microhabitat use by the long-footed potoroo, *Potorous longipes*. *Wildlife Research*, **25**, 357–72.

Greenwood, P. J. (1980). Mating systems, philopatry and dispersal in birds and mammals. *Animal Behaviour*, **28**, 1140–62.

Gubernick, D. J. (1994). Biparental care and male–female relations in mammals. In *Infanticide and Parental Care*, ed. S. Parmigiani & F. vom Saal, pp. 427–63. London: Harwood Academic Press.

Gubernick, D. J. & Teferi, T. (2000). Adaptive significance of male parental care in a monogamous mammal. *Proceedings of the Royal Society of London, Series B*, **267**, 147–50.

Haimoff, E. H. (1986). Convergence in the duetting of monogamous Old World primates. *Journal of Human Evolution*, **15**, 51–9.

Hannon, S. J. (1983). Spacing and breeding density of willow ptarmigan in response to an experimental alteration of sex ratio. *Journal of Animal Ecology*, **52**, 807–20.

Hannon, S. J. & Dobush, G. (1997). Pairing status of male willow ptarmigan: is polygyny costly to males? *Animal Behaviour*, **53**, 369–80.

Hannon, S. J. & Martin, K. (1992). Monogamy in willow ptarmigan: is male vigilance important for reproductive success and survival of females? *Animal Behaviour*, **43**, 747–57.

Hartley, I. R., Davies, N. B., Hatchwell, B. J., Desrouchers, A., Nebel, D. & Burke, T. A. (1995). The polyandrous mating system of the alpine accentor, *Prunella collaris*. II. Multiple paternity and parental effort. *Animal Behaviour*, **49**, 789–803.

Hasselquist, D. S. & Sherman, P. W. (2001). Social mating systems and extrapair fertilizations in passerine birds. *Behavioral Ecology*, **12**, 457–66.

Hatchwell, B. J., Russell, A. F. & Ross, D. J. (2000). Divorce in cooperatively breeding long-tailed tits: a consequence of

inbreeding avoidance? *Proceedings of the Royal Society of London, Series B*, **267**, 813–19.
Hausfater, G. & Hrdy, S. B. (1984). *Infanticide. Comparative and Evolutionary Perspectives*. New York: Aldine.
Heller, K.-G., Achmann, R. & Witt, K. (1993). Monogamy in the bat *Rinolophus sedulus*? *Zeitschrift für Säugetierkunde*, **58**, 376–7.
Hendrichs, H. (1975). Changes in a population of Dikdik, *Madoqua* (*Rhynchotragus*) *kirki* (Gunther 1880). *Zeitschrift für Tierpsychologie*, **38**, 55–69.
Heymann, E. W. (2000). The number of adult males in callitrichine groups and its implications for callitrichine social evolution. In *Primate Males. Causes and Consequences of Variation in Group Composition*, ed. P. M. Kappeler, pp. 159–68. Cambridge: Cambridge University Press.
Hoi-Leitner, M. H., Hoi, H., Romero-Pujante, M. & Valera, F. (1999). Female extra-pair behaviour and environmental quality in the serin (*Serinus serinus*): a test of the 'constrained female hypothesis'. *Proceedings of the Royal Society of London, Series B*, **266**, 1021–6.
Hrdy, S. B. (1979). Infanticide among animals: a review, classification, and examination of the implications for the reproductive strategies of females. *Ethology and Sociobiology*, **1**, 13–40.
(1986). Empathy, polyandry, and the myth of the coy female. In *Feminist Approaches to Science*, ed. R. Bleier, pp. 119–46. New York: Pergamon Press.
Hrdy, S. B., Janson, C. & van Schaik, C. P. (1995). Infanticide: let's not throw out the baby with the bath water. *Evolutionary Anthropology*, **3**, 151–4.
Hughes, C. (1998). Integrating molecular techniques with field methods in studies of social behavior: a revolution results. *Ecology*, **79**, 383–99.
Isbell, L. A. (1994). Predation on primates: ecological patterns and evolutionary consequences. *Evolutionary Anthropology*, **3**, 61–71.
Jamieson, I. G. (1997). Testing reproductive skew models in a communally breeding bird, the pukeko, *Porphyrio porphyrio*. *Proceedings of the Royal Society of London, Series B*, **264**, 335–40.
Janson, C. H. (1992). Evolutionary ecology of primate social structure. In *Evolutionary Ecology and Human Behavior*, ed. E. A. Smith & B. Winterhalder, pp. 95–130. New York: Aldine de Gruyter.
Jeffreys, A. J., Wilson, V. & Thein, S. L. (1985). Hypervariable "minisatellite" regions in human DNA. *Nature*, **314**, 67–73.
Johnstone, R. A. & Keller, L. (2000). How males can gain by harming their mates: sexual conflict, seminal toxins, and the cost of mating. *American Naturalist*, **156**, 368–77.
Keane, B., Waser, P. M., Creel, S. R., Creel, N. M., Elliott, L. F. & Minchella, D. J. (1994). Subordinate reproduction in dwarf mongooses. *Animal Behaviour*, **47**, 65–75.
Kempenaers, B. (1997). Does reproductive synchrony limit male opportunities or enhance female choice for extra-pair paternity. *Animal Behaviour*, **134**, 551–62.
Kempenaers, B., Everding, S., Bishop, C., Boag, P. & Robertson, R. J. (2001). Extra-pair paternity and the role of male floaters in the tree swallow (*Tachycineta bicolor*). *Behavioral Ecology and Sociobiology*, **49**, 251–9.
Kempenaers, B., Verheyen, G. R., van den Broeck, M., Burke, T., van Broeckhoven, C. & Dhondt, A. A. (1992). Extra-pair paternity results from female preferences for high-quality males in the blue tit. *Nature*, **357**, 494–6.
Kishimoto, R. & Kawamichi, T. (1996). Territoriality and monogamous pairs in a solitary ungulate, the Japanese serow, *Capricornis crispus*. *Animal Behaviour*, **52**, 673–82.
Kleiman, D. G. (1977). Monogamy in mammals. *Quarterly Review of Biology*, **52**, 39–69.
Kleiman, D. G. & Malcolm, J. R. (1981). The evolution of male paternal investment in mammals. In *Paternal Care in Mammals*, ed. D. J. Gubernick & P. H. Klopfer, pp. 347–87. New York: Plenum Press.
Kokita, T. (2002). The role of the female behavior in maintaining monogamy of a coral-reef fish. *Ethology*, **108**, 157–68.
Komers, P. E. & Brotherton, P. N. M. (1997). Female space use is the best predictor of monogamy in mammals. *Proceedings of the Royal Society of London, Series B*, **264**, 1261–70.
Lack, D. (1968). *Ecological Adaptations for Breeding in Birds*. London: Methuen.
Leighton, D. R. (1987). Gibbons: territoriality and monogamy. In *Primate Societies*, ed. B. B. Smuts, D. L. Cheney, R. M. Seyfarth, R. W. Wrangham & T. T. Struhsaker, pp. 135–45. Chicago: University of Chicago Press.
Lessells, C. M. (1999). Sexual conflict in animals. In *Levels of Selection in Evolution*, ed. L. Keller, pp. 75–99. Princeton, New Jersey: Princeton University Press.
Ligon, J. D. (1999). *The Evolution of Avian Breeding Systems*. Oxford: Oxford University Press.
Malcolm, J. R. (1985). Paternal care in canids. *American Zoologist*, **25**, 853–9.
Marlowe, F. (2000). Paternal investment and the human mating system. *Behavioural Processes*, **51**, 45–61.
McKinney, F. (1986). Ecological factors influencing the social systems of migratory dabbling ducks. In *Ecological Aspects of Social Evolution: Birds and Mammals*, ed. D. Rubenstein & R. Wrangham, pp. 153–71. Princeton, New Jersey: Princeton University Press.
McKinney, F. & Stolen, P. (1982). Extra-pair-bond courtship and forced copulation among captive green-winged teal (*Anas crecca caolinensis*). *Animal Behaviour*, **30**, 461–74.

McKinney, F., Cheng, K. M. & Bruggers, D. J. (1984). Sperm competition in apparently monogamous birds. In *Sperm Competition and the Evolution of Animal Mating Systems*, ed. R. L. Smith, pp. 523–45. London: Academic Press.

Mitani, J. C. (1984). The behavioural regulation of monogamy in gibbons (*Hylobates muelleri*). *Behavioral Ecology and Sociobiology*, **15**, 225–9.

Mitani, J., Gros-Louis, J. & Manson, J. H. (1996). Number of males in primate groups: comparative tests of competing hypotheses. *American Journal of Primatology*, **38**, 315–32.

Mock, D. W. (1985). An introduction to the neglected mating system. In *Avian Monogamy, Ornithological Monographs*, Volume 37, ed. P. A. Gowaty & D. W. Mock, pp. 1–10, Washington, DC: American Ornithologists' Union.

Mock, D. W. & Fujioka, M. (1990). Monogamy and long-term pair bonding in vertebrates. *Trends in Ecology and Evolution*, **5**, 39–43.

Moehlman, P. D. (1989). Instraspecific variation in canid social systems. In *Carnivore Behavior, Ecology, and Evolution*, ed. J. L. Gittleman, pp. 143–63. Ithaca, New York: Cornell University Press.

Møller, A. P. (1988*a*). Infanticidal and anti-infanticidal strategies in the swallow *Hirundo rustica*. *Behavioral Ecology and Sociobiology*, **22**, 365–71.

(1988*b*). Female choice selects for male sexual tail ornaments in the monogamous swallow. *Nature*, **332**, 640–2.

(1994). *Sexual Selection and the Barn Swallow*. Oxford: Oxford University Press.

(1998). Sperm competition and sexual selection. In *Sperm Competition and Sexual Selection*, ed. T. R. Birkhead & A. P. Møller, pp. 55–90. San Diego: Academic Press.

(2000). Male parental care, female reproductive success and extra-pair paternity. *Behavioral Ecology*, **11**, 161–8.

Murray, B. G. Jr (1984). A demographic theory on the evolution of mating systems as exemplified by birds. In *Evolutionary Biology*, Volume 18, ed. M. K. Hecht, B. Wallace & G. T. Prance, pp. 71–140. New York: Plenum Press.

Orians, G. H. (1969). On the evolution of mating systems in birds and mammals. *American Naturalist*, **103**, 589–603.

Otter, K., Ratcliffe, L., Michaud, D. & Boag, P. T. (1998). Do female black-capped chickadees prefer high-ranking males as extra-pair partners. *Behavioral Ecology and Sociobiology*, **43**, 25–36.

Parker, G. A. (1974). Courtship persistence and female-guarding as male time investment strategies. *Behaviour*, **48**, 157–84.

(1979). Sexual selection and sexual conflict. In *Sexual Selection and Reproductive Competition in Insects*, ed. M. S. Blum & N. A. Blum, pp. 123–66. New York: Academic Press.

(1984). Sperm competition and the evolution of animal mating strategies. In *Sperm Competition and the Evolution of Animal Mating Systems*, ed. R. L. Smith, pp. 1–59. London: Academic Press.

Parker, G. A., Baker, R. R. & Smith, V. G. F. (1972). The origin and evolution of gamete dimorphism and the male–female phenomenon. *Journal of Theoretical Biology*, **36**, 529–33.

Petrie, M. & Kempenaers, B. (1998). Why does the proportion of extra-pair paternity in birds vary between species and between populations? *Trends in Ecology and Evolution*, **13**, 52–8.

Petrie, M., Doums, C. & Møller, A. P. (1998). The degree of extra-pair paternity increases with genetic variability. *Proceedings of the National Academy of Sciences of the USA*, **95**, 9390–5.

Pizzari, T. & Birkhead, T. R. (2000). Female feral fowl eject sperm of subdominant males. *Nature*, **405**, 787–9.

Pribil, S. & Picman, J. (1996). Polygyny in the red-winged blackbird: do females prefer monogamy or polygamy? *Behavioral Ecology and Sociobiology*, **38**, 183–90.

Pribil, S. & Searcy, W. A. (2001). Experimental confirmation of the polygyny threshold model for red-winged blackbirds. *Proceedings of the Royal Society of London, Series B*, **268**, 1643–6.

Rees, E. C., Lievesley P., Pettifor, R. A. & Perrins, C. (1996). Mate fidelity in swans: an interspecific comparison. In *Partnerships in Birds: The Study of Monogamy*, ed. J. M. Black, pp. 118–37. Oxford: Oxford University Press.

Reichard, U. (1995). Extra-pair copulations in a monogamous gibbon (*Hylobates lar*). *Ethology*, **100**, 99–112.

Reichard, U. & Sommer, V. (1997). Group encounters in wild gibbons (*Hylobates lar*): agonism, affiliation, and the concept of infanticide. *Behaviour*, **134**, 1135–74.

Ribble, D. O. (1991). The monogamous mating system of *Peromyscus californicus* as revealed by DNS fingerprinting. *Behavioral Ecology and Sociobiology*, **29**, 161–6.

Roberts, S. C. & Dunbar, R. I. M. (2000). Female territoriality and the function of scent-marking in a monogamous antelope (*Oreotragus oreotragus*). *Behavioral Ecology and Sociobiology*, **47**, 417–23.

Royle, N. J., Hartley, I. R. & Parker, G. A. (2002). Sexual conflict reduces offspring fitness in zebra finches. *Nature*, **416**, 733–6.

Rutberg, A. T. (1983). The evolution of monogamy in primates. *Journal of Theoretical Biology*, **104**, 93–112.

Sandell, M. & Liberg, O. (1992). Roamers and stayers: a model on male mating tactics and mating systems. *American Naturalist*, **139**, 177–89.

Sandell, M. & Smith, H. G. (1996). Already mated females constrain male mating success in the European starling.

Proceedings of the Royal Society of London, Series B, **263**, 743–7.

Sheldon, B. C. (1994). Male phenotype, fertility, and the pursuit of extra-pair copulations by female birds. *Proceedings of the Royal Society of London, Series B*, **257**, 25–30.

Shellman-Reeve, J. S. & Reeve, H. K. (2000). Extra-pair paternity as the result of reproductive transactions between paired mates. *Proceedings of the Royal Society of London, Series B*, **267**, 2543–6.

Sillero-Zubiri, C., Gottelli, D. & Macdonald, D. W. (1996). Male philopatry, extra-pack copulations and inbreeding avoidance in Ethiopian wolves (*Canis simensis*). *Behavioral Ecology and Sociobiology*, **38**, 331–40.

Slagsvold, T. (1983). Female–female aggression and monogamy in great tits *Parus major*. *Ornis Scandinavica*, **24**, 155–8.

Slagsvold, T. & Drevon, T. (1999). Female pied flycatchers trade between male quality and mating status in mate choice. *Proceedings of the Royal Society of London, Series B*, **266**, 917–21.

Slagsvold, T. & Lifjeld, J. T. (1994). Polygyny in birds: the role of competition between females for male parental care. *American Naturalist*, **143**, 59–94.

Slagsvold, T., Amundsen, S., Dale, S. & Lampe, H. (1992). Female–female aggression explains polyterritoriality in male pied flycatchers. *Animal Behaviour*, **43**, 397–407.

Smiseth, P. T. & Amundsen, T. (1995). Female bluethroats (*Luscinias svecica*) regularly visit territories of extrapair males before egg laying. *Auk*, **112**, 1049–53.

Snowdon, C. T. (1997). The "nature" of sex differences: myths of male and female. In *Feminism and Evolutionary Biology. Boundaries, Intersections, and Frontiers*, ed. P. A. Gowaty, pp. 276–93. New York: Chapman & Hall.

Sommer, S. (2000). Sex-specific predation on a monogamous rat, *Hypogeomys antimena* (Muridae: Nesomyinae). *Animal Behaviour*, **59**, 1087–94.

Sommer, S. & Tichy, H. (1999). Major histocompatibility complex (MHC) class II polymorphism and paternity in the monogamous *Hypogeomys antimena*, the endangered, largest endemic Malagasy rodent. *Molecular Ecology*, **8**, 1259–72.

Sommer, V. (1994). Infanticide among the langurs of Jodhpur – testing the sexual selection hypothesis with a long-term record. In *Infanticide and Parental Care*, ed. S. Parmigiani & F. vom Saal, pp. 155–93. London: Harwood Academic Press.

Sommer, V. & Reichard, U. (2000). Rethinking monogamy: the gibbon case. In *Primate Males. Causes and Consequences of Variation in Group Composition*, ed. P. M. Kappeler, pp. 159–68. Cambridge: Cambridge University Press.

Southgate, V. R., Jourdane, J. & Tchuemtchuenté, L. A. (1998). Recent studies on the reproductive biology of the schistosomes and their relevance to speciation in the Digenea. *International Journal of Parasitology*, **28**, 1159–72.

Spencer, P., Horsup, A. & Marsh, H. (1998). Enhancement of reproductive success through mate choice in a social rock-wallaby, *Pterogale assimilis* (Macropodidae) as revealed by microsatellite markers. *Behavioral Ecology and Sociobiology*, **43**, 1–9.

Sterck, E. H. M. & Korstjens, A. H. (2000). Female dispersal and infanticide avoidance in primates. In *Infanticide by Males and its Implications*, ed. C. P. van Schaik & C. H. Janson, pp. 293–321. Cambridge: Cambridge University Press.

Sterling, E. J. & Richard, A. F. (1995). Social organization in the aye-aye (*Daubentonia madagascariensis*) and the perceived distinctiveness of nocturnal primates. In *Creatures of the Dark. The Nocturnal Prosimians*, ed. L. Alterman, G. A. Doyle & M. K. Izard, pp. 439–51. New York: Plenum Press.

Taber, A. B. & Macdonald, D. W. (1992). Spatial organization and monogamy in the mara, *Dolichotis patagonum*. *Journal of Zoology*, **227**, 417–38.

Tardif, S. D. (1994). Relative energetic costs of infant care in small bodied Neotropical primates and its relation to infant-care patterns. *American Journal of Primatology*, **34**, 133–43.

Temrin, H. & Arak, A. (1989). Polyterritoriality and deception in passerine birds. *Trends in Ecology and Evolution*, **4**, 106–9.

Travis, S. E., Slobodchikoff, C. N. & Keim, P. (1995). Ecological and demographic effects on intraspecific variation in the social system of prairie dogs. *Ecology*, **76**, 1794–1803.

(1996). Social assemblages and mating relationships in prairie dogs: a DNA fingerprint analysis. *Behavioral Ecology*, **7**, 95–100.

Treesucon, U. & Raemaekers, J. J. (1984). Group formation in gibbon through displacement of an adult. *International Journal of Primatology*, **5**, 387.

Trivers, R. L. (1972). Parental investment and sexual selection. In *Sexual Selection and the Descent of Man 1871–1971*, ed. B. Campbell, pp. 136–79. Chicago: Aldine Press.

Tullberg, B. S., Ah-King, M. & Temrin, H. (2002). Phylogenetic reconstruction of parental-care systems in the ancestors of birds. *Philosophical Transactions of the Royal Society of London, Series B*, **357**, 251–7.

van Rhijn, J. G. (1991). Mate guarding as a key factor in the evolution of parental care in birds. *Animal Behaviour*, **41**, 963–70.

van Schaik, C. P. (1983). Why are diurnal primates living in groups. *Behaviour*, **87**, 120–44.

(2000*a*). Infanticide by male primates: the sexual selection hypothesis revisited. In *Infanticide by Males and its*

Implications, ed. C. P. van Schaik & C. H. Janson, pp. 27–60. Cambridge: Cambridge University Press.

(2000*b*). Vulnerability to infanticide by males: patterns among mammals. In *Infanticide by Males and its Implications*, ed. C. P. van Schaik & C. H. Janson, pp. 61–72. Cambridge: Cambridge University Press.

van Schaik, C. P. & Dunbar, R. I. M. (1990). The evolution of monogamy in large primates: a new hypothesis and some crucial tests. *Behaviour*, **115**, 30–62.

van Schaik, C. P. & Hörstermann, M. (1994). Predation risk and the number of adult males in a primate group: a comparative test. *Behavioral Ecology and Sociobiology*, **35**, 261–72.

van Schaik, C. P. & Janson, C. H. (ed.) (2000). *Infanticide by Males and its Implications*. Cambridge: Cambridge University Press.

van Schaik, C. P. & Kappeler, P. M. (1993). Life history, activity period and lemur social systems. In *Lemur Social Systems and their Ecological Basis*, ed. P. M. Kappeler & J. Ganzhorn, pp. 241–60. New York: Plenum Press.

(1997). Infanticide risk and the evolution of male–female association in primates. *Proceedings of the Royal Society of London, Series B*, **264**, 1687–94.

van Schaik, C. P. & van Hooff, J. A. R. A. M. (1996). Toward an understanding of the orangutan's social system. In *Great Ape Societies*, ed. W. C. McGrew, L. Marchandt & T. Nishida, pp. 3–15. Cambridge: Cambridge University Press.

van Schaik, C. P. & van Noordwijk, M. A. (1989). The special role of male *Cebus* monkeys in predation avoidance and its effect on group composition. *Behavioral Ecology and Sociobiology*, **24**, 265–76.

Veiga, J. P. (1992). Why are house sparrows predominantly monogamous: a test of hypotheses. *Animal Behaviour*, **43**, 361–70.

(2000). Infanticide by male birds. In *Infanticide by Males and its Implications*, ed. C. P. van Schaik & C. H. Janson, pp. 189–220. Cambridge: Cambridge University Press.

Verner, J. (1964). Evolution of polygyny in the long-billed marsh wren. *Evolution*, **18**, 252–61.

Verner, J. & Willson, M. L. (1966). The influence of habitats on mating systems of North American passerine birds. *Ecology*, **47**, 143–7.

Vincent, A. C. J. (1995). A role for daily greetings in maintaining seahorse pair bonds. *Animal Behaviour*, **49**, 258–60.

Wagner, R. H. (1992). The pursuit of extra-pair copulations by monogamous female razorbills: how do females benefit? *Behavioral Ecology and Sociobiology*, **29**, 455–64.

Wickler, W. & Seibt, U. (1981). Monogamy in Crustacea and Man. *Zeitschrift für Tierpsychologie*, **57**, 215–34.

(1983). Monogamy: an ambiguous concept. In *Mate Choice*, ed. P. Bateson, pp. 33–50. Cambridge: Cambridge University Press.

Williams, M. & McKinney, F. (1996). Long-term monogamy in a river specialist – the Blue duck. In *Partnerships in Birds: The Study of Monogamy*, ed. J. M. Black, pp. 73-90. Oxford: Oxford University Press.

Williams, T. D. (1996). Mate fidelity in penguins. In *Partnerships in Birds: The Study of Monogamy*, ed. J. M. Black, pp. 268–85. Oxford: Oxford University Press.

Wittenberger, J. F. & Tilson, R. L. (1980). The evolution of monogamy: hypothesis and evidence. *Annual Review of Ecology and Systematics*, **11**, 197–232.

Wolf, L., Kettterson, E. D. & Nolan, V. Jr. (1988). Paternal influence on growth and survival of dark-eyed junco young: do parental males benefit? *Animal Behaviour*, **36**, 1601–18.

Woodroffe, R. & Vincent, A. (1994). Mother's little helpers: patterns of male care in mammals. *Trends in Ecology and Evolution*, **9**, 294–7.

Wrangham, R. W. (1980). An ecological model of female-bonded primate groups. *Behaviour*, **75**, 262–300.

Wright, P. C. (1990). Patterns of paternal care in primates. *International Journal of Primatology*, **11**, 89–102.

Wundt, W. (1894). *Lectures on Human and Animal Psychology*. London: Swan Sonnenschein.

Wynne-Edwards, K. E. (1987). Evidence for obligate monogamy in the Djungarian hamster, *Phodopus campbelli*: pup survival under different parenting conditions. *Behavioral Ecology and Sociobiology*, **20**, 427–37.

PART I

Evolution of social monogamy

CHAPTER 2

The evolution of monogamy: mating relationships, parental care and sexual selection

Anders Pape Møller

INTRODUCTION

Monogamy is defined as a unique social relationship between one adult female and one adult male for the purpose of reproduction. Monogamy is a mating system particularly common in birds, but also occurring infrequently in invertebrates, fish, amphibians, reptiles, and mammals. Polygyny or promiscuous mating systems appear to be predominant in animals. Why should that be the case? And why is monogamy so rare?

Using birds as a model system, in this chapter I explore a number of different selection pressures that may shape monogamy. The reason for restricting the analyses to birds is that comparative data of a similar extent are simply unavailable for any other class of animals. However, whenever possible, I refer to other taxa (i.e., when data are available).

While the close spatial co-occurrence of a single male and a single female during an extended period of time is the pre-eminent criterion for defining monogamy, a number of other features characterize some, but not necessarily all, monogamous mating systems: (i) a social relationship between a male and a female; (ii) male provisioning of parental care; and (iii) a reduced sexual dimorphism compared with polygynous mating systems. The second feature involves parental investment by an adult male in offspring at the expense of mating effort, suggesting that male care, as opposed to pursuit of additional females, is a superior way of creating fitness returns. The third feature suggests that if males indeed differ less from females in socially monogamous mating systems, then the intensity of sexual selection is less under monogamy compared with polygyny. While many cases of extravagant sexual dimorphism in species such as pheasants, grouse, hummingbirds, and birds of paradise are associated with extreme polygyny, there are also cases of extreme dimorphism in monogamous species, such as the quetzal (*Pharomachrus mocinno*) and many other species. Indeed, Møller (1986), in a preliminary analysis of sexual dichromatism in European passerine birds, found little evidence of differences between monogamous and polygynous species.

Recently, the selection pressures that shape mating systems and the phenotypes of males and females involved in these mating systems have been clearly identified. While the social mating system is but one component of sexual selection, later stages of the reproductive events of males and females may exert just as much or even more important selection pressures (Figure 2.1; Møller, 1994). Thus, independent of the social mating system, sexual selection associated with fertilization, abortion, infanticide, parental investment, and differential parental investment may shape the evolution of sexual dimorphism and comprise additional layers of complexity on top of the social mating system. These sequential selection events may also interact with each other (in the statistical sense), implying that, for example, the relationship between a male trait and social mating success may become even stronger after fertilization. Hence, an understanding of the selection pressures that shape individuals within a given mating system depends upon estimates of these component parts and their interactions.

The aims of this chapter are to explore the role of extra-pair paternity (EPP) in the evolution of social monogamy, where social monogamy (henceforth monogamy) refers to the social, but not necessarily the sexual relationship between a male and a female. While neither the number of social nor of sexual partners may have any fitness consequences if they do not result in fertilizations, EPP adds a second layer of complexity to the analysis of selection pressures because it comprises the level closest to fitness. Before entering this

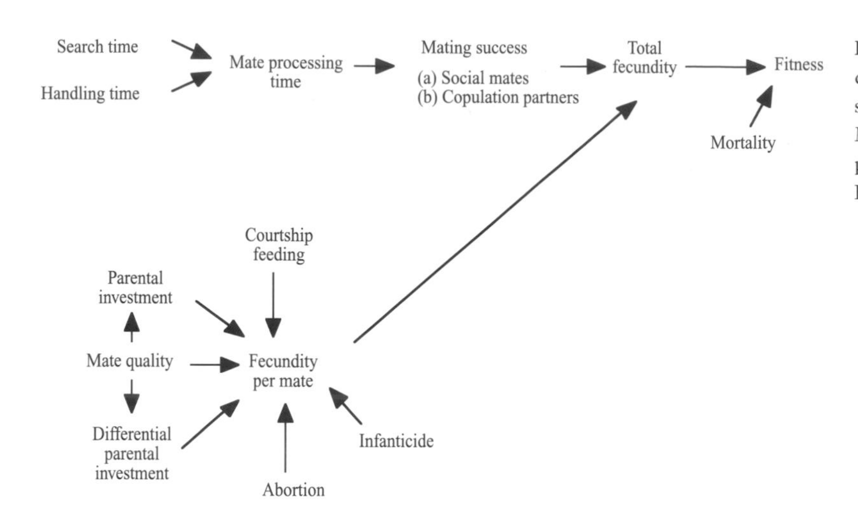

Figure 2.1. Mating system described as a sequence of sexual selection events. Adapted from Møller (1994). Reprinted by permission of Oxford University Press.

exploration, I will briefly review different hypotheses for the evolution of monogamy and discuss these in relation to EPP.

KINDS OF MONOGAMY

Evolution of a mating system

Monogamy has been the focus of intensive research and numerous hypotheses have been suggested. Wittenberger and Tilson (1980), in their influential review of monogamy, identified three main categories of hypotheses explaining the evolution of this mating system: (i) male care is essential and only the presence of a male and a female can ensure the survival of offspring; (ii) failed polygynists because the polygyny threshold has not been reached; and (iii) enforced monogamy by males monopolizing females (alternatively females may use aggression to prevent other females from settling: Gowaty, 1996).

Previous comparative work on birds by Birkhead and Møller (1996) suggested that many species may have a monogamous mating system because male care is essential. Species that are monogamous because males are failed polygynists are also common, while enforced monogamy was not identified. Birkhead and Møller (1996) also showed that extra-pair paternity was uncommon in species where male care is essential, although this analysis was not corrected for similarity among species due to common ancestry.

Extra-pair paternity

What is extra-pair paternity (EPP)? As defined by Birkhead and Møller (1992), EPP is the proportion of offspring fathered by males other than the primary male in a reproductive unit (the status of a male was defined based on his dominance rank). Although this was an attempt to devise a broad definition based on sperm competition theory, it disregards any obscuring effects of the social mating system. Thus, in a socially monogamous species, EPP would simply be the proportion of all offspring that are fertilized by males from other reproductive units. In a cooperative breeder, EPP would be the proportion of offspring fathered by males other than the alpha male, as members either of the reproductive unit or of other units. Similarly, in a polyandrous species, EPP would be the proportion of offspring fertilized by males other than the focal male.

This definition works well when the focus is on sperm competition. However, when the focus shifts to sexual selection, things may change. In species with cooperative breeding systems, EPP should be defined as the percentage of offspring sired by males other than the attending males in the reproductive unit. These local males will obviously compete for fertilizations, but this competition is not necessarily related to sexual selection. The reason is that the variance in reproductive success among males in a social group is constrained by a single female (or, rarely, a couple of females) laying the eggs. In contrast, if males are competing for reproductive success among a large number of females, irrespective of membership of particular social groups, there is virtually no upper bound to the variance in success. Hence, males competing for fertilizations within social groups will not increase the variance in success, or increase it by only a very small amount, resulting in no or only very weak sexual selection. Using this definition, dunnocks (*Prunella modularis*) have a high degree of shared paternity in polygynandrous groups, but no EPP in monogamous groups (Burke *et al.*, 1989), suggesting that mixed paternity in groups of this cooperative breeder is mainly caused by male competition for fertilizations within groups and hence is unrelated or weakly related to sexual selection. The same applies to many other species of 'cooperative' breeders. However, there are exceptions such as the superb fairy wren (*Malurus cyaneus*) in which almost all paternity is contributed by extra-group males with a phenotype preferred by females, while within group males hardly account for any paternity (Dunn & Cockburn, 1999). If sexual selection due to a female preference for a particular male phenotype was driving multiple matings by females, we should also expect female dunnocks in monogamous groups to engage in extra-pair copulations. They do not. However, female superb fairy wrens engage in extra-pair copulations almost independent of group composition.

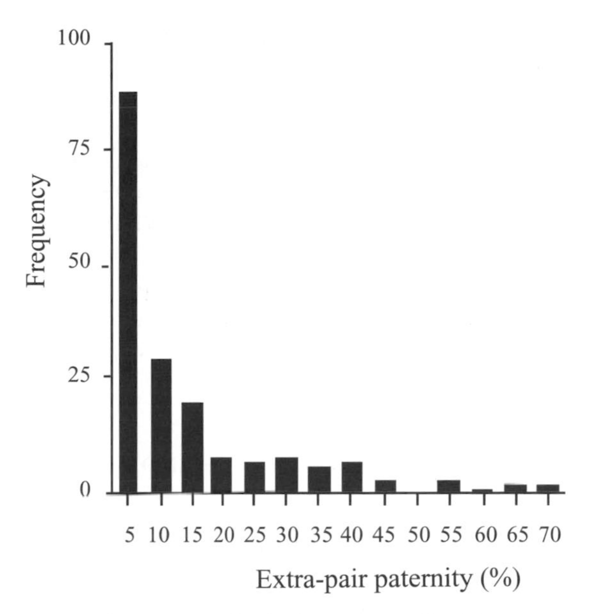

Figure 2.2. Frequency distribution of EPP in 185 bird species.

Extra-pair paternity, social competition, and sexual selection

While many cooperative species may seem marginally relevant for a chapter on social monogamy, many cooperative breeders do have social units that consist of a single male and a single female. Even in this situation, EPP can be clearly categorized, as described above. Thus, there appear to be at least two different kinds of EPP. This distinction has important implications for studies of paternity in primates and other taxa. First, any intraspecific study should attempt to determine whether EPP is due to social competition within or among groups. Second, any attempt to understand interspecific variation in paternity in birds as well as in other taxa will have to distinguish between these two kinds of EPP. In the following I will only consider the second kind, which seems to be more clearly related to sexual selection in socially monogamous species.

Extra-pair paternity variability

EPP is extremely variable within and among species. For example, studies of paternity in the willow warbler (*Phylloscopus trochilus*) have revealed estimates of 0.0%, 27.94%, and 33.03% in three studies with reasonably large sample sizes (Gyllensten *et al.*, 1990; Bjørnstad & Lifjeld, 1997; Fridolfsson *et al.*, 1997). In the species where at least two estimates have been obtained, we can assess the reliability of single estimates by calculating the repeatability of multiple estimates. EPP estimates had a highly significant repeatability (Falconer & Mackay, 1996)of 0.68 (Petrie *et al.*, 1998), even though several species, including the willow warbler, showed marked intraspecific variation in EPP. Thus, these estimates provide reliable information about species-specific features, although this clearly does not imply that intraspecific variation is uninteresting since several hypotheses have been put forward (Griffith, 2000; Møller, 2001).

During the last 15 years, EPP has been emphasized as an important component of mating success. The frequency ranges from complete absence in 53 out of 185 species (28.6% of species) to 67% in the superb fairy wren and 65% in the tree swallow (*Tachycineta bicolor*). A staggering 21.1% have more than 20% EPP among their offspring (Figure 2.2). The mean value among all 185 was 11.51% (SE = 1.12), and among the species with any EPP it was 16.14% (SE = 1.37, $N = 132$). What determines this tremendous variation?

Variation in EPP among males causes variation in reproductive success, and this variance may provide an estimate of the opportunity for sexual selection. This opportunity for sexual selection can simply be estimated as

$$\left(s_1^2/\text{mean}_1^2\right)/\left(s_2^2/\text{mean}_2^2\right) \quad (2.1)$$

where s_1^2 is the variance in male success taking EPP into account (i.e., real reproductive success) and $mean_1$ is the mean value, while s_2^2 is the variance in male success without taking EPP into account (i.e., apparent reproductive success) and $mean_2$ is mean male apparent success. When there is no EPP, this variance ratio is 1. As the variance in male success due to EPP increases, so does the variance ratio. This measure of increase in variance due to EPP was on average 3.593 (SE = 0.611) in 27 species with non-zero EPP (author's unpublished data). This implies that EPP caused a 3.6-fold increase in variance in success.

RELATIONSHIP BETWEEN EPP AND SOCIAL MATING SYSTEMS

Why should EPP be related to mating systems? EPP may increase with the evolution of polygyny because there is a trade-off between male effort spent on attracting additional females and male defence of paternity with already attracted females (Hasselquist & Sherman, 2001). Alternatively, we may expect a reduction in EPP among polygynously mated males if females prefer such males due to the indirect fitness benefits that they provide. Hence, it seems unlikely that females mated to polygynous males would seek extra-pair copulations with additional males for indirect benefits (Hasselquist & Sherman, 2001).

Intraspecific studies have generally shown increased EPP in polygynously compared with monogamously mated males, with an overall small effect size (Cohen (1984) defined small effects as having a mean Pearson $r = 0.10$, intermediate effects a mean $r = 0.30$, and large effects a mean $r = 0.50$) (mean Pearson correlation coefficient adjusted for sample size between paternity and polygyny across 12 studies $r = -0.119$: Møller & Ninni, 1998). The empirical evidence at the interspecific level suggests that species with a high frequency of polygyny have less EPP than sister species with a monogamous mating system (11% vs. 23%: Hasselquist & Sherman, 2001). The analysis in the Hasselquist and Sherman study was based on only 40 species. However, the more extensive analysis conducted for 185 species in the study presented here showed a different pattern: a regression based on the raw data revealed no significant relationship between EPP and the frequency of polygyny (Figure 2.3; $F = 0.95$, d.f. = 1, 183, $r^2 = 0.005$, $P = 0.332$). An analysis based on statistically independent contrasts revealed a very similar result.

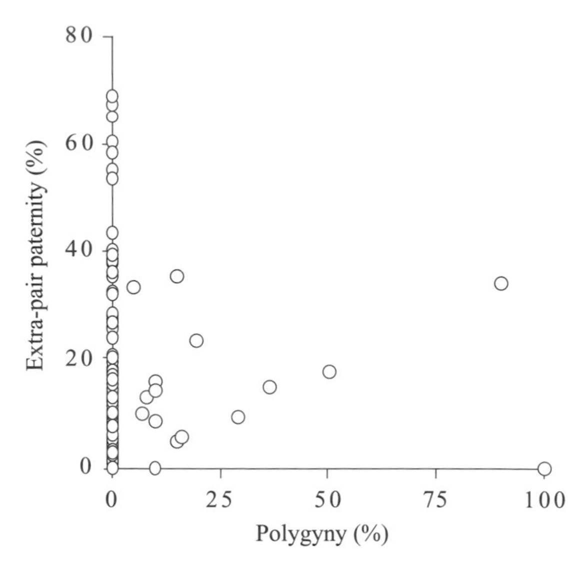

Figure 2.3. Frequency of EPP (%) in relation to frequency of polygyny (% males polygynous) in 185 species of birds.

In conclusion, there is very little evidence to suggest that EPP is more common in highly polygynous (and hence less common in monogamous) species of birds.

RELATIONSHIP BETWEEN MONOGAMY, EPP, AND SEXUAL DISPLAYS

Females are often constrained in their mate choice because a very attractive male cannot mate with all females. Such constraints are particularly severe in monogamous mating systems, but can also be significant even in lek mating systems (Møller, 1992). This effect of restrictions on the level of mate choice was demonstrated experimentally by Møller (1988), who showed that female barn swallows (*Hirundo rustica*) mated to males that had become attractive by elongation of their outermost tail feathers were less likely to engage in extra-pair copulations than control females, while females mated to males that had their tail feathers shortened were more likely to participate in extra-pair copulations. A subsequent experiment showed that this actually resulted in differences in paternity among treatments (Saino *et al.*, 1997*a*).

What is the evidence that secondary sexual characters of males affect the level of extra-pair paternity? At the intraspecific level, Møller and Ninni (1998) found a mean correlation coefficient adjusted for sample size between paternity and male ornamentation of 0.341 across 17 studies. Thus, EPP is more common when males have small secondary sexual characters.

At the interspecific level, Møller and Birkhead (1994) investigated the relationship between EPP and sexual dichromatism, under the assumption that the expression of secondary sexual characters would reflect past pressures of sexual selection. They found a significant positive correlation, even when controlling for similarity due to common descent. A multiple regression with EPP and polygyny as independent predictors of sexual dichromatism showed that EPP explained 14.8% of the variance in sexual dichromatism, while polygyny only accounted for 0.0004% of the variance. In a subsequent analysis, Dunn *et al.* (2001) concluded that the social mating system was more important than the sexual mating system. They based their calculations on residual testes mass as an indicator of the level of sperm competition. However, this is a poor predictor of sperm competition since residual testes mass explains only 12.8–22.8% of the interspecific variance in EPP (Møller & Briskie, 1995). Using such a poor predictor to make conclusions about the relative importance of selection pressures seems unwarranted. In addition, testes produce more than sperm. A comparative analysis of circulating testosterone levels in birds has shown that relative testes mass is a reasonable predictor of testosterone concentration, particularly in relationship to the degree of EPP, but also to polygyny (Møller *et al.*, unpublished data). Hence, correlations between sexual dichromatism and relative testes mass and polygyny, respectively, may arise for reasons other than sperm competition.

RELATIONSHIP BETWEEN MONOGAMY, EPP, AND MALE PARENTAL CARE

Organisms have been selected for differential allocation of parental effort to their own offspring and discrimination against the offspring of other individuals. Theoretical models suggest that under certain circumstances males may be selected to discriminate against extra-pair offspring and provide care only for their own descendants. This requires that males adjust their level of care, and that they somehow assess any extra-pair sexual activity of their own mate.

Many studies have tested whether males discriminate against broods with high levels of EPP and most have found no significant evidence for such an effect. However, this conclusion should be drawn with care because sample sizes are often very small, thus rendering very low the validity of the statistical tests. Two studies have investigated whether males adjust their level of care when the level of EPP changes between broods. Males are usually consistent in their level of EPP among broods, but sometimes levels change. Dixon *et al.* (1994) showed that for the reed bunting (*Emberiza schoeniclus*), males that experienced an increase in EPP actually decreased their feeding rate, suggesting that they were able to discriminate between broods with high and low levels of EPP. Møller and Tegelström (1997) provided similar data for the barn swallow. Further studies are needed to determine whether this is a general finding.

Interspecific relationships between male parental care and EPP have shown a negative covariation for some kinds of care, but not for others. Møller and Cuervo (2000) analysed a large data set for birds and showed that male feeding of offspring was reduced in response to an evolutionary increase in EPP. However, this was not the case for male nest building, male courtship feeding, or male share of incubation. In a second comparative analysis, Møller (2000) provided evidence that in species where female reproductive success was significantly reduced when her partner was removed, EPP was infrequent. When male presence had little effect on female reproductive success, EPP was common. The findings of these comparative analyses do not imply that extensive paternal care and high levels of EPP cannot co-occur in monogamous species. They do! Examples among birds include the reed bunting, the superb fairy wren and many other species with extensive male care; similar examples can be found in other taxa (e.g., Fietz *et al.*, 2000). However, the general trend is for EPP to be rare in monogamous species with extensive male parental care.

Direct fitness benefits are the simplest explanation for female mate preferences for particular male phenotypes. Although theoreticians have assumed this hypothesis to be self-evident, the empirical support is far from clear. An extensive review of direct fitness benefits by Møller and Jennions (2001) found that for male feeding of offspring, male secondary sexual characters only explained 1.3% of the variance, while male traits explained 6.3% of the variance in fertility, 2.3% of the variance in fecundity, and 23.6% of the variance in hatching rate in male-guarding ectotherms. Thus, direct benefits seem weak and perhaps unimportant when assessment is based on male secondary sexual characters, the single exception being hatching rate in guarding ectotherms.

Should males signal their direct benefits? Males could either signal their parental care ability honestly, or, if they are particularly attractive to females, they could reduce their level of parental care since females invest differentially in reproduction in cases of exceptionally attractive males (Kokko, 1998). Differential parental investment occurs when a female's investment depends on the phenotype of her partner (Burley, 1986). The optimal strategy for a male depends on the marginal gains from EPP (or polygyny). Males should honestly advertise their parenting ability when the opportunities for extra-pair copulations are minimal. When males have ample opportunities for extra-pair copulations, differential parental investment should evolve. Møller and Thornhill (1998) showed that the most attractive males provided the most parental care in some species of birds, while they provided the least in others. The intraspecific relationship between male attractiveness and male care was significantly correlated with EPP, so that the most attractive males in species with little EPP provided the most care, while the situation was reversed in species with frequent EPP.

In conclusion, intraspecific and interspecific patterns of EPP and paternal care suggest that male care interacts with sexual selection, allowing males to reduce their care and increase EPP when females invest differentially, but forcing males to invest in offspring and forgo extra-pair copulations when male care is essential.

RELATIONSHIP BETWEEN EPP, MONOGAMY, AND SYNCHRONY

Which ecological conditions favour EPP? Stutchbury and Morton (1995) suggested that temporal synchrony in reproduction would facilitate females seeking extra-pair copulations because more potential copulation partners would be available. This hypothesis was extended to explain geographical patterns of paternity in birds, including apparently low levels of EPP in the tropics (Stutchbury & Morton, 2001).

What are the intraspecific relationships between EPP and synchrony? Møller and Ninni (1998) reviewed 12 studies and found that the degree of EPP actually decreased significantly when synchrony increased (mean correlation coefficient adjusted for sample size between EPP and synchrony across 12 studies $r = -0.086$). This finding is counter to that predicted by Stutchbury and Morton (1995).

What are the interspecific relationships between EPP and synchrony? Stutchbury (1998; Stutchbury & Morton, 2001) analysed interspecific patterns of EPP in relation to a synchrony index and found that species with higher degrees of synchrony had more EPP. Tropical species were less synchronous than temperate zone species and their levels of EPP were low. However, these analyses do not consider causation. EPP could drive the evolution of synchrony, or synchrony could promote the evolution of EPP (the latter hypothesis is that favoured by Stutchbury and Morton, 2001). In a comparative analysis, Spottiswoode and Møller (2003) found that EPP changed in association with synchrony. Which alternative explanations could account for this finding? One possibility is that females choose extra-pair partners when the quality of those partners can be readily distinguished. A simple mechanism for distinguishing high quality from low quality males is subjecting the males to severe physical tests: for example, bird migration. A male that can successfully complete a long-distance migration of thousands of kilometres ahead of his competitors is bound to be of better quality. Spottiswoode and Møller (2003) tested this hypothesis by including migration distance in the comparative analyses and found that once this variable was controlled statistically, there was no additional effect of breeding synchrony. Thus, migration appears to be a better predictor of EPP than breeding synchrony.

In conclusion, according to Stutchbury and Morton (2001), there is a positive correlation between breeding synchrony and EPP in birds. However, their analyses do not control for potentially confounding variables, neither do they take into account the order of evolutionary events. The intraspecific patterns of EPP and synchrony, and the analyses based on their relationships (after accounting for migratory status) suggest that synchrony does not drive extra-pair paternity in birds.

RELATIONSHIP BETWEEN MONOGAMY, EPP, AND LIFE HISTORY

Longevity may be an important predictor of EPP for several reasons. Mauck *et al.* (1999) used a simple dynamic programming model to investigate the relationship between EPP and life history variables under the assumption that adult survival should influence trade-offs in reproductive effort. Level of paternity, ability to assess paternity, and adult survival rate were investigated to quantify male tolerance of reduced paternity of a brood. Adult survival had the greatest influence on male decisions, so that for any given level of cost of reproduction and value of male parental care, tolerance of EPP decreased with increasing male survival rate. The optimal male response also depended on the ability of males to assess reliably the probability of paternity.

However, survival and longevity may also be important for a second reason: accumulation of mutations should be more important in long-lived species (e.g., Medawar, 1952), and the negative effects of mutations may be particularly costly in monogamous pairs because the failure of a partner to provide her or his share of parental care may be impossible for the other partner to compensate. Thus, particularly in species that were long-lived, we could expect that females would seek extra-pair copulations to ensure that their eggs were fertilized by the most viable sperm. Hence, different predictions can be made concerning EPP and survival, opening up the possibility of conducting tests that can distinguish between alternative explanations for extra-pair paternity. However, as just described, we may also consider the possibility that females are subject to conflicting selection pressures from different sources, with no clear prediction arising.

The empirical evidence testing this hypothesis is as follows. Using a database with 71 bird species, Wink and Dyrcz (1999) showed that EPP was negatively related to the probability of survival of both pair members. They suggested that if the probability of both pair members surviving is high, then they are likely to gain mutual benefits from breeding together, thus reducing the incentive for seeking extra-pair copulations. Using a much more extensive data set, I found that a regression of EPP on survival rate revealed a negative relationship, explaining more than 14% of the variance (Figure 2.4*a*; $F = 24.70$, d.f. = 1,146, $r^2 = 0.145$, $P < 0.001$, slope (SE) = -0.445 (0.090)). However, life history traits show strong effects of allometry, large-sized species having small clutch sizes and high survival rates. What we want to test is really a question of survival rate after accounting for interspecific variation in other confounding variables. After removing the allometric effect of body mass from the survival estimate, this relationship between survival rate and EPP completely vanished (Figure 2.4*b*; $F = 0.74$, d.f. = 1,134, $r^2 = 0.005$, $P = 0.393$, slope (SE) = -0.113 (0.131)). Hence, there is little evidence consistent with the hypothesis that there is a positive link between longevity and EPP.

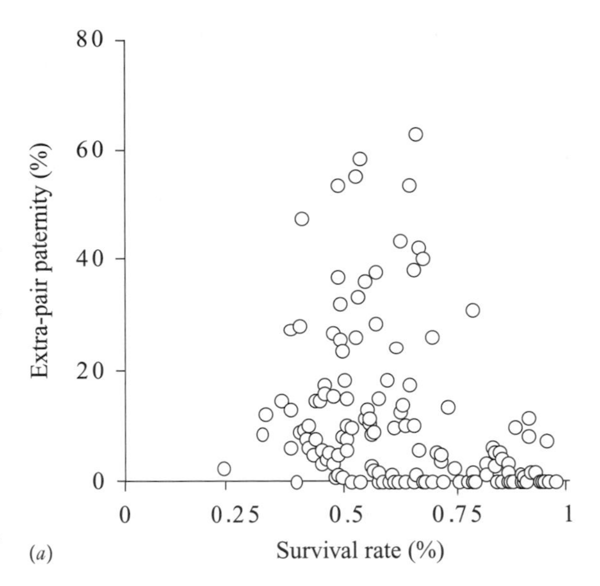

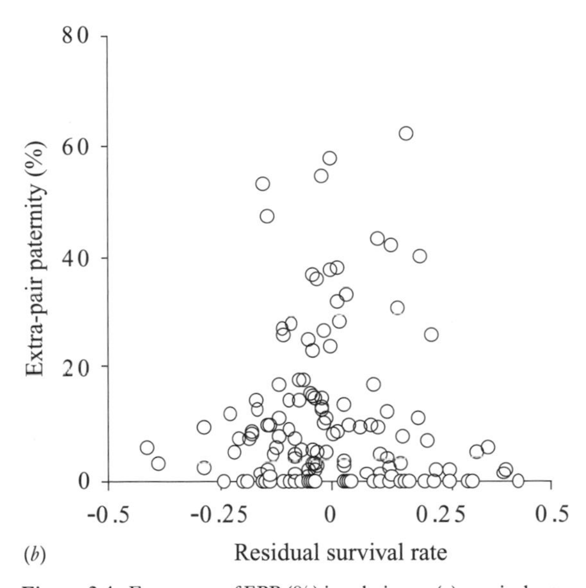

Figure 2.4. Frequency of EPP (%) in relation to (a) survival rate (%) and (b) residual survival rate from a regression of arcsine-transformed survival rate on log-transformed body mass. Data for 136 species of birds.

All life history traits are intercorrelated (Roff, 1992; Stearns, 1992). Hence, we may even have to control for covariation between life history traits in order to test adequately the association between EPP and survival. Which variable is the most important in explaining interspecific variation in EPP? A multiple stepwise regression analysis revealed that only the relative duration of the nestling period and the period until independence explained significant amounts of variation in EPP, species with high levels of EPP having relatively short nestling periods and periods until independence. In addition, survival rate did not even begin to enter the picture as a significant predictor (A. P. Møller and C. Spottiswoode, unpublished data). This was also the case when the analyses were based on statistically independent contrasts that control for similarity among taxa due to common ancestry.

In conclusion, there is little evidence to suggest that survival or longevity affects the evolution of EPP in birds.

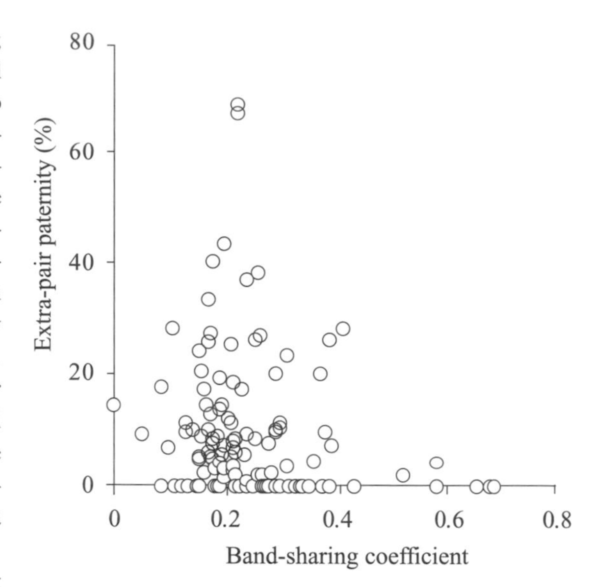

Figure 2.5. Frequency of EPP (%) in relation to band-sharing coefficients among unrelated adults. Data for 117 species of birds.

RELATIONSHIP BETWEEN MONOGAMY, EPP, AND GENETIC VARIATION

Why is genetic variation important in discussions of EPP? Females may obtain indirect fitness benefits from their mate choice in terms of genetic advantages among offspring in the following generation. However, such benefits may be illusive if strong directional preferences by females cause beneficial genetic variants to go to fixation. Hence the lek paradox which is based on the puzzling observation of unanimous female mate preferences for genetic benefits when there should not be much genetic variance in fitness (Taylor & Williams, 1982; Charlesworth, 1987). Surprisingly, there is a small, but highly significant additive genetic component of fitness (Burt, 1995). How is this possible? A comparative study of birds showed that species with more intense sexual selection as reflected by EPP tended to have more heterozygosity in RFLPs (Restriction Fragment Length Polymorphism) than closely related species with less intense sexual selection. EPP was also positively correlated with allozyme polymorphism across a large number of species (Petrie *et al.*, 1998). A second data set on genetic variation, measured as the band-sharing coefficient among supposedly unrelated adult individuals in populations of birds likewise suggested genetic variation and EPP to be positively correlated, although this relationship was no longer significant after controlling for similarity due to common descent (Figure 2.5; Spottiswoode & Møller, unpublished data). How can genetic variation be maintained in the presence of strong directional selection imposed by females? Møller and Cuervo (unpublished data) suggested that a mechanism of male-biased mutations associated with fidelity of replication, repair, or recombination could increase genetic variation when sexual selection through sperm competition was the evolutionary process. Consistent with this prediction, Møller and Cuervo (unpublished data) found that minisatellite mutation rates were positively correlated with EPP across birds. In addition, the order of evolutionary events for genetic variation and EPP suggested that EPP changed before genetic variability, implying that sexual selection through sperm competition is the evolutionary motor generating genetic variation.

The geographical patterns of genetic variation also have important implications for the interpretation of EPP in relation to monogamy and breeding synchrony. In a comparative analysis of birds, genetic variation was significantly lower in the tropics than in the temperate zones, even when controlling for similarity among species due to common descent (Spottiswoode & Møller, unpublished data). Migration and migratory distance covaried positively with genetic variation, implying that migrants have more genetic variation than residents

(Spottiswoode & Møller, unpublished data). This is as expected since migrants generally have longer dispersal distances than residents, resulting in more long-distance gene flow in migrants than in residents (Belliure *et al.*, 2000). Thus, the lower frequency of EPP in the tropics observed by Stutchbury and Morton (2001) could be a consequence of lower genetic variation rather than a lower degree of breeding synchrony.

DISCUSSION

Extra-pair paternity has traditionally been interpreted as a homogeneous phenomenon arising from sperm competition (competition among males for the fertilization of eggs of a single female during a single reproductive event). However, the data presented here suggest that EPP may in fact include at least two different phenomena; one arising from within-group dynamics relating to social competition among multiple males in a single breeding unit (so-called cooperative breeding) and a second arising from among-group dynamics resulting from sexual selection. Here, I will consider only the second category, which seems to be the most common in socially monogamous birds. This does not imply that the first category is uninteresting. For example, Hartley and Davies (1994) have discussed factors determining partitioning of paternity within social groups, and Shellman-Reeve and Reeve (2000) have used reproductive skew theory as an explanation for EPP in birds. Thus, reproductive skew theory might be a unifying concept for analysing the two categories of EPP, their relative importance, and their causes and consequences.

Extra-pair paternity and sexual selection

Extra-pair paternity is an important component of sexual selection because the variance in reproductive success among males is considerably increased over the variance in apparent reproductive success. Although EPP may arise for random reasons (Reyer *et al.*, 1997), this seems to be an unlikely explanation for the vast amount of variation found among populations and species. Furthermore, species are highly consistent in levels of EPP across populations. Hence, in the following I will assume that EPP arises for biological reasons.

Sexual selection may, among other factors, arise from variance in social mating success (Figure 2.1). There has been controversy over the relative importance of polygyny and EPP as evolutionary correlates of secondary sexual characters. The only study addressing this issue directly is that of Møller and Birkhead (1994), who showed that EPP accounted for *c.* 15% of the variance in sexual dichromatism, while polygyny accounted for hardly any. Dunn *et al.* (2001) used relative testes mass as an indirect measure of the importance of sperm competition and concluded that the social mating system was much more important than the sexual mating system. However, this conclusion seems premature given that relative testes mass explains only a very small amount of the variance in EPP (Møller & Briskie, 1995). Thus, we cannot make very reliable inferences about the relative importance of EPP based on such a measure. Hasselquist and Sherman (2001) used a small sample of passerine birds to investigate the difference in EPP between monogamous and polygynous species of birds and found evidence of a higher frequency in monogamous species. An analysis based on the much larger sample of 185 species revealed no clear relationship (Figure 2.3). Hence, the social and sexual mating systems appear to be unrelated. This raises the question of which factors determine interspecific variation in monogamous social and sexual mating systems.

Clearly, both the degree of polygyny and EPP are important components of sexual selection. Numerous studies of EPP have found evidence for strong positive correlations between secondary sexual characters and the proportion of offspring being sired by a focal male in his own nest(s), but also of the number of offspring sired elsewhere (reviews in Møller, 1998; Møller & Ninni, 1998). Experimental manipulations of secondary sexual characters demonstrate that these effects are causal (Møller, 1988; Burley *et al.*, 1996; Saino *et al.*, 1997*a*). Why should females care about the identity of the sire of their offspring? Sexual selection theory suggests that females may gain direct, material, or indirect, genetic benefits from their mate choice (Andersson, 1994).

Interspecific variations

Males should be selected to provide parental care only for offspring that they have sired, since otherwise they would be altruists. Still, males of many species with high frequencies of EPP provide paternal care. Why should that be the case? To address this question we need to quantify direct and indirect fitness benefits of sexual selection. Møller and Jennions (2001) made an extensive meta-analysis of the literature on direct fitness benefits, suggesting that the mean effect size adjusted for sample size for male care in relation to their attractiveness was small and not significantly different from zero. However, there was highly significant heterogeneity among studies, with some showing strong positive relationships and others showing strong negative relationships. This can best be understood by quantifying the relationship between the importance of male parental care and EPP. When experimental removal of males resulted in a dramatic reduction in female reproductive success, the frequency of EPP was in general very low, while species exhibiting little or no effect in response to male removal had consistently high frequencies of EPP (Møller, 2000). Thus, male care is essential for successful reproduction in some species, and in these species females are sexually faithful to their mates. In other species, where male parental care is of little importance, females often pursue an extra-pair breeding strategy and invest differentially in reproduction. In the first group of species (male care essential), this is clearly demonstrated by males signalling their parenting ability through their secondary sexual characters, while in the second category of species (male care not essential) females invest differentially in reproduction (Møller & Thornhill, 1998). It seems likely that indirect, genetic benefits accrue to choosy females in the second category. These findings are consistent with optimality models of male care in relation to EPP (Kokko, 1999). Overall, EPP is associated with a reduction in male parental care, mainly in terms of offspring provisioning (Møller & Cuervo, 2000), and this even seems to be the case when considering the decisions of individual males in particular species (Dixon *et al.*, 1994; Møller & Tegelström, 1997).

Ecology and evolution

The ecological and evolutionary factors associated with high levels of EPP must be identified and investigated to understand the basis for interspecific differences in sexual selection. Breeding synchrony has been proposed as one such factor (Stutchbury & Morton, 1995, 2001). Synchrony may be important because it allows females to choose among a larger number of potential sires. Although interspecific and geographic patterns of EPP are consistent with this hypothesis, there are also several problems. First of all, females seem to benefit from breeding asynchrony when making a mate choice because a desired sire will then be free to copulate with an additional female without risking the loss of paternity in his own nest(s) (Saino *et al.*, 1997*b*). In the barn swallow, females seem to breed asynchronously with desired sires as a strategy to facilitate EPP (Saino *et al.*, 1997*b*). A positive association between EPP and asynchrony is also the general intraspecific pattern in birds (Møller & Ninni, 1998). Second, the geographical association between breeding asynchrony and latitude does not hold when potentially confounding variables are controlled (Spottiswoode & Møller, 2003). The association between breeding synchrony and EPP disappears when migratory behaviour is controlled statistically. Migration may be important for EPP because males are clearly sorted with respect to quality when forced to deal with the obstacles presented by long-distance migration.

Life history

Life history is the second factor that has been proposed to explain interspecific variation in EPP (Mauck *et al.*, 1999). Wink and Dyrcz (1999) found that EPP was negatively associated with survival rate, as predicted by a dynamic programming model (Mauck *et al.*, 1999). Again, the interpretations and predictions are not as simple as they may seem. Life history traits are generally intercorrelated, and relationships with a single variable do not provide strong evidence when other potentially confounding variables are not controlled. With respect to survival and EPP, the relationship vanishes when the association between survival rate and body mass is controlled statistically (Figure 2.4). This conclusion is substantiated when controlling for the effects of other life history variables.

Genetic variation

Genetic benefits cannot persist without the maintenance of genetic variation. This has traditionally been seen as the major obstacle to 'good genes' models of sexual selection (Taylor & Williams, 1982). Recent studies suggest that, in fact, genetic variation is greater in species with high levels of EPP (Petrie *et al.*, 1998; Spottiswoode & Møller, unpublished data), and that this genetic variation is maintained by greater mutation rates (Møller & Cuervo, unpublished data). Hence, sexual selection arising from EPP may tend to maintain genetic variation, thereby sustaining the evolutionary consequences of sexual selection. Good genes effects are important in sexual selection (Møller & Alatalo, 1999). The exact nature of these genetic benefits remains to be determined, but resistance to parasites appears to be a prime candidate (review in Møller, 1997; Møller *et al.*, 1999).

Social monogamy, extra-pair paternity, and paternal care

The general model developed from the analyses that relate social monogamy to EPP and paternal care is described in detail in Møller and Thornhill (1998), Petrie *et al.* (1998), Møller (2000), Møller and Cuervo (2000), and Møller and Cuervo (unpublished data) (see also Kokko, 1999). Following is a brief summary of this evolutionary scenario. Males may evolve reliable signals of phenotypic quality, particularly in monogamous mating systems because such mating systems are associated with high costs of reproduction. Initially, such signals will provide reliable information about direct fitness benefits provided to choosy females by males. However, as attractive males may increasingly benefit from copulations with other females (and hence increased mating effort) rather than providing parental care, EPP may become increasingly common. This applies particularly to species where nestling periods and periods until offspring are independent are relatively short, since these females may be more likely to raise their offspring with reduced male parental care. The ecological conditions associated with these life history traits must still be identified. The increase in EPP can have the consequence of rendering social monogamy evolutionarily unstable (Kokko, 1999). Increased variance in male reproductive success associated with EPP may also increase the germ-line mutation rate, due to an increased number of germ-line cell divisions. Since this will increase the additive genetic variance in such species, intense sexual selection due to female preferences for particular male phenotypes can be maintained for extended evolutionary periods. This scenario also helps solve the lek paradox (Taylor & Williams, 1982), which poses the question of how females can continue to choose mates, apparently entirely for genetic fitness benefits, when such strong directional selection will tend to eliminate the remaining genetic variation.

SUMMARY

The emphasis of this chapter has been strongly biased towards birds for the simple reason that comparable data do not exist for any other class of animals. However, we do know that the evolution of male parental care is associated with certainty of paternity in insects, fish, and other taxa (review in Clutton-Brock, 1991). Likewise, the ecological and social factors associated with intense sperm competition are generally known across the animal kingdom (reviews in Birkhead & Møller, 1996). However, it remains to be determined whether monogamy in, for example, primates is associated with EPP in the same way as it is in birds. Likewise, the ecological, life history, and genetic patterns described for birds in this chapter also need to be explored in other taxa. A priori, there is no reason to believe that these patterns are unique to birds. For example, the remarkable convergence between socially monogamous birds and humans in terms of sexual behaviour and sperm competition (Baker & Bellis, 1995) suggests that the patterns found in birds can be extended to other taxa.

DIRECTIONS FOR FUTURE STUDY

What are the main outstanding questions? Although great progress has been made in understanding the causes and consequences of extra-pair paternity and its relationship to monogamy, considerable work still needs to be done. The following three questions seem particularly pertinent.

1 What are the ecological conditions promoting polygyny and EPP?
2 What are the ecological conditions associated with indispensable male parental care and hence monogamy?
3 And finally, since sexual selection may increase rates of speciation, do increased mutation rates associated with EPP result in increased rates of divergence and speciation?

Acknowledgements

I would like to thank C. Boesch and U. Reichard for inviting me to make this contribution.

References

Andersson, M. (1994). *Sexual Selection*. Princeton, New Jersey: Princeton University Press.

Baker, R. R. & Bellis, M. (1995). *Human Sperm Competition*. London: Chapman & Hall.

Belliure, J., Sorci, G., Møller, A. P. & Clobert, J. (2000). Dispersal distances predict subspecies richness in birds. *Journal of Evolutionary Biology*, **13**, 480–7.

Birkhead, T. R. & Møller, A. P. (1992). *Sperm Competition in Birds: Evolutionary Causes and Consequences*. London: Academic Press.

(1996). Monogamy and sperm competition in birds. In *Partnerships in Birds: The Study of Monogamy*, ed. J. M. Black, pp. 323–43. Oxford: Oxford University Press.

Bjørnstad, G. & Lifjeld, J. T. (1997). High frequency of extra-pair paternity in a dense and synchronous population of Willow Warblers, *Phylloscopus trochilus*. *Journal of Avian Biology*, **28**, 319–24.

Burke, T. A., Davies, N. B., Bruford, M. W. & Hatchwell, B. J. (1989). Parental care and mating behaviour of polygynandrous dunnocks, *Prunella modularis,* related to paternity by DNA fingerprinting. *Nature*, **338**, 249–51.

Burley, N. (1986). Sexual selection for aesthetic traits in species with biparental care. *American Naturalist*, **127**, 415–45.

Burley, N. T., Parker, P. G. & Lundy, K. (1996). Sexual selection and extrapair fertilization in a socially monogamous passerine, the zebra finch (*Taeniopygia guttata*). *Behavioral Ecology*, **7**, 218–26.

Burt, A. (1995). The evolution of fitness. *Evolution*, **49**, 1–8.

Charlesworth, B. (1987). The heritability of fitness. In *Sexual Selection: Testing the Alternatives*, ed. J. W. Bradbury & M. B. Andersson, pp. 21–40. New York: John Wiley.

Clutton-Brock, T. H. (1991). *The Evolution of Parental Care*. Princeton, New Jersey: Princeton University Press.

Cohen, J. (1984). *Statistical Power Analysis for the Behavioral Sciences*, 2nd Edition. Hillsdale, New Jersey: Lawrence Erlbaum.

Dixon, A., Ross, D., O'Malley, S. L. C. & Burke, T. (1994). Paternal investment inversely related to degree of extra-pair paternity in the reed bunting. *Nature*, **371**, 698–700.

Dunn, P. O. & Cockburn, A. (1999). Extrapair mate choice and honest signaling in cooperatively breeding superb fairy-wrens. *Evolution*, **53**, 938–46.

Dunn, P. O., Whittingham, L. A. & Pitcher, T. E. (2001). Mating systems, sperm competition, and the evolution of sexual dimorphism in birds. *Evolution*, **55**, 161–75.

Falconer, D. S. & Mackay, T. F. C. (1996). *Introduction to Quantitative Genetics*, 4th Edition. New York: Longman.

Fietz, J., Zischler, H., Schwiegk, C., Tomiuk, J., Danzman, K. H. & Ganzhorn, J. U. (2000). High rates of extra-pair young in the pair-living fat-tailed dwarf lemur, *Cheirogaleus medius*. *Behavioral Ecology and Sociobiology*, **49**, 8–17.

Fridolfsson, A. K., Gyllensten, U. B. & Jakobsson, S. (1997). Microsatellite markers for paternity testing in the willow warbler, *Phylloscopus trochilus*: high frequency of extra-pair young in an island population. *Hereditas*, **126**, 127–32.

Gowaty, P. A. (1996). Battles of the sexes and origins of monogamy. In *Partnerships in Birds: The Study of Monogamy*, ed. J. M. Black, pp. 21–52. Oxford: Oxford University Press.

Griffith, S. C. (2000). High fidelity on islands: a comparative study of extrapair paternity in passerine birds. *Behavioral Ecology*, **11**, 265–73.

Gyllensten, U. B., Jacobsson, S. & Temrin, H. (1990). No evidence for illegitimate young in monogamous and polygynous warblers. *Nature*, **343**, 168–70.

Hartley, I. R. & Davies, N. B. (1994). Limits to cooperative polyandry in birds. *Proceedings of the Royal Society of London, Series B*, **257**, 67–73.

Hasselquist, D. & Sherman, P. W. (2001). Social mating systems and extrapair fertilizations in passerine birds. *Behavioral Ecology*, **12**, 457–66.

Kokko, H. (1998). Should advertising parental care be honest? *Proceedings of the Royal Society of London, Series B*, **265**, 1871–8.

(1999). Cuckoldry and the stability of biparental care. *Ecology Letters*, **2**, 247–55.

Mauck, R. A., Marschall, E. A. & Parker, P. G. (1999). Adult survival and imperfect assessment of parentage: effects on male parenting decisions. *American Naturalist*, **154**, 99–109.

Medawar, P. B. (1952) *An Unsolved Problem in Biology*. London: H. K. Lewis.

Møller, A. P. (1986). Mating systems among European passerines. *Ibis*, **128**, 234–50.

(1988). Female choice selects for male sexual tail ornaments in the monogamous swallow. *Nature*, **322**, 640–2.

(1992). Frequency of female copulations with multiple males and sexual selection. *American Naturalist*, **139**, 1089–101.

(1994). *Sexual Selection and the Barn Swallow*. Oxford: Oxford University Press.

(1997). Immune defence, extra-pair paternity and sexual selection in birds. *Proceedings of the Royal Society of London, Series B*, **264**, 561–6.

(1998). Sperm competition and sexual selection. In *Sperm Competition and Sexual Selection*, ed. T. R. Birkhead & A. P. Møller, pp. 55–90. London: Academic Press.

(2000). Male parental care, female reproductive success and extra-pair paternity. *Behavioral Ecology*, **11**, 161–8.

(2001). Sexual selection, extra-pair paternity, genetic variability and conservation. *Acta Zoologica Sinica*, **47**, 2–12.

Møller, A. P. & Alatalo, R. V. (1999). Good genes effects in sexual selection. *Proceedings of the Royal Society of London, Series B*, **266**, 85–91.

Møller, A. P. & Birkhead, T. R. (1994). The evolution of plumage brightness in birds is related to extra-pair paternity. *Evolution*, **48**, 1089–100.

Møller, A. P. & Briskie, J. V. (1995). Extra-pair paternity, sperm competition and the evolution of testis size in birds. *Behavioral Ecology and Sociobiology*, **36**, 357–65.

Møller, A. P. & Cuervo, J. J. (2000). The evolution of paternity and paternal care. *Behavioral Ecology*, **11**, 472–85.

(2003). Sexual selection, germline mutation rate and sperm competition. *BioMed Central Evolutionary Biology* (in press).

Møller, A. P. & Jennions, M. D. (2001). How important are direct fitness benefits of sexual selection? *Naturwissenschaften*, **88**, 401–15.

Møller, A. P. & Ninni, P. (1998). Sperm competition and sexual selection: a meta-analysis of paternity studies of birds. *Behavioral Ecology and Sociobiology*, **43**, 345–58.

Møller, A. P. & Tegelström, H. (1997). Extra-pair paternity and tail ornamentation in the barn swallow, *Hirundo rustica*. *Behavioral Ecology and Sociobiology*, **41**, 353–60.

Møller, A. P. & Thornhill, R. (1998). Male parental care, differential parental investment by females, and sexual selection. *Animal Behaviour*, **55**, 1507–15.

Møller, A. P., Christe, P. & Lux, E. (1999). Parasite-mediated sexual selection: effects of parasites and host immune function. *Quarterly Review of Biology*, **74**, 3–20.

Petrie, M., Doums, C. & Møller, A. P. (1998). The degree of extra-pair paternity increases with genetic variability. *Proceedings of the National Academy of Sciences of the USA*, **95**, 9390–5.

Reyer, H.-U., Bollmann, K., Schläpfer, A. R., Schymainda, A. & Klecack, G. (1997). Ecological determinants of extra-pair fertilizations and egg dumping in Alpine water pipits (*Anthus spinoletta*). *Behavioral Ecology*, **8**, 534–43.

Roff, D. (1992). *Life History Evolution*. New York: Chapman & Hall.

Saino, N., Primmer, C. R., Ellegren, H. & Møller, A. P. (1997*a*). An experimental study of paternity and tail ornamentation in the barn swallow (*Hirundo rustica*). *Evolution*, **51**, 562–70.

(1997*b*). Breeding synchrony and paternity in the barn swallow (*Hirundo rustica*). *Behavioral Ecology and Sociobiology*, **45**, 211–18.

Shellman-Reeve, J. S. & Reeve, H. K. (2000). Extra-pair paternity as the result of reproductive transactions between paired mates. *Proceedings of the Royal Society of London, Series B*, **267**, 2543–6.

Spottiswoode, C. & Møller, A. P. (2003). Extra-pair paternity, migration and breeding synchrony in birds. *Behavioral Ecology* (in press).

Stearns, S. C. (1992). *Life History Evolution*. Oxford: Oxford University Press.

Stutchbury, B. J. M. (1998). Breeding synchrony best explains variation in extra-pair mating system among avian species. *Behavioral Ecology and Sociobiology*, **43**, 221–2.

Stutchbury, B. J. M. & Morton, E. S. (1995). The effect of breeding synchrony on extra-pair mating. *Behaviour*, **132**, 675–90.

(2001). *Behavioral Ecology of Tropical Birds*. London: Academic Press.

Taylor, P. D. & Williams, G. C. (1982). The lek paradox is not resolved. *Theoretical Population Biology*, **22**, 392–409.

Wink, M. & Dyrcz, A. (1999). Mating systems in birds: a review of molecular studies. *Acta Ornithologica*, **34**, 91–109.

Wittenberger, J. F. & Tilson, R. L. (1980). The evolution of monogamy: hypotheses and evidence. *Annual Review of Ecology and Systematics*, **11**, 197–232.

CHAPTER 3

Mate guarding and the evolution of social monogamy in mammals

Peter N. M. Brotherton & Petr E. Komers

INTRODUCTION

The majority of hypotheses proposed to explain the evolution of social monogamy focus on the benefits of biparental care to offspring (e.g., Kleiman, 1977; Wittenberger & Tilson, 1980). This is probably for two reasons. First, most socially monogamous fish, birds and mammals show some form of biparental care; indeed, this association is so close that many early authors included biparental care in their definitions of monogamy (e.g., Lorenz, 1963, p. 167; Lack, 1968, pp. 4–5; Brown, 1975, p. 168; Wilson, 1975, p. 589). Second, given the enormous difference that normally exists between the potential reproductive rates of the sexes (especially in mammals because of female gestation and lactation: Clutton-Brock & Vincent, 1991), it is difficult to explain why the most competitive males are unable to monopolize more than one mate, unless males are constrained by paternal commitments.

Although understandable, this emphasis on the benefits of biparental care has, we believe, shifted the focus of the debate on social and/or genetic monogamy away from its original evolution and onto its current function and maintenance. In this chapter we address this imbalance by considering the origins of social monogamy in mammals. First, we attempt to reconstruct the evolutionary routes by which social monogamy could have evolved, by considering the social and parental care systems that may have been exhibited by non-monogamous ancestors. We go on to review the hypotheses that have been advanced to account for the evolution of social monogamy, and consider their applicability to a monogamous dwarf antelope, Kirk's dik-dik (*Madoqua kirkii*), as a case study. Finally, we consider how our conclusions regarding the dik-dik may be generalized to help explain the evolution of social and/or sociogenetic monogamy in other species. Throughout our discussion and analysis, we focus on mammals.

THREE ROUTES TO MONOGAMY

What were the social and parental care systems of the ancestors of socially monogamous individuals likely to have been? In other words, where did social monogamy come from? Phylogenetic analyses indicate that social monogamy is likely to have evolved either from situations where females were solitary (Komers & Brotherton, 1997) or where females lived in groups accompanied by one or more males (van Schaik & Kappeler, chapter 4). From these starting points, we identify three possible evolutionary routes to social monogamy (Figure 3.1).

Paternal investment route

In the paternal investment route, the ancestral state (Figure 3.1: 1A) comprises stable groups that contain several females accompanied by one or more males, and the parental care system is female-only. Because the groups are stable, males associate with their young and males exhibiting paternal activities that increase the number or quality of surviving young have a selective advantage over males that do not. As a result of this selective pressure, male care evolves (Figure 3.1: 1B). If the paternal investment is depreciable (i.e., has less value if it is shared by more than one female), selection may then occur that excludes other females until only one is the sole beneficiary of the male's investment. As female group size decreases, the group will become easier for a single male to defend, and so we also envision a transition from multi-male to one-male units (alternatively, the group may have contained a single male in the first place). This reduction in female group size ultimately

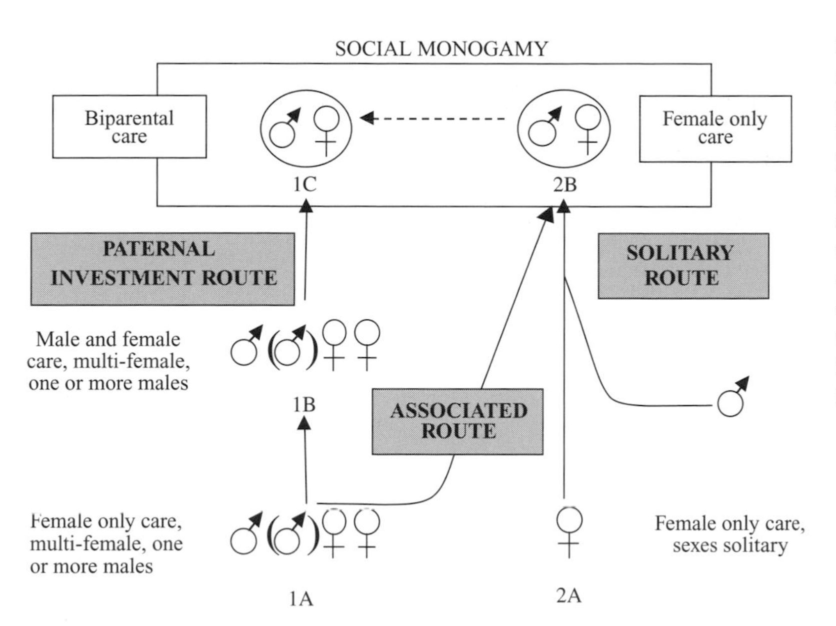

Figure 3.1. Three evolutionary routes to social monogamy. Full details of each pathway (shaded boxes) are given in the text. Arrows indicate potential evolutionary transitions from ancestral states to social monogamy (upper box). Once social monogamy with female-only care has evolved, biparental care may or may not evolve, as shown by the dashed arrow. The parentheses indicate that the social group may have contained one or more males.

leads to the situation where social monogamy evolves as a result of the benefits of biparental care (Figure 3.1: 1C).

If the ancestral state comprised groups containing more than one male, there could still be a selective advantage to paternal investment: if males are related to each other (through kin-selected benefits); if males preferentially help their own offspring; or if females reward males for their help through increased mating access.

Associated route

Here we assume that the ancestral state (Figure 3.1: 1A) comprises groups containing several females accompanied by one male, the parental care system is female-only, and males are associated with females primarily to secure access to mates. In the associated route, just as in the parental investment route, males remain associated with females throughout the course of evolutionary events. However, in this case female group size decreases due to changing ecological pressures (i.e., a shift in the relative costs and benefits of living in a group), not as a result of competition for male help. Ultimately, if female group size declines to one, social monogamy will have evolved (Figure 3.1: 2B). Later, we will discuss why males might remain associated with a single female instead of attempting to defend several female ranges (see also van Schaik & Kappeler, chapter 4).

Solitary route

For the solitary route, the ancestral state (Figure 3.1: 2A) is one comprising solitary females with whom males associate only in order to mate. Again, the initial parental care system is female-only. For social monogamy to evolve from this starting point there needs to be some reason for males and females to form a stable, long-term association (possible causes of this association are discussed later). Once this pair bond has become established we have social monogamy without biparental care (Figure 3.1: 2B). This state may be stable, or alternatively there may then be selection pressure for paternal investment, as discussed above, potentially leading to biparental care within a socially monogamous mating system (Figure 3.1: 1C).

DISTINGUISHING BETWEEN THE EVOLUTIONARY ROUTES TO SOCIAL MONOGAMY

Which of these routes is the most likely to account for social monogamy in mammals? Both of the 'ancestral states' (Figure 3.1: 1A and 2A) are found in mammals (examples of the 1A state include several social ungulates such as the gemsbok (*Oryx gazella*) and some monkeys belonging to the genus *Cercopithecus* sp., while examples of the 2A state include polar bears (*Thalarctos maritimus*), orangutans (*Pongo pygmaeus*), and many microtine rodents and smaller ungulates: see Gosling, 1986; Clutton-Brock, 1989; Komers & Brotherton, 1997). Similarly, there are a number of examples of the intermediate station along the paternal investment route (Figure 3.1: 1B), where male care exists in polygynous or multi-male, multi-female societies (e.g., in some social carnivores and primates such as lions (*Panthera leo*), banded mongooses (*Mungos mungo*), and chimpanzees (*Pan troglodytes*): see Kleiman & Malcolm, 1981; Woodroffe & Vincent, 1994). Moreover, van Schaik and Kappeler (chapter 4) present phylogenetic evidence that the evolution of social monogamy via an associated route is likely to have occurred in several primates; while Ribble (chapter 5) proposes a paternal investment route for the California mouse (*Peromyscus californicus*). Thus, at first inspection, any of the routes would appear possible. In the following sections we attempt to distinguish between these possibilities by reviewing the existing hypotheses that have been advanced to account for the evolution of social monogamy in mammals. Our intention is to highlight the relative merits of the rationale underlying these hypotheses rather than attempt an exhaustive review. In each case, we focus on the applicability of the hypothesis to our study animal, Kirk's dik-dik.

SITE, TIME, AND SPECIES

Kirk's dik-dik inhabits scrub and open woodland in semi-arid regions across Southwest and East Africa (Tinley, 1969; Hendrichs & Hendrichs, 1971). Weighing under 6 kg (Kellas, 1955), dik-diks are among the smallest antelopes. They are both socially and genetically monogamous (Brotherton *et al.*, 1997), and pair members have been shown to remain together for at least four years, though they probably do so until one pair member dies (Hendrichs, 1975).

We collected data between February 1991 and April 1993 in Etosha National Park, Namibia (18° 50′ S, 16° 57′ E) and from September 1993 to July 1994 in Tsavo East National Park, Kenya (2° 60′ S, 35° 70′ E). The Namibian site contained 23 dik-dik territories and the average territory size was 3.5 ha (range 2.1–7.0 ha), while the Kenyan site contained 21 territories averaging 2.4 ha (range: 0.8–4.6 ha). In Namibia 38 individuals were ear-tagged (18 males and 20 females) and 22 were radio-collared (10 males and 12 females) (Brotherton, 1994), while in Kenya 57 individuals were ear-tagged (20 males, 21 females and 16 juveniles), of which 32 adults were collared (17 males and 15 females) (Komers, 1996*a*). Our results are based on over 1200 hours of behavioural observations. Further details of the Namibian and Kenyan study sites and methods are given in Brotherton and Manser (1997) and Komers (1996*a*).

HYPOTHESES FOR THE EVOLUTION OF MONOGAMY IN MAMMALS

A review of the literature reveals that the hypotheses that have been advanced to explain the evolution of social monogamy broadly fall into five categories (Table 3.1). Two of these, joint territory defence and kin selection, do not appear to be relevant to our discussion of the evolution of social monogamy in mammals. Joint territory defence has been proposed to account for social monogamy in butterfly fishes (family Chaetodontidae) because the loss of one pair member causes a reduction in territory size and intrusion rates increase by up to 10 times (Fricke, 1986). We know of no comparable situation in any socially monogamous mammal. Indeed, even within butterfly fish there are theoretical reasons to question this hypothesis: if group size is so important for territory defence it might be expected that butterfly fish would benefit from forming larger groups, or that defence would be achieved just as effectively by homosexual pairs; however, neither of these groupings occurs (Roberts & Ormond, 1992). As to kin selection, in species where females have clutches or litters containing more than one young, offspring that are full-sibs may survive better than those that are half-sibs because of decreased sibling competition (Peck & Feldman, 1988). Therefore, within a given breeding event there may be selection for females to be genetically monogamous. However, a female could obtain these benefits by fertilizing all her offspring with sperm from a single mating (which may require sperm storage in birds: Gomendio & Roldan, 1993). Thus, this does not appear to provide a satisfactory explanation for prolonged associations between males and females. In the discussion that follows we focus on the three remaining explanations for the evolution of social monogamy.

Biparental care

Biparental care appears to be important for offspring survival in most socially monogamous fish (Barlow,

Table 3.1. *Hypotheses for the evolution of social (and genetic) monogamy. Here it has been assumed that males are the sex with the highest potential reproductive rate (Clutton-Brock & Vincent, 1991); however, the same principles are likely to apply in sex-role reversed species*

Explanation for monogamy	References
1 Biparental care	Kleiman, 1977; Wittenberger & Tilson, 1980
(a) Male care is essential	Bart & Tornes, 1989
(b) Polygyny threshold is not reached	Orians, 1969; Davies, 1989
(c) Female–female aggression prevents polygyny	Yasukawa & Searcy, 1982; Veiga, 1992; Slagsvold & Lifjeld, 1994
2 Female dispersion: males are unable to monopolize more than one female range	Jarman, 1974; Emlen & Oring, 1977; Rutberg, 1983; Barlow, 1988
3 Mate guarding: males have the highest reproductive rate by remaining with one female	Parker, 1974; Wittenberger & Tilson, 1980; Wickler & Seibt, 1981
4 Joint territory defence is essential	Wilson, 1975; Fricke, 1986
5 Kin selection	Peck & Feldman, 1988

Table 3.2. *Forms of male care in socially monogamous mammals. For further examples of species showing each type of care see Kleiman & Malcolm, 1981; Clutton-Brock, 1989; Woodroffe & Vincent, 1994*

Type of male help	Examples	References
Providing food	Most canids	Moehlman, 1989
	Prairie vole (*Microtus ochrogaster*)	Thomas & Birney, 1979
Incubation	Djungarian hamster (*Phodopus campbelli*)	Wynne-Edwards, 1987
	California mouse (*Peromyscus californicus*)	Ribble, chapter 5
Den/refuge construction	European beaver (*Castor fiber*)	Wilsson, 1971
Infant carrying	Titi monkey (*Callicebus moloch*)	Mendoza & Mason, 1986
	Siamang (*Hylobates syndactylus*)	Chivers, 1974
Protection from predators	Klipspringer (*Oreotragus oreotragus*)	Dunbar & Dunbar, 1980
	Aardwolf (*Proteles cristatus*)	Richardson, 1987
Protection from male infanticide	Possibly gibbons (*Hylobates* sp.)	van Schaik & Dunbar, 1990
Resource defence	Possibly gibbons (*Hylobates* sp.)	Rutberg, 1983; van Schaik & van Hooff, 1983

1984), birds (Lack, 1968; Wittenberger & Tilson, 1980), and mammals (Kleiman, 1977; Clutton-Brock, 1989). In birds, for example, both sexes are equally capable of incubating eggs and feeding chicks and this probably explains why *c.* 91% of bird species are socially monogamous (Lack, 1968). Although male mammals are unable to gestate or lactate (but see Francis *et al.*, 1994), paternal activities are also thought to contribute substantially to offspring survival in many socially monogamous mammals (Table 3.2). For example, Wynne-Edwards (1987) found that 95% of Djungarian hamster (*Phodopus campbelli*) pups survived to weaning when both parents were present, but only 47% survived when fathers were removed. In species such as these, it is clear that males play an important parental role. Of relevance to this discussion is whether this role evolved before or after social monogamy. In the sections that follow, we consider whether the importance of biparental care is likely to explain the origins of social monogamy in dik-diks and other mammals.

Biparental care and sociogenetic monogamy in dik-dik

What types of paternal care might occur in dik-diks (Table 3.2)? Dik-diks are browsers, and infants hide alone after they are born (Hendrichs, 1975); hence, males can have no paternal role in providing food or incubation for their young. Similarly, den construction does not occur and infant antelopes are not carried. However, there are three forms of paternal care that might be provided by male dik-diks: (i) defending resources (Rutberg, 1983); (ii) reducing predation risk (Dunbar & Dunbar, 1980); and (iii) defending against infanticide by other males (Cockburn, 1988, pp. 90–1; van Schaik & Dunbar, 1990).

Each of these forms generates testable predictions. (i) If males are defending resources for their mate and offspring, we would expect males to defend the territory against intruders of both sexes since intruding females (and males) will use the food resources of the territory. If defence by males is sex-specific, monogamous females would have no more resources available to them than solitary females. (ii) If males reduce the risk of predation, we would expect: females to benefit from the presence of their mate by generally having to be less vigilant when they are together than when she is alone; females to be alerted to approaching predators sooner when with their mates; and males to respond when their young are threatened by predators. (iii) If males are required to defend against infanticide, we would expect new resident males to kill any infants in their territory.

Using a combination of observational and experimental techniques, we showed that none of these predictions hold in dik-diks (Brotherton & Rhodes, 1996). (i) Territory defence is largely sex-specific, with females performing much of the territory defence themselves; indeed males sometimes prevent their mates from chasing intruding females away (Brotherton, 1994). (ii) Females are no less vigilant in the presence of their mates than when they are alone; they are not alerted sooner to the presence of an approaching predator when males are present; and fathers, unlike mothers, did not respond to the playback of the call of an avian predator that preys on young dik-diks but not adults. (iii) Males did not kill infants that they had not sired. Thus, there is convincing evidence that biparental care is not important in dik-diks (Brotherton & Rhodes, 1996).

Biparental care and the evolution of social monogamy in mammals

While there are undoubtedly socially monogamous mammals in which biparental care is now important (Kleiman & Malcolm, 1981; Woodroffe & Vincent, 1994), the key question is, did paternal investment evolve after social monogamy or before? A recent phylogenetic analysis of mammalian social and parental care systems shows clearly that social monogamy has evolved more frequently in the absence of paternal care than in its presence (Komers & Brotherton, 1997). This effect was highly significant and robust across alternative mammalian phylogenies: in 61 out of 64 combinations of alternative phylogenies, social monogamy evolved significantly more often in the absence of paternal care than in its presence, and in the three non-significant combinations the trend was also in this direction (Komers & Brotherton, 1997). This is the opposite of what we would expect if the most likely scenario for the evolution of social monogamy is via the paternal investment route. Consequently, we conclude that in most monogamous mammals exhibiting biparental care, paternal investment probably developed secondarily in species that were for other reasons already monogamous (but see Ribble, chapter 5).

Infanticide: a special case?

As discussed above, phylogenetic analyses suggest that if social monogamy evolves via the associated or solitary route, biparental care is likely to have evolved secondarily. This also makes intuitive sense because in species with internal fertilization there would appear to be no opportunity for paternal investment to evolve unless males and females remain in association after copulation (Elwood, 1983; van Rhijn, 1991). The one exception to this may be male defence against infanticide (Table 3.2; Cockburn, 1988, pp. 90–1; van Schaik & Dunbar, 1990; van Schaik & Kappeler, chapter 4). In several mammal species, males that kill a female's current offspring benefit by reducing the interval before she enters her next fertile period (Hausfater & Hrdy, 1984); this may also be true in some birds (Veiga, 1990). In many rodents, females reduce the reproductive cost of infanticide by aborting fetuses in the presence of a new male ('the Bruce Effect': e.g., Schwagmeyer, 1979) and may show a pregnancy block in response to multiple matings or the presence of more than one male (Dewsbury, 1982; Wynne-Edwards & Lisk, 1984; Labov *et al.*, 1985). In these circumstances there would appear to be clear benefits to a male remaining with a female after mating until the risk of infanticide is removed. If this guarding period is prolonged or represents a significant part of the breeding season, a male may achieve a greater number of matings by remaining with his partner until her next fertile period than by searching for additional mates.

At first inspection, then, male defence against infanticide by other males could explain the evolution of social monogamy, and this hypothesis has specifically been proposed to account for social monogamy in rodents (Cockburn, 1988, pp. 90–1) and primates (van Schaik & Dunbar, 1990). However, this theory presents several problems. First, there is a lack of empirical evidence. Most reports of infanticide in wild populations of rodents have been from polygynous species, and it is predominantly females, not males, that are infanticidal (Labov *et al.*, 1985). In addition, male removal experiments have shown that there is little or no risk of infanticide by new males in the sociogenetically monogamous rodent *Peromyscus californicus* (Ribble, chapter 5). Moreover, in many of the rodent species in which infanticide occurs, the females are solitary and so the threat of infanticide does not appear sufficient to cause social monogamy to evolve on its own, and even where prolonged male–female associations and infanticide co-occur, there is no apparent adaptive link between the two (Palombit, 2000).

Infanticide by new resident males has been observed in at least ten species of primates, and is most common in species living in single-male groups where male tenure is short (Hiraiwa-Hasegawa, 1988). However, infanticide has not been confirmed in any socially monogamous primate (Hiraiwa-Hasegawa, 1988; van Schaik & Dunbar, 1990). In addition, although the infanticide hypothesis was specifically formulated to account for social monogamy in the gibbons (*Hylobates* spp.) (van Schaik & Dunbar 1990), there have been several reports of adult male gibbons being replaced in groups containing infants and in no case did infanticide occur (Palombit, 1994*a*). It may be difficult to separate cause and effect here (i.e., are low rates of infanticide in socially monogamous species a *consequence* of reduced infanticide risk?), but proponents of this hypothesis must at the very least show that infanticide consistently occurs following male takeovers. More data on the effect of male replacement on infant survival are required.

In addition to this lack of empirical evidence, comparative studies offer little support for the idea that the risk of male infanticide favours the evolution of social monogamy. Nunn and van Schaik (2000) showed that, compared with ecological factors such as predation risk, infanticide risk is a very poor predictor of primate social systems. Furthermore, in an analysis of the costs and benefits of male–female bonding for both sexes, Palombit (2000) demonstrated that the ecological and social conditions and behaviours usually found in species where male infanticide risk is important are largely absent in gibbons and lemurs (though the evidence for lemurs was somewhat equivocal).

Finally, there are at least three theoretical reasons why infanticide by males would not be expected in socially monogamous primates. First, the tenure of socially monogamous males is usually relatively long compared with that of their polygynous counterparts, and so the benefits to males of committing infanticide are likely to be less in socially monogamous species. Second, socially monogamous primates are usually monomorphic (Kleiman, 1977) and so females are probably capable of defending their offspring from infanticidal males, or at least of making infanticide costly. There is evidence that female aggression may indeed deter infanticide in some rodents (Wolff, 1985). Third, and perhaps most important, just as reducing the risk of infanticide may benefit males who stay with females after mating, infanticidal acts are similarly only beneficial to males who are going to stay with the females whose young they kill. It is important to take this into account when we consider the order of the evolutionary events that may have occurred. If, in the ancestral state, females were solitary (Figure 3.1: 2A), males would not benefit from committing infanticide unless they remained in the vicinity of the female and had priority of access the next time she was receptive. This would require males to be territorial, and so the ancestral state for males, if infanticidal behaviour is to evolve, would have needed to have been polygyny, with males defending more than one female range. Given this polygynous starting point, it would appear extremely unlikely that it would ever pay males to forgo guarding several females in favour of social monogamy.

Van Schaik & Kappeler (chapter 4) raise the intriguing possibility that in primates, as a result of infanticide avoidance, the habit of living in 'associated pairs' may have evolved directly from societies in which females are group-living, not from those containing solitary females (cf. the associated route in Figure 3.1). They argue that if infanticide is associated with male take-overs and these are more likely to occur in groups containing more than one female, then there may be selective pressure on females to live apart and bond with a guarding male who protects her young from other males. They also provide phylogenetic evidence in support of the transition from social polygyny to social monogamy, and this interesting idea warrants further investigation. Whether or not this hypothesis proves robust will depend upon the answers to a number of questions, including: (i) Why have females been selected to live apart in some species at risk for infanticide but not in others? (ii) What prevents the most competitive males from defending two or more female ranges from infanticidal males (recall our argument above, that infanticidal behaviour is only beneficial if the male increases his chances of mating with the female whose infant he kills, i.e., he must take over the territory)? (iii) Why do males in socially monogamous species such as gibbons spend their time so closely bonded to females even when there are no infants present, if the primary role of the bond is to reduce infanticide?

We believe there is little evidence so far to support the hypothesis that the primary reason for monogamous males to remain with their mates is to reduce the risk of infanticide. How does this compare with our understanding of the role of infanticide in other mating systems? In several polygynous species, including lions and langurs, infanticide frequently occurs following male take-overs (Hrdy, 1979; van Schaik & Janson, 2000). As a consequence, both infant survival and male mating success commonly correlate positively with the duration of male tenure in these species. Is one of these the primary evolutionary cause of the social patterns we observe or do both play a role? This question would be a useful topic for future research. Our suspicion is that males in these species remain with their mates in order to gain future mating opportunities and to avoid the costs associated with attempts to take over additional females, not to prevent infanticide. We see no reason to interpret prolonged associations in socially monogamous species any differently.

In summary, the paternal investment route is unlikely to provide a general explanation for the evolution of social monogamy in mammals (but see Ribble, chapter 5), and we should instead focus on the alternative routes (Figure 3.1). In species where biparental care is now important, it probably evolved secondarily. Consequently, we should be cautious when studying biparental, socially monogamous species because the social and ecological conditions they now experience may be very different from those experienced by their ancestors when monogamy evolved. While the maintenance and function of monogamy in these circumstances can be fascinating, we should be clear that this is what we are studying. This problem is perhaps best exemplified by the apparent state of flux in the mating systems of many socially monogamous birds. Chicks may survive best under conditions of biparental care, but both sexes may attempt to parasitize the caring efforts of their mates and neighbours: females may mate with extra-pair males and attempt to dump eggs in the nests of conspecifics, while males often search for EPC partners, and in some cases allow females to lay in their nests in return for mating (Emlen & Wrege, 1986; Rohwer & Freeman, 1989; Birkhead & Møller, 1992). Biparental care may have been key to the success of birds as an order, but it may be the case that monogamy would not evolve in the current social and ecological environment experienced by many bird species today (see also van Rhijn, 1991).

Female dispersion

In socially monogamous mammals that do not exhibit biparental care (e.g., elephant shrews: Rathbun, 1979; Kirk's dik-dik: Brotherton & Rhodes, 1996; mara: Pankhurst, 1998), monogamy is typically thought to have evolved because female ranges are defensible, but too large or too dispersed for males to be able to defend more than one (Emlen & Oring, 1977; Rutberg, 1983; van Schaik & van Hooff, 1983; Barlow, 1988). The assumption underlying this hypothesis is that males would defend more than one female if they could.

Female dispersion and sociogenetic monogamy in dik-dik

Since male dik-diks show no paternal care (see above), we would expect the main determinant of female dispersion to be the distribution of resources rather than the dispersion of males (Emlen & Oring, 1977; Davies, 1991). Thus, we would expect little variation in the amount of resources contained in a territory: socially monogamous males should defend enough resources for one female and her young, but not two. If male dik-diks are socially monogamous due to female dispersion, we would expect to see a flexible social system where males take advantage of any opportunity to defend additional space, especially if this extra space was occupied by a female.

At both study sites, we determined the amount of resources available in a sample of territories and found no negative relationships between territory size and food availability, shrub density, or total cover (Komers, 1996*a*; Brotherton & Manser, 1997). This led us to conclude that territory size was a good indicator of resource availability. Despite this, there was substantial variation in territory size. In Namibia, territory size ranged from 2.1 to 7.0 ha ($N = 14$), the modal territory size being 2.01–2.50 ha (based on a frequency distribution at 0.5 ha intervals: Brotherton & Manser, 1997). Similarly, in the Kenyan study site there was substantial variation in territory size and quality. Again, the modal territory size was 2.01–2.50 ha, with the size ranging from 0.8 to 4.6 ha, although the males defending the three smallest territories (<1.5 ha) did not appear to be able to form stable bonds with females. Males and females consume similar amounts of browse (Manser & Brotherton, 1995), and if we assume that their offspring consume the same, then a bigamous male would require a territory with 1.67 times the resources of that of a monogamous male. It seems likely that territories of the modal size (<2.5 ha) contain sufficient resources to support a pair and their offspring. Consequently, territories that are at least 4.2 ha (1.67 × 2.5 ha) should contain enough resources for an extra female and her young. In the Namibian study, three males defended territories larger than this, and in Kenya two males did so, and yet they were socially monogamous (Komers, 1996*b*; Brotherton & Manser, 1997).

In the Kenyan population, we investigated the basis of dik-dik social monogamy further by temporarily removing eight males from their territories. If males are constrained into monogamy by female dispersion, we would expect paired males to respond to these available mates by annexing their territories. However, none did so, even though in eight cases the possible expansion would have resulted in a combined territory size smaller than the largest existing territory (Komers, 1996*a*). These observations are inconsistent with the hypothesis that males are constrained into social monogamy by female dispersion.

FEMALE DISPERSION AND THE EVOLUTION OF SOCIAL MONOGAMY IN MAMMALS

Just as dik-dik males do not appear to be constrained by female dispersion, there is no good evidence that female ranges are unusually large or dispersed in most other socially monogamous mammals. Comparative studies show that female range size is similar in polygynous and socially monogamous rodents (Cockburn, 1988, p. 89) and primates (Dunbar, 1988, pp. 273–88). In addition, mathematical models of the maximum possible territory size of socially monogamous male primates, based on daily range, indicate that males of many species should be able to defend the ranges of several females (Dunbar, 1988, p. 278; van Schaik & Dunbar, 1990).

Further evidence against the female dispersion hypothesis comes from phylogenetic comparisons (Komers & Brotherton, 1997). In this analysis we found that female range size (standardized to control for differences in body mass and trophic level: herbivore, omnivore, and carnivore) was actually smaller in socially monogamous species than in non-monogamous ones. Thus, although there may be some examples of socially monogamous species in which males are constrained by female dispersion (e.g., the Japanese serow: Kishimoto & Kawamichi, 1996; Kishimoto, chapter 10), it does not provide a general explanation of the evolution of social monogamy in mammals.

Mate guarding

The foregoing analysis and discussion has concluded that neither the benefits of biparental care, nor constraints associated with female dispersion, can adequately explain why some 5% of mammal species are socially monogamous. If this conclusion is correct, then we believe there are two key questions concerning the evolution of social monogamy in mammals. (i) Why, in socially monogamous species, do the most competitive males not monopolize more than one mate? (ii) Why, since male care is unimportant, do females accept social or, in particular, sociogenetic monogamy? In the discussion that follows, we consider each of these questions in turn and attempt to answer them for the dik-dik and other mammals. The exact same mate-guarding arguments apply when considering either the 'associated' or 'solitary' routes to monogamy (Figure 3.1): it makes no difference whether we are trying to explain why one male joins a solitary female and then remains associated with her (solitary route), or why a male has remained associated with a female who has chosen to live apart from other females (associated route).

Why do the most competitive males not monopolize more than one mate?

One possible answer to this question is mate guarding. It has been suggested that social monogamy may evolve as a male mate-guarding strategy in two situations. First, when there is a male-biased sex ratio in the population, social monogamy may evolve because the average male will achieve higher reproductive success by sequestering one female than by competing for matings in a promiscuous population (Wittenberger & Tilson, 1980). However, in birds and mammals, the operational sex ratio is almost invariably male-biased and this is likely to have a greater effect on the intensity of mating competition than the population sex ratio (Emlen & Oring, 1977). Thus, it is not clear why, even in the presence of a male-biased sex ratio, the most competitive males are unable to monopolize additional females. Second, it has been suggested that social and sociogenetic monogamy may evolve if males achieve a higher reproductive rate by mating with a single female during successive reproductive events than by leaving females after copulation and searching for additional mates (Parker, 1974; Wickler & Seibt, 1981). This may occur if breeding is highly synchronized within the population or if the costs of searching for or defending additional females are high. Socially monogamous mammals typically do not have extremely synchronized breeding, so we focus on this final possibility here, again using Kirk's dik-dik as our case study.

Why do high quality males neither rove nor defend multiple females in socially monogamous species? Both of these strategies occur in other mammals. A roving male strategy is adopted by many male carnivores, especially mustelids (Sandell, 1989), and roving also typically occurs in species of microtine rodents in which females are territorial (Ostfeld, 1985; Ribble, chapter 5). Similarly, males in several ungulate species do defend several female ranges (Barrette, 1987). Why do high quality socially monogamous males not adopt either of these strategies? Or, to put this question another way, what enables relatively poor quality males to secure mates in socially monogamous species? To answer this, we must consider the costs and benefits of alternative strategies.

Roving costs

Obvious costs of roving include the increased predation risk and decreased feeding efficiency that is likely to be associated with being in unfamiliar areas, as well as the costs of repeated fights to gain access to females. If fights are costly, then it may pay males to fight for access to a guarded female only if there is a high probability that she is in oestrus. The information that males have regarding the oestrous state of females is variable for two reasons. First, females advertise oestrus in different ways. In many non-monogamous species, females advertise oestrus unambiguously. For example, female African elephants attract males over long distances by emitting ultrasonic rumbles (Poole *et al.*, 1988), and many female primates undergo obvious physical changes during oestrus (Hrdy & Whitten, 1987). Second, the length of oestrus varies in relation to the duration of the mating season. If oestrus is synchronized or relatively long compared to the duration of the breeding season, it may still pay males to fight for access to guarded females at times when they are likely to be receptive.

Is roving likely to pay off as a strategy in the dik-dik? Dik-diks are vulnerable to a wide range of terrestrial and aerial predators (Hendrichs & Hendrichs, 1971) and the increased predation risk from roving is likely to be significant. In addition, fighting costs are likely to be high because, unlike the weapons of many larger male ungulates (Estes, 1991), dik-dik horns are not adapted for prolonged tests of strength. Rather, they are stabbing weapons with which either contestant could do considerable harm. There is also evidence that the pay-off of winning any fight is likely to be low. Oestrus is asynchronous (the breeding season lasts several months) and lasts about one day (Hendrichs, 1975; Dittrich & Böer, 1980), and so the chances of a female being receptive on any randomly selected day is likely to be low. Additionally, guarding males control the movements of females (Komers, 1996*a*) and appear able to conceal the oestrous period of their mates by over-marking their scent (Brotherton *et al.*, 1997). Males over-mark both the preorbital secretions and the urine and dung of females (Hendrichs, 1975; Tilson & Tilson, 1986), especially during oestrus (Brotherton, 1994) and, in the case of urine and dung deposits, males precede their over-marking by scraping several centimetres of dirt on top of the female deposit. Other males may therefore have little or no information regarding the oestrous state of a guarded female and they will have to fight just to test whether or not she is in oestrus. This rarely happens: few matings are contested and males do father the offspring of their social partner (Brotherton *et al.*, 1997).

Taken together, the costs would appear to outweigh the benefits of roving for male dik-diks. Similar factors may also apply in other socially monogamous mammals. In many species (both monogamous and non-monogamous), roving may incur increased predation risk or reduced feeding efficiency. In addition, in other socially monogamous species, guarding males typically show high levels of attendance and/or have specially adapted guarding behaviours such as over-marking (Roberts & Dunbar, 2000). Further, we have found no evidence to suggest that the duration of oestrus in dik-diks (24 hours) is unusually short for a socially monogamous mammal (van Tienhoven & van Tienhoven, 1993). Consequently, it may often be the case that in socially monogamous species only guarding males are certain when their pair mates are in oestrus, unless the females overtly advertise their state to other males, which has only rarely been reported in socially monogamous mammals (Richardson, 1987). There is likely, therefore, to be an asymmetry of information regarding the oestrous state of females, with guarding males having a significant advantage. And this may be the key to explaining why males are able to secure access to their mates in monogamous mammals (Brotherton & Manser, 1997; Pankhurst, 1998; Roberts & Dunbar, 2000; Kishimoto, chapter 10). Achieving the asymmetry does not require oestrus to be completely unpredictable or guarding to be perfect, it simply requires the guard to have a significant advantage relative to other males.

Although roving may not be an option for socially monogamous males, they could potentially be polygynous by defending the ranges of more than one female. As discussed above, males in many socially monogamous mammals should be able to do so. Why don't male dik-diks defend more than one female? Again the answer appears to be mate guarding. Males do not just guard their mates during oestrus. Year-round, males maintain proximity within pairs and follow female changes in activity (Brotherton *et al.*, 1997). Males also spend *c.* 65% of their time with their partners and over-mark most of their preorbital secretions and urine and faeces, even outside the oestrous period (Hendrichs, 1975; Tilson & Tilson, 1986; Brotherton *et al.*, 1997). This intensive mate guarding appears to be essential if males are to retain their territory and mate. Although territorial challenges are very uncommon, males are typically replaced within three days if they die (Brotherton, 1994). Experiments using female scent suggest that this rapid response by non-resident males is due to the uncovered female urine and faecal deposits that occur in territories when no male is present to over-mark them (Brotherton, 1994). Furthermore, removal experiments show that males are usually unable to regain territories if they lose them: six out of eight removed males permanently lost their territories (Komers, 1996*a*). Consequently, any decrease in the efficacy of advertising territorial occupancy is likely to be costly, and we suggest that it is these costs that prevent males from attempting to defend more than one female territory, hence constraining them into social monogamy. Extending this argument, if two females were to share the same territory (and dung piles) and move together, we would expect a single male to be able to defend them; this is precisely what occurred in four temporarily polygynous groups in the Namibian study population (Brotherton, 1994; Brotherton & Manser, 1997).

Similarly, mate-guarding constraints may explain why males in other socially monogamous mammals defend only a single female, and why pair bonds are often maintained outside breeding seasons. Fitzgibbon (1997) found that male golden-rumped elephant shrews (*Rhynchocyon chrysopygus*) that attempted to defend an extra female territory, following the experimental removal of males, lost weight and could only temporarily hold the extra territory. This was apparently attributable to the costs of defending an extra female, not the extra space, because some monogamous males had territories more than twice the size of others (Fitzgibbon, 1997). In many monogamous mammals, males almost continually attend their partners (e.g., klipspringer (*Oreotragus oreotragus*): Dunbar & Dunbar, 1980; gibbons (*Hylobates* spp.): van Schaik & Dunbar, 1990; mara (*Dolichotis patagonium*): Pankhurst, 1998), and advertise their 'ownership' through, for example, scent marking or singing duets, activities which would be impossible to carry out for more than one female. Instead of competing for guarded females, unpaired males respect these advertisements of ownership: to our knowledge there are few examples of territory take-overs in socially monogamous species. Thus the primary function of year-round pair bonding seems to be to avoid the costs of conflicts over females. In this respect, monogamous male mammals may be viewed as risk-averse (cf. McNamara & Houston, 1992, who used this term for risk-averse foraging behaviour), maximizing their reproductive success by avoiding conflicts and living for a long time, rather than fighting repeatedly over females.

As a result of mate-guarding constraints, males are thus unable to defend more than one female range; but why do they not adopt a mixed strategy: pair with a single female, and attempt to mate with neighbours? Males adopting this strategy would not incur the increased mate-guarding costs that would be associated with permanently defending more than one female, nor the increased predation risk of roving. However, they would be subject to the information asymmetry discussed above, and so would be unlikely to be successful in gaining access to receptive females, unless the females themselves chose to mate with them. This is the subject of the next section.

Why do females accept social or sociogenetic monogamy?

Extra-pair copulations (EPCs) occur in over 57% of socially monogamous bird species (Møller & Birkhead, 1993). Far from being passive or unwilling EPC partners, female birds may actively solicit EPCs (Smith, 1988; Kempenaers *et al.*, 1992), and females that do mate with multiple partners are likely to be able to control which partner fathers their offspring (Birkhead & Møller, 1993). EPCs may also represent an alternative reproductive strategy for socially monogamous female mammals. For example, female titi monkeys and Mongolian gerbils have been reported to leave their territories and solicit extra-pair matings (Mason, 1966; Ågren *et al.*, 1989), while oestrous female aardwolves apparently attract neighbouring males to their territory by scent marking (Richardson, 1987). In addition, a confirmed case of extra-pair paternity (EPP) was found in a sample of ten mara families (Pankhurst, 1998). More genetic studies are needed to confirm how common EPP is in monogamous species; however, behavioural data strongly suggest that most matings are within-pair (with the possible exception of the aardwolf (*Proteles cristatus*): Richardson, 1987). Since EPCs occur in so many socially monogamous bird species it is important to consider why this is not more common in socially monogamous mammals.

Possible benefits and costs of EPCs to females are summarized in Table 3.3 (after Westneat *et al.*, 1990; Birkhead & Møller, 1992). The most obvious benefit that would seem to apply to female dik-diks and females in other socially monogamous species is good genes. Choosing high quality EPC partners might be expected to be particularly important for female dik-diks because they may have little control over the choice of male to which they are pair-bonded. There are three reasons for this. First, searching for a good territory with a high quality mate may be costly, and so female dik-diks may have no choice but to settle with the first unpaired male that they find. Second, when a resident male dies he is usually replaced by a young male on a 'first-come/first-served' basis (Brotherton, 1994), and the female does not appear to have any choice over which male becomes her new mate. And third, because males are not known to change territory, even if a higher quality one becomes available (Hendrichs, 1975), there may be no correlation between male quality and territory quality, and so females may be forced to choose either a good mate or a good territory. Females are therefore unlikely to be pair-bonded to the 'best' male in the vicinity; however, they do have ample opportunity to assess neighbouring males, and should therefore be in a good position to select a high quality father for their offspring. Females in other socially monogamous species are likely to be faced with a similar situation.

Why, then, do females not seek EPCs (Brotherton *et al.*, 1997)? We believe the answer for dik-diks, and probably for females in many socially monogamous mammals, is the costs of harassment by extra-pair males. Male harassment has been shown to be costly to oestrous females in several ungulates (Geist, 1971, pp. 208–11; Rubenstein, 1986; Komers *et al.*, 1999), and may even cause females to leave mix-sex herds to obtain matings at leks (Wrangham, 1980; Clutton-Brock & Parker, 1995). Harassment may be taken to the extreme in waterfowl, where females may be drowned during forced copulation attempts by multiple males (McKinney *et al.*, 1983). Oestrous female dik-diks will always be attended by their mates, and so females are unlikely to be able to mate with extra-pair males without instigating a prolonged and potentially dangerous conflict. On the two occasions when we observed disputed matings, the females responded

Table 3.3. *Potential benefits and costs of extra-pair copulations to females (after Westneat* et al., *1990; Birkhead & Møller, 1992). Species whose names are in parentheses are not monogamous, and so the benefits/costs are associated with multiple matings, rather than EPCs*

Benefit/cost	Examples	References
Benefits		
1 Infertility insurance	Siamang	Birkhead & Møller, 1992; Palombit, 1994*b*
2 Courtship feeding	(Orthopteran insects)	Gwynne, 1983
	(Orange-rumped honey-guides)	Cronin & Sherman, 1976
3 Paternal care	Saddle-back tamarin	Terborgh & Goldizen, 1985
	Dunnock	Burke *et al.*, 1989
4 Avoidance of infanticide	(Hanuman langur)	Hrdy, 1977
5 Avoidance of rejection costs	No examples provided	Parker, 1974
6 Genetic diversity	No examples provided	Williams, 1975
7 Good genes	Barn swallow	Møller, 1988*a*
	Zebra finch	Houtman, 1992
	Blue tit	Kempenaers *et al.*, 1992
Costs		
1 Bad genes/unsexy sons	No examples provided	Birkhead & Møller, 1992
2 Retaliation by paired male	Mountain bluebird	Barash, 1976
	Barn swallow	Møller, 1988*b*
3 Harassment by EPC males	Ducks	McKinney *et al.*, 1983
	(Mountain sheep)	Geist, 1971
4 Increased predation risk during mating	No examples provided	Daly, 1978
5 Transfer of parasites or disease	(Rat)	Hart *et al.*, 1987
		Read, 1990

either by hiding, or running away with the males in pursuit. Since females suckle their young at least four times every 24 hours (Hendrichs & Hendrichs, 1971), this may be particularly costly for females that have a postpartum oestrus, as usually occurs in East Africa (Hendrichs & Hendrichs, 1971) and occasionally occurs in Namibia (Brotherton, 1994). Oestrous females who are attended by multiple males, all competing for mating, may be prevented from suckling their infants, or their offspring may be injured in the process.

Thus, there may be high reproductive and phenotypic costs of EPCs, probably explaining why female dik-diks, and possibly females of other species, mate monogamously. If there are phenotypic benefits to females of mating with guarding males, female choice may also enhance the advantage that guarders have over intruders and so actually facilitate the evolution of male social monogamy.

FOUR CONDITIONS FAVOURING THE EVOLUTION OF SOCIAL MONOGAMY IN MAMMALS

This discussion leads us to conclude that the following four conditions favour the evolution of social monogamy in mammals.

Females range independently of each other. This is usually because females occupy small and exclusive territories (Komers & Brotherton, 1997); however, in some cases females may have overlapping ranges but still move independently (e.g., the mara, Pankhurst, 1998).

Guarding males have an advantage in conflicts over their females. This may be because (i) there is an asymmetry of information regarding the female's oestrous state; and/or (ii) the costs of fighting for guarded females is high, relative to the likely benefits.

Male–male competition for unguarded females is high. This may be because there are many non-guarding males in the population. If this condition is not met, the best quality males will be expected to monopolize more than one mate, at least occasionally.

Females are prepared to be mated by their guarding male. This may be because the costs of EPCs are too high, perhaps due to the costs of harassment by multiple males, or because there are parental care benefits in allowing partners at least some mating access (Richardson, 1987). If this condition is not met, females would be expected to evolve strategies to overcome males' guarding efforts.

When these conditions are met, social monogamy may evolve along either the solitary route or the associated route described at the start of this chapter (Figure 3.1). Interpreting social monogamy as a mate-guarding strategy is entirely consistent with our understanding of other mammalian mating systems. In species where females form small stable groups, males adopt a variety of strategies, depending on the prevailing ecological conditions and on the extent to which the female group is defensible (Clutton-Brock, 1989). Understanding the relative costs and benefits of defending single females (or single groups of females) will, we believe, help explain why some species in which females live apart from each other are socially monogamous while other species are polygynous.

Once the pair bond has formed, biparental care may be expected to evolve if suitable opportunities for paternal investment exist. However, we should not assume that the current association between male care and social monogamy is evidence that it was important in its evolution: the more likely explanation for the evolution of social monogamy via either the 'associated' or 'solitary' routes (Figure 3.1) is mate guarding. Of these, phylogenetic evidence (Komers & Brotherton, 1997) suggests that the solitary route is the more common pathway.

SUMMARY

On the basis of empirical and phylogenetic evidence, we conclude that neither the benefits of biparental care nor the constraints imposed by female range size are likely to provide a general explanation for the evolution of social monogamy in mammals. Instead by considering the alternative strategies open to males – roving, and defending multiple females – we have explored the possibility that social monogamy may have evolved as a result of male mate guarding. We suggest that male roving does not occur in socially monogamous mammals because intruding males typically experience a high cost:benefit ratio in fighting for guarded females. In many species, this appears to be heightened by an asymmetry of information regarding the oestrous state of guarded females, giving residents an advantage in conflicts. Further, the intensity of guarding required to achieve this advantage means that resident males are unable to defend multiple females. Instead, we suggest that males in socially monogamous species typically adopt a risk-averse strategy, maximizing their reproductive success by avoiding conflicts, living for a long time, and mating with females during successive breeding events, rather than fighting repeatedly over females. We further suggest that females may accept being mated by guarding males when there are high phenotypic costs of attempting to mate with extra-pair males. Thus, social and sociogenetic monogamy can evolve as an extreme form of mate guarding. Socially monogamous male mammals effectively adopt the same strategy as males in harem-holding species (i.e., sequestering females), except that the female group size in socially monogamous species is equal to one.

Acknowledgements

We thank U. Reichard, C. Boesch, D. Ribble and two anonymous referees for their comments on this chapter. Our ideas have been developed through valuable discussions with T. H. Clutton-Brock, R. Woodroffe, R. I. M. Dunbar, J. Deutsch, S. Balshine Earn, N. B. Davies, and C. van Schaik, together with all the other participants of the monogamy workshop. For help and support in the field we thank A. Rhodes, M. Manser, E. Komers, J. Mwanzia, K. Roth, the staff of the Namibian Ministry of Wildlife and Tourism, and the Kenyan Wildlife Service.

References

Ågren, G., Zhou, Q. & Zhong, W. (1989). Ecology and social behavior of Mongolian gerbils, *Meriones unguiculatus*, at Xilinhot, Inner Mongolia, China. *Animal Behaviour*, **37**, 11–27.

Barash, D. P. (1976). Male response to apparent female adultery in the Mountain Bluebird (*Sialia currucoides*): an evolutionary interpretation. *American Naturalist*, **110**, 1097–101.

Barlow, G. W. (1984). Patterns of monogamy among teleost fishes. *Archiv für Fischereiwissenschaft*, **35** (supplement 1), 75–123.

(1988). Monogamy in relation to resources. In *The Ecology of Social Behavior*, ed. C. N. Slobodchikoff, pp. 55–79. London: Academic Press.

Barrette, C. (1987). The comparative behavior and ecology of chevrotains, musk deer, and morphologically conservative deer. In *Biology and Management of the Cervidae*, ed. C. W. Wemmer, pp. 200–13. Washington, DC: Smithsonian Institute.

Bart, J. & Tornes, A. (1989). Importance of monogamous male birds in determining reproductive success. *Behavioral Ecology and Sociobiology*, **24**, 109–16.

Birkhead, T. R. & Møller, A. P. (1992). *Sperm Competition in Birds: Evolutionary Causes and Consequences*. London: Academic Press.

(1993). Female control of paternity. *Trends in Ecology and Evolution*, **8**, 100–3.

Brotherton, P. N. M. (1994). The Evolution of Monogamy in the Dik-dik. Ph.D. thesis, University of Cambridge.

Brotherton, P. N. M. & Manser, M. B. (1997). Female dispersion and the evolution of monogamy in the dik-dik. *Animal Behaviour*, **54**, 1413–24.

Brotherton, P. N. M. & Rhodes, A. (1996). Monogamy without biparental care in a dwarf antelope. *Proceedings of the Royal Society of London, Series B*, **263**, 23–9.

Brotherton, P. N. M., Pemberton, J. M., Komers, P. E. & Malarky, G. (1997). Genetic and behavioural evidence of monogamy in a mammal, Kirk's dik-dik (*Madoqua kirkii*). *Proceedings of the Royal Society of London, Series B*, **264**, 675–81.

Brown, J. L. (1975). *The Evolution of Behavior*. New York: Norton.

Burke, T., Davies, N. B., Bruford, M. W. & Hatchwell, B. J. (1989). Parental care and mating behaviour of polyandrous dunnocks *Prunella modularis* related to paternity by DNA fingerprinting. *Nature*, **338**, 249–51.

Chivers, D. J. (1974). The siamang in Malaya: a field study of a primate in a tropical forest. In *Contributions to Primatology*, Volume 4, ed. H. Kuhn, C. R. Luckett, C. R. Noback, A. H. Schultz, D. Starck & F. S. Szalay, pp. 1–335. Basel: Karger.

Clutton-Brock, T. H. (1989). Mammalian mating systems. *Proceedings of the Royal Society of London, Series B*, **236**, 339–72.

Clutton-Brock, T. H. & Parker, G. A. (1995). Sexual coercion in animal societies. *Animal Behaviour*, **49**, 1345–65.

Clutton-Brock, T. H. & Vincent, A. C. J. (1991). Sexual selection and the potential reproductive rates of males and females. *Nature*, **351**, 58–60.

Cockburn, A. (1988). *Social Behaviour in Fluctuating Populations*. London: Croom Helm.

Cronin, J. W. J. & Sherman, P. W. (1976). A resource-based mating system: the orange-rumped honeyguide. *Living Bird*, **15**, 5–32.

Daly, M. (1978). The cost of mating. *American Naturalist*, **112**, 771–4.

Davies, N. B. (1989). Sexual selection and the polygamy threshold. *Animal Behaviour*, **38**, 226–34.

(1991). Mating systems. In *Behavioural Ecology: an Evolutionary Approach*, 3rd Edition, ed. J. R. Krebs & N. B. Davies, pp. 263–94. Oxford: Blackwell Scientific Publications.

Dewsbury, D. A. (1982). Pregnancy blockage following multiple-male copulation or exposure at the time of mating in deer mice, *Peromyscus maniculatus*. *Behavioral Ecology and Sociobiology*, **11**, 37–42.

Dittrich, L. & Böer, M. (1980). *Verhalten und Fortpflanzung von Kirks Rüssel-Dikdiks (Madoqua (Rhynchotragus) kirkii) im Zoologischen Garten*. Hannover: Freimann & Fuchs.

Dunbar, R. I. M. (1988). *Primate Social Systems*. London: Croom Helm.

Dunbar, R. I. M. & Dunbar, E. P. (1980). The pairbond in klipspringer. *Animal Behaviour*, **28**, 219–29.

Elwood, R. W. (1983). Paternal care in rodents. In *Parental Behaviour of Rodents*, ed. R.W. Elwood, pp. 235–57. Chichester: John Wiley.

Emlen, S. T. & Oring, L. W. (1977). Ecology, sexual selection, and the evolution of mating systems. *Science*, **197**, 215–23.

Emlen, S. T. & Wrege, P. H. (1986). Forced copulations and intra-specific parasitism: two costs of social living in the white-fronted bee-eater. *Ethology*, **71**, 2–29.

Estes, R. D. (1991). *The Behavior Guide to African Mammals*. Berkeley: University of California Press.

Fitzgibbon, C. D. (1997). The adaptive significance of monogamy in the golden-rumped elephant-shrew. *Journal of Zoology, London*, **242**, 167–77.

Francis, C. M., Anthony, E. L. P., Brunton, J. A. & Kunz, T. H. (1994). Lactation in male fruit bats. *Nature*, **367**, 691–2.

Fricke, H. W. (1986). Pair swimming and mutual partner guarding in monogamous butterflyfish (Pisces, Chaetodontidae): a joint advertisement for territory. *Ethology*, **73**, 307–33.

Geist, V. (1971). *Mountain Sheep: A Study in Behavior and Evolution*. Chicago: University of Chicago Press.

Gomendio, M. & Roldan, E. R. S. (1993). Mechanisms of sperm competition: linking physiology and behavioural ecology. *Trends in Ecology and Evolution*, **8**, 95–100.

Gosling, L. M. (1986). The evolution of mating strategies in male antelopes. In *Ecological Aspects of Social Evolution*, ed. D. I. Rubenstein & R. W. Wrangham, pp. 244–81. Princeton, New Jersey: Princeton University Press.

Gwynne, D. T. (1983). Male nutritional investment and the evolution of sexual differences in Tettigoniidae and other Orthoptera. In *Orthopteran Mating Systems: Sexual Competition in a Diverse Group of Insects*, ed. D. T. Gwynne & G. K. Morris, pp. 337–66. Boulder, Colorado: Westview Press.

Hart, B. J., Korinek, E. & Brennan, P. (1987). Postcopulatory genital grooming in male rats: prevention of sexually transmitted infections. *Physiology & Behavior*, **41**, 321–5.

Hausfater, G. & Hrdy, S. B. (1984). *Infanticide: Comparative and Evolutionary Perspectives*. New York: Aldine de Gruyter.

Hendrichs, H. (1975). Changes in a population of Dikdik *Madoqua* (*Rhynchotragus*) *kirkii* (Gunther 1880). *Zeitschrift für Tierpsychologie*, **38**, 55–69.

Hendrichs, H. & Hendrichs, U. (1971). *Dikdik und Elefanten*. Munich: Piper Verlag.

Hiraiwa-Hasegawa, M. (1988). Adaptive significance of infanticide in primates. *Trends in Ecology and Evolution*, **3**, 102–5.

Houtman, A. M. (1992). Female zebra finches choose extra-pair copulations with genetically attractive males. *Proceedings of the Royal Society of London, Series B*, **249**, 3–6.

Hrdy, S. B. (1977). *The Langurs of Abu*. Cambridge, Massachusetts: Harvard University Press.

(1979). Infanticide among animals: a review, classification, and examination of the implications for the reproductive strategies of females. *Ethology and Sociobiology*, **1**, 13–40.

Hrdy, S. B. & Whitten, P. L. (1987). Patterning of sexual activity. In *Primate Societies*, ed. B. B. Smuts, D. L. Cheney, R. M. Seyfarth, R. W. Wrangham & T. T. Struhsaker, pp. 370–84. Chicago: University of Chicago Press.

Jarman, P. J. (1974). The social organisation of antelope in relation to their ecology. *Behaviour*, **48**, 215–67.

Kellas, L. M. (1955). Observations on the reproductive activities, measurements, and growth rate of Dikdik (*Rhynchotragus kirkii thomasi* Neumann). *Proceedings of the Zoological Society of London*, **124**, 751–84.

Kempenaers, B., Verheyen, G. R., van den Broeck, M., Burke, T., van Broeckhoven, C. & Dhondt, A. A. (1992). Extra-pair paternity results from female preference for high-quality males in blue tit. *Nature*, **357**, 494–6.

Kishimoto, R. & Kawamichi, T. (1996). Territoriality and monogamous pairs in a solitary ungulate, the Japanese serow, *Capricornis crispus*. *Animal Behaviour*, **52**, 673–82.

Kleiman, D. G. (1977). Monogamy in mammals. *Quarterly Review of Biology*, **52**, 39–69.

Kleiman, D. G. & Malcolm, J. R. (1981). The evolution of male parental investment in mammals. In *Parental Care in Mammals*, ed. D. Gubernick & P. Klopfer, pp. 347–87. New York: Plenum Press.

Komers, P. E. (1996*a*). Obligate monogamy without paternal care in Kirk's dikdik. *Animal Behaviour*, **51**, 131–40.

(1996*b*). Conflicting territory use in males and females of a monogamous ungulate, the Kirk's dikdik. *Ethology*, **102**, 568–79.

Komers, P. E. & Brotherton, P. N. M. (1997). Female space use is the best predictor of monogamy in mammals. *Proceedings of the Royal Society of London, Series B*, **264**, 1261–70.

Komers, P. E., Birgersson, B. & Ekvall, K. (1999). Timing of estrus influenced by male age in fallow deer. *American Naturalist*, **153**, 431–6.

Labov, J. B., Huck, U. W., Elwood, R. W. & Brooks, R. J. (1985) Current problems in the study of infanticidal behavior in rodents. *Quarterly Review of Biology*, **60**, 1–20.

Lack, D. (1968). *Ecological Adaptations for Breeding in Birds*. London: Methuen.

Lorenz, K. (1963). *On Aggression*. London: Methuen.

Manser, M. B. & Brotherton, P. N. M. (1995). Environmental constraints on the foraging behaviour of a dwarf antelope (*Madoqua kirkii*). *Oecologia*, **102**, 404–12.

Mason, W. A. (1966). Social organisation of the South American monkey, *Callicebus moloch:* a preliminary report. *Tulane Studies in Zoology*, **13**, 23–8.

McKinney, F., Derrickson, S. R. & Mineau, P. (1983). Forced copulation in waterfowl. *Animal Behaviour*, **86**, 250–94.

McNamara, J. M. & Houston, A. I. (1992). Risk-sensitive foraging: a review of the theory. *Bulletin of Mathematical Biology*, **54**, 355–78.

Mendoza, S. P. & Mason, W. A. (1986). Parental division of labour and differentiation of attachments in a monogamous primate (*Callicebus moloch*). *Animal Behaviour*, **34**, 1336–47.

Moehlman, P. D. (1989). Intraspecific variation in canid social systems. In *Carnivore Behavior, Ecology, and Evolution*, ed. J. L. Gittleman, pp. 143–63. London: Chapman & Hall.

Møller, A. P. (1988*a*). Female choice selects for male sexual tail ornaments in the monogamous swallow. *Nature*, **322**, 640–2.

(1988*b*). Paternity and paternal care in the swallow (*Hirundo rustica*). *Animal Behaviour*, **36**, 996–1005.

Møller, A. P. & Birkhead, T. R. (1993). Cuckoldry and sociality: a comparative study of birds. *American Naturalist*, **142**, 118–40.

Nunn, C. L. & van Schaik, C. P. (2000) Social evolution in primates: relative roles of ecology and intersexual conflict. In *Infanticide by Males and its Implications*, ed. C. P. van Schaik & C. H. Janson, pp. 388–412. Cambridge: Cambridge University Press.

Orians, G. H. (1969). On the evolution of mating systems in birds and mammals. *American Naturalist*, **103**, 589–603.

Ostfeld, R. S. (1985). Limiting resources and territoriality in microtine rodents. *American Naturalist*, **126**, 1–15.

Palombit, R. A. (1994*a*). Extra-pair copulations in a monogamous ape. *Animal Behaviour*, **47**, 721–3.

(1994*b*). Dynamic pair bonds in hylobatids: implications regarding monogamous social systems. *Behaviour*, **128**, 65–101.

(2000). Infanticide and the evolution of male–female bonds in animals. In *Infanticide by Males and its Implications*, ed. C.P. van Schaik & C. H. Janson, pp. 239–50. Cambridge: Cambridge University Press.

Pankhurst, S. J. (1998). The Social Organisation of the Mara at Whipsnade Wild Animal Park. Ph.D. thesis, University of Cambridge.

Parker, G. A. (1974). Courtship persistence and female guarding as male time investment strategies. *Behaviour*, **48**, 157–84.

Peck, J. R. & Feldman, M. W. (1988). Kin selection and the evolution of monogamy. *Science*, **240**, 1672–4.

Poole, J. H., Payne, K., Langbauer, W. R. Jr & Moss, C. R. (1988). The social context of some very low frequency calls of African elephants. *Behavioral Ecology and Sociobiology*, **22**, 385–92.

Rathbun, G. B. (1979). The social structure and ecology of elephant-shrews. *Zeitschrift für Tierpsychologie*, **20** (supplement), 1–76.

Read, A. F. (1990). Parasites and the evolution of host sexual behaviour. In *Parasitism and Host Behaviour*, ed. C. J. Barnard & J. M. Behnke, pp. 117–57. London: Taylor and Francis.

Richardson, P. R. K. (1987). Aardwolf mating system: overt cuckoldry in an apparently monogamous mammal. *South African Journal of Science*, **83**, 405–10.

Roberts, S. C. & Dunbar, R. I. M. (2000). Female territoriality and the function of scent-marking in a monogamous antelope (*Oreotragus oreotragus*). *Behavioral Ecology and Sociobiology*, **47**, 417–23.

Roberts, C. M. & Ormond, R. F. G. (1992). Butterflyfish social behavior, with special reference to the incidence of territoriality – a review. *Environmental Biology of Fishes*, **34**, 79–93.

Rohwer, F. C. & Freeman, S. (1989). The distribution of conspecific nest parasitism in birds. *Canadian Journal of Zoology*, **67**, 239–57.

Rubenstein, D. I. (1986). Ecology and sociality in horses and zebras. In *Ecological Aspects of Social Evolution*, ed. D. I. Rubenstein & R. W. Wrangham, pp. 282–302. Princeton, New Jersey: Princeton University Press.

Rutberg, A. T. (1983). The evolution of monogamy in primates. *Journal of Theoretical Biology*, **104**, 93–112.

Sandell, M. (1989). The mating tactics and spacing patterns of solitary carnivores. In *Carnivore Behavior, Ecology, and Evolution*, ed. J. L. Gittleman, pp. 164–82. London: Chapman & Hall.

Schwagmeyer, P. L. (1979). The Bruce Effect: an evaluation of male/female advantages. *American Naturalist*, **114**, 932–38.

Slagsvold, T. & Lifjeld, J. T. (1994). Polygyny in birds: the role of competition between females for male parental care. *American Naturalist*, **143**, 59–94.

Smith, S. M. (1988). Extra-pair copulations in black-capped chickadees: the role of the female. *Behaviour*, **129**, 99–111.

Terborgh, J. & Goldizen, A. W. (1985). On the mating system of the cooperatively breeding saddle-backed tamarin (*Saguinus fuscicollis*). *Behavioral Ecology and Sociobiology*, **16**, 293–9.

Thomas, J. A. & Birney, E. C. (1979). Parental care and mating system of the prairie vole, *Microtus ochrogaster*. *Behavioral Ecology and Sociobiology*, **5**, 171–86.

Tilson, R. L. & Tilson, J. W. (1986). Population turnover in a monogamous antelope (*Madoqua kirki*) in Namibia. *Journal of Mammlogy*, **67**, 610–13.

Tinley, K. L. (1969). Dikdik *Madoqua kirki* in South West Africa: notes on distribution, ecology and behaviour. *Madoqua*, **1**, 7–33.

van Rhijn, J. G. (1991). Mate guarding as a key factor in the evolution of parental care in birds. *Animal Behaviour*, **41**, 963–70.

van Schaik, C. P. & Dunbar, R. I. M. (1990). The evolution of monogamy in large primates – a new hypothesis and some crucial tests. *Behaviour*, **115**, 30–62.

van Schaik, C. P. & Janson, C. H. (ed.) (2000). *Infanticide by Males and its Implications*. Cambridge: Cambridge University Press.

van Schaik, C. P. & van Hooff, J. A. R. A. M. (1983). On the ultimate causes of primate social systems. *Behaviour*, **85**, 91–117.

van Tienhoven, A. & van Tienhoven, A. (1993). *Asdell's Patterns of Mammalian Reproduction: A Compendium of Species-specific Data*. Ithaca, New York: Cornell University Press.

Veiga, J. P. (1990). Infanticide by male and female house sparrows. *Animal Behaviour*, **39**, 496–502.

(1992). Why are house sparrows predominantly monogamous? A test of hypotheses. *Animal Behaviour*, **43**, 361–70.

Westneat, D. F., Sherman, P. W. & Morton, M. L. (1990). The ecology and evolution of extra-pair copulations in birds. In *Current Ornithology*, ed. D. W. Power, pp. 331–69. New York: Plenum Press.

Wickler, W. & Seibt, U. (1981). Monogamy in crustacea and man. *Zeitschrift für Tierpsychologie*, **57**, 215–34.

Williams, G. C. (1975). *Sex and Evolution*. Princeton: Princeton University Press.

Wilson, E. O. (1975). *Sociobiology*. Cambridge, Massachusetts: Harvard University Press.

Wilsson, L. (1971). Observations and experiments on the ethology of the European beaver (*Castor fiber* L.). *Viltrevy*, **8**, 115–266.

Wittenberger, J. F. & Tilson, R. L. (1980). The evolution of monogamy: hypotheses and evidence. *Annual Review of Ecology and Systematics*, **11**, 197–232.

Wolff, J. O. (1985). Maternal aggression as a deterent to infanticide in *Peromyscus leucopus* and *P. maniculatus*. *Animal Behaviour*, **34**, 1568.

Woodroffe, R. & Vincent, A. (1994). Mother's little helpers: patterns of males care in mammals. *Trends in Ecology and Evolution*, **9**, 294–7.

Wrangham, R. W. (1980). Female choice of least costly mates: a possible factor in the evolution of leks. *Zeitschrift für Tierpsychologie*, **54**, 352–67.

Wynne-Edwards, K. E. (1987). Evidence for obligate monogamy in the Djungarian hamster, *Phodopus campbelli*: pup survival under different parenting conditions. *Behavioral Ecology and Sociobiology*, **20**, 427–37.

Wynne-Edwards, K. E. & Lisk, R. D. (1984). Djungarian hamsters fail to conceive in the presence of multiple males. *Animal Behaviour*, **32**, 626–8.

Yasukawa, K. & Searcy, W. A. (1982). Aggression in female red-winged blackbirds: a strategy to ensure male parental investment. *Behavioral Ecology and Sociobiology*, **11**, 13–17.

CHAPTER 4

The evolution of social monogamy in primates

Carel P. van Schaik & Peter M. Kappeler

INTRODUCTION

Empathy probably guides most hypotheses about social behaviour (Hrdy, 1986), or at least the amount of attention given to particular problems. For several decades now, behavioural biologists have been fascinated with sexual monogamy, or its absence, in both birds and mammals. However, the question of the evolution of social monogamy has also been a topic of sustained interest (e.g., Kleiman, 1977; Barlow, 1988; Komers & Brotherton, 1997; Fuentes, 2000). Unfortunately, despite this interest, a comprehensive model of the evolution of social monogamy, at least among mammals, has so far eluded us, quite possibly because there are both multiple pathways leading up to it, and multiple benefits favouring its maintenance in different lineages. There is, accordingly, little hope for a simple unitary model for the evolution of social monogamy.

In this chapter, we examine the evolution of social monogamy in one mammalian lineage: primates. Primates are a very suitable group for a study of the evolution of social monogamy. Pairs are remarkably common among them (Kleiman, 1977; Müller & Thalmann, 2000), and range from dispersed to associated, and from variable to uniform. In the following paragraphs we outline our approach to this question.

Our focus is on the evolution of pairs as a social system, i.e., the associations and social interactions among individual animals. The social system encompasses the social organization (its size, composition, and spatiotemporal cohesion), the mating system, and social structure (patterning of social interactions and relationships among members of the social unit) (Kappeler & van Schaik, 2002). More specifically, our focus is on the social organization. There has been a tendency in earlier work to focus on the mating system and assume that other aspects of the social system are a direct consequence. However, mating system and social organization do not necessarily correspond, as amply illustrated in the case of monogamy (Reichard, 1995*b*, chapter 13), and we also need to consider factors other than just those that affect the mating system, for instance defence against predation or infanticide, that may affect social organization.

Hence, our goal is to examine the evolution of *social monogamy*, where social units consisting of one adult male and one adult female make up the predominant social organization. These units are not only characterized by the association of a pair, but also by the fact that reproduction is limited to a single female, which is unusual compared with most other types of primate social units. This insight helps us to define the trait so that we can reconstruct its evolution. For phylogenetic reconstructions to be meaningful, the trait's definition must encompass common variation within populations or variation experienced by the same individual over time. Variation at this level includes groups with a variable number of adult males, or groups with normally only a single breeding female and additional adult females that are reproductively inhibited (as in many callitrichids: Goldizen, 1990; Dunbar, 1995; Reichard, chapter 13). These cooperative breeders actually represent derived changes reflecting a long history of monogamy (discussed under Direct male care, below). Thus, in order to include these social units in our trait, we must focus on *social units in which only a single female breeds* (Single Breeding Female Unit or SBFU), although for convenience we will continue to refer to the unit of interest as *pairs*. Note, however, that this term may occasionally include units with multiple adult males or even multiple adult females, and that the mating system of these 'pairs' often involves polyandry.

The proportion of pairs (SBFUs) within a species contains information on the evolution of the trait because all species showing pair living today are likely to have started out with variable social organization. We distinguish between variable and uniform pair living.

For a species to have *variable* pairs, we decided that SBFUs are the modal, i.e., the most common, social unit but characterize less than 90% of the social units. The other units will generally have multiple breeding females. We designated a species as having *uniform* pairs when more than 90% of the social units are pairs. The point of demarcation between variable and uniform is of course somewhat arbitrary. Our statistical usage of variable versus uniform differs from the facultative–obligate dichotomy developed by Kleiman (1977, 1981) because that dichotomy conflates what are now known to be independently varying components (frequency of pairs, degree of spatial association, and extent of direct male help).

Another aspect of social organization, in addition to demographic composition of social units, is its spatial structuring (Kappeler & van Schaik, 2002). During the period of activity, pairs can show clear spatial association or be dispersed (while still sharing a common range and interacting more with each partner than with others). Hence, we should distinguish between *associated* and *dispersed pairs*.[1] Most species can be characterized unambiguously as associated or dispersed, although judgement is needed where males and females are not in close proximity at all times, but do seem to coordinate their travel and come together multiple times during a single activity period, as in *Tarsius* (MacKinnon & MacKinnon, 1980).

The evolution of pairs

A study of the evolution of any biological feature, including pair living, can involve several different steps. First, one needs to reconstruct the historical pathways leading up to the trait. This is a phylogenetic analysis, usually based on parsimony-based reconstructions in a phylogeny, ideally directed by information from the fossil record. But even the most accurate historical reconstruction does not tell us why some taxa changed toward pair living. Thus, the second step is to explain why the trait originated. This is a difficult task, bound to remain largely speculative, but comparative analysis is one of the best tools for it. The third and final step is to explain why the trait is maintained, i.e., why it is adaptive in the sense of current function or utility. This latter task is more feasible because we can turn to extant pair-living species and to the various ways of testing adaptations (Futuyma, 1998).

Our approach, then, will be to reconstruct the historical pathways leading to pair living in each taxon that currently shows variable or uniform pair living, to identify the general preconditions for the transition towards variable pair living, to demonstrate that uniform pairs followed on variable pairs, to identify the selective benefits that favoured the transition from variable to uniform pairs, and to identify the forces that currently maintain pair living.

THE HISTORICAL EVOLUTION OF PRIMATE PAIRS

Historical reconstructions

Reconstruction of the social system of the stem-primates suggests that they were solitary foragers, with a mating system in which males were predominantly polygynous and females were polyandrous or at least not strictly monandrous (Kappeler, 1999; Müller & Thalmann, 2000). Accordingly, most primates do not live in pairs. Pairs are derived social systems; indeed, as we shall show here, they tend to be found near the tips of the phylogenetic tree.

If uniform pairs can only derive from variable pairs (see the following section), it makes sense to consider the female social arrangement as ordered states (i.e., making it impossible to leap from non-pairs straight to uniform pairs). However, because the length of this variable phase may vary considerably as a function of the benefits of pair living, it is not clear whether the switch to uniform always involved speciation. Hence, we also defined the female social arrangements as unordered states, and performed our reconstructions using both assumptions.

In the cladogram of Figure 4.1, we consider the female social arrangements' states ordered and, so as to resolve states on all branches, reconstruct their evolution using the rule that delays ambiguous transitions as much as possible (DELTRAN option in MacClade: see

Figure 4.1. Primate phylogeny, indicating the origins of pair living as defined in the text. To maximize the estimated number of origins, we used the DELTRAN reconstruction option, which forces ambiguous changes closer to the tips of the tree. The phylogeny is based on Purvis (1995), but modified according to Fleagle (1999) and Purvis and Webster (1999), where relevant.

Otolemur
Galagoides zanzibaricus
Galagoides
Galago
Euoticus
Nycticebus
Loris
Perodicticus
Arctocebus
Phaner
Cheirogaleus
Cheirogaleus medius
Mirza
Microcebus
Allocebus
Avahi
Indri
Propithecus
Hapalemur
Eulemur
Eulemur rubriventer
Varecia
Lemur
Lepilemur
Daubentonia
Tarsius spectrum
Tarsius pumilus
Tarsius bancanus & other Tarsius
Cebuella
Callithrix
Callimico
Leontopithecus
Saguinus
Aotus
Saimiri
Cebus
Callicebus
Pithecia
Chiropotes
Cacajao
Alouatta
Lagothrix
Ateles
Brachyteles
Presbytis
Presbytis potenziani
Semnopithecus
Trachypithecus
Nasalis
Simias
Pygathrix
Rhinopithecus
Colobus
Procolobus
Cercopithecus
Chlorocebus
Erythrocebus
Miopithecus
Allenopithecus
Macaca
Cercocebus
Mandrillus
Lophocebus
Theropithecus
Papio
Hylobates
Pongo
Gorilla
Pan paniscus
Pan troglodytes
Pairs ordered
Other
Variable pairs
Uniform pairs

Table 4.1. *Number of independent origins of variable and uniform pairs, considering the different female grouping character states ordered or unordered, and using three different rules to reconstruct character evolution (parsimony, DELTRAN [resolving states that remain ambiguous when using parsimony so as to delay changes] and ACCTRAN [resolving ambiguous states so as to accelerate changes])*

Kinds of pairs	Unordered states			Ordered states		
	Parsimony	DELTRAN	ACCTRAN	Parsimony	DELTRAN	ACCTRAN
Variable	7	7	7	6	5	6
Uniform	10	10	7	10	10	8
Any pairs	11	10	8	11	10	8

Maddison & Maddison, 1992). Table 4.1 gives the number of origins of variable and uniform pairs using several different criteria.

Both the most parsimonious and the DELTRAN reconstructions of the evolution of uniform pairs suggest ten independent origins of uniform pair living in primates, whereas the reconstruction that assumes the fewest origins, forcing equivocal reconstructed states towards the root of the tree (ACCTRAN), produces seven or eight independent origins. The origins identified are listed in Table 4.2.

We also find five or six origins of variable pairs, rising to seven if we consider the social arrangements unordered states (i.e., make no assumptions about the order of change: Table 4.2). The nocturnal *Tarsius spectrum*, for instance, forms mainly pairs, but there are enough exceptions to suggest variable pairs (MacKinnon & MacKinnon, 1980; Gursky, 2000*a*, *b*). Other nocturnal species, *Galagoides zanzibaricus* and *Lepilemur edwardsi* (and possibly other congeners), often live in pairs (Harcourt & Nash, 1986; Müller & Thalmann, 2000; Rasoloharijaona *et al.*, 2000). Among the infant-carrying lemurs, most genera show variable pairs, and some (e.g., *Eulemur mongoz*, *Hapalemur aureus*) may even show a strong tendency towards pair living (e.g., Kappeler, 1997*a*; see Table 1 in Kappeler, 2000). Among monkeys, *Pithecia* spp. are usually found in pairs, but in some areas larger groups are common (Lehman *et al.*, 2001) although, intriguingly, adult sex ratios remain approximately even. Pairs are less common in *Simias* (Newton & Dunbar, 1994). Not included in Figure 4.1 are species where some groups may be so small that they contain only a single female, as in *Cercopithecus neglectus* and *Presbytis comata*, because they clearly are at the multi-female end of the spectrum (Cords, 1987; Newton & Dunbar, 1994).

Several of the taxa with variable pairs show sexual dimorphism in body or canine size, or both (the exceptions being the lemurs and *Pithecia*). Sexual dimorphism is consistent with the common occurrence of multi-female groups, and thus polygynous mating systems. Moreover, there is a tendency for species with variable pairs to show female-biased adult sex ratios at the population level, and those with uniform pairs to show the opposite sex bias (van Schaik & Kappeler, unpublished data), again suggesting different evolutionary histories.

Did variable pairs precede uniform pairs?

It is reasonable to assume that variable pair living arose more or less by accident when the more typical social and mating system failed to materialize (often referred to, from the male perspective, as 'polygyny failed' e.g., Barlow, 1988). In some taxa, such variable systems would subsequently have turned into uniform pair living, whereas they would have remained variable in others, and disappeared again in yet others. However, it is important to test this assumption of compulsory evolutionary directionality by formulating and testing additional predictions. First, variable pair living should give rise to uniform pair living, but the reverse transition should be much less common. This prediction is tentative because it is not known how long it takes to acquire uniform pair living, or whether speciation needs to be involved. In Figure 4.1, when states of female social arrangements are considered unordered, uniform pairs evolved once from variable pairs, whereas the reverse never did. However, when we assume the states are

Table 4.2. *Kinds of pairs: origins of uniform and variable pair living among primates and various correlates (using non-pairs, variable, and uniform pairs as unordered states)*

Clade	Uniform/ variable pairs	Activity period	Body size (kg)	Associated/ dispersed	Mode of infant care	Litter size	Male care (1–4)	Pairs (unif./ var.) among sister taxon	Female party size in sister taxon
Hylobates spp.	U	D	5–10	Associated	Carry	1	1, 3	No	<2
Presbytis potenziani	U	D	6	Associated	Carry	1	[a]	Some	2–5
Callicebus	U	D	1	Associated	Carry	1	4	Yes	<2–5
Aotus	U	N	1	Associated	Carry	1	4	No	>5
Callitrichids	U	D	<1	Associated	Carry	2	3–4	Yes	<2
Eulemur rubriventer	U	C	2	Associated	Carry	1	2?	Yes	2–5
Avahi	U	N	1	Associated	Carry	1	1	Yes	<2
Indri	U	D	7	Associated	Carry	1	1?	Yes	2–5
Cheirogaleus medius	U	N	<1	Dispersed	Nest + park	2	3–4	No?	Sol
Phaner	U	N	<1	Dispersed (mainly)	Freq. park	1	1	Yes	Sol
Simias concolor	V	D	8	Associated	Carry	1	[a]	No	2–5
Pithecia spp.	V	D	1	Associated	Carry	1	1–2	No	2–5
Tarsius spectrum	V	N	<1	Associated (mainly)	Nest + freq. park	1	1 (–2?)	No?	Sol
Hapalemur, *Eulemur*, *Varecia*	V	C (D)	1–3	Associated	Mainly carry	1–2	1 (*Varecia*: 3)	Yes	<2
Lepilemur p.p.	V	N	<1	Dispersed	Nest + park	1	1	Yes	Sol
Propithecus spp.	V	D	3–6	Associated	Carry	1	1	Yes	<2
Galagoides zanzibaricus	V	N	<1	Dispersed	Nest + park	1–2	1	No?	Sol

Abbreviations: U, uniform; V, variable; C, cathemeral; D, diurnal; N, nocturnal; Sol, solitary.

[a] Unknown.

Female party size in sister taxon refers to reconstructed value of the root of the sister clade using parsimony. Female party size of ancestor is the reconstructed party size on branch below the pair-living taxon in the tree. Levels of male care follow the categories of Ross & MacLarnon, 2000.

Sources: Kappeler, 1997*a*, 1998; van Schaik & Kappeler, 1997; Ross & Jones, 1999; Ross & MacLarnon, 2000; Gursky, 2000*a*, *b*; Müller & Thalmann, 2000; Nunn & van Schaik, 2000.

ordered, uniform pairs evolved from variable pairs five times, whereas the reverse never happened (the ordering of states does not constrain the direction of change). Thus, the reconstructions are consistent with the first prediction.

A second prediction is that uniform pairs should only rarely revert to other social systems. The degree of irreversibility may depend on the degree to which female and offspring biology (infant development rate, litter size) have changed to rely on male care: uniform pairs with uniform male care can probably not escape from their state. (Note that the obvious exception – a change toward polyandrous helper systems – is subsumed under pairs in this analysis.) Indeed, in Figure 4.1 there are no cases of uniform pairs giving way to other social arrangements, regardless of whether we consider the states ordered or unordered, or trace character states parsimoniously or by delaying changes (DELTRAN). On the other hand, there are one or two transitions from variable pairs towards other states: to larger groups, in *Lemur*; and (assuming ordered states and DELTRAN) to a solitary social organization, in *Cheirogaleus*. Thus, the second prediction is also supported.

Finally, extensive male involvement in infant care, considered to be a result of long periods of uniform pair living (see below), should only be seen in uniform pairs, whereas we expect no direct male care in the variable pairs. This is indeed the case (Table 4.1), although evidently not all uniform pairs show strong male involvement.

Although in none of these cases could we make a quantitative assessment of the strength of support, the consistency of all three tests with the predictions make it parsimonious to accept the assumption that variable pairs indeed precede uniform pairs.

Pathways to pairs

An important task is the reconstruction of the pathways that have led to pairs. The ancestral states can be solitary females or females living in (mixed-sex) groups. For each origin of associated or dispersed pairs we can ask whether the state at its origin was solitary or mixed-sex groups, which is reconstructed separately from the current distribution of these two states. Table 4.3 shows that there are two main reconstructed pathways (Fisher exact test, $P = 0.0021$): from solitary females to dispersed pairs, and from mixed-sex groups to associated pairs. The existence of these two major pathways is not surprising, but it points out the need for separate analyses for the origin of dispersed and associated pairs.

Table 4.3. *Pathways to pair living, as reconstructed from information presented in the cladogram of Figure 4.1, and in Table 4.1.*

	To:	
From:	Dispersed	Associated
Solitary	4	1
Mixed-sex groups	0	12

Fisher exact test, $P = 0.0021$.

The evolution of dispersed pairs from solitary ancestors

Among nocturnal, infant-caching primates, solitary foraging is the rule (Bearder, 1987; Kappeler, 1998; Müller & Thalmann, 2000). Due to the scarcity of data on spatial coordination and interactions during the active phase, their social systems are poorly described, but perhaps the best approximation comes from daytime sleeping associations and from ranging data. A plausible hypothesis for the evolution of variable pairs among these taxa is that they arise when a male is unable to defend mating access to more than one female. Among solitary foragers, the precondition for this outcome is more or less exclusive female ranges, so that within the range of a female a roving male cannot find other oestrous females (cf. Komers & Brotherton, 1997). This precondition can give rise to uniform pair living when these exclusive ranges are so large that a male cannot cover an area that includes more than one female's range without risking losing exclusive or main access to any single female, or when despite moderate female home range size,

Figure 4.2. Phylogeny of strepsirrhine primates, with a reconstruction of the evolution of female territoriality. Phylogeny as in Figure 4.1; ranging data from Charles-Dominique & Hladik, 1971; Charles-Dominique, 1977; Harcourt & Nash, 1986; Bearder, 1987; Sterling, 1993; Kappeler, 1997*b*; Warren & Crompton, 1997; Müller, 1999; Atsalis, 2000; Gursky, 2000*a*, *b*; Müller & Thalmann, 2000; Radespiel, 2000; personal observation.

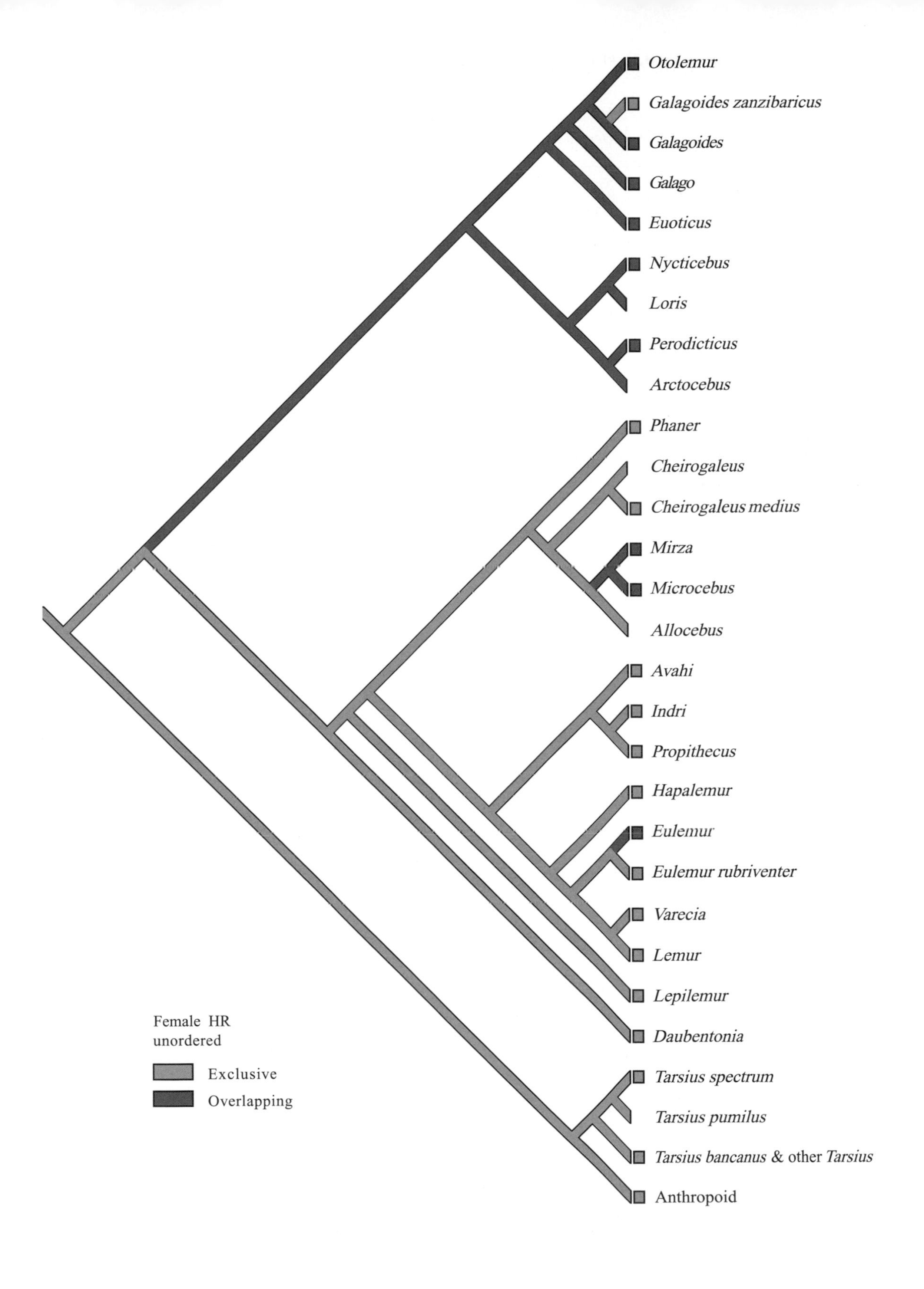

Otolemur
Galagoides zanzibaricus
Galagoides
Galago
Euoticus
Nycticebus
Loris
Perodicticus
Arctocebus
Phaner
Cheirogaleus
Cheirogaleus medius
Mirza
Microcebus
Allocebus
Avahi
Indri
Propithecus
Hapalemur
Eulemur
Eulemur rubriventer
Varecia
Lemur
Lepilemur
Daubentonia
Tarsius spectrum
Tarsius pumilus
Tarsius bancanus & other Tarsius
Anthropoid
Female HR
unordered
Exclusive
Overlapping

Table 4.4. *Proportion of basal branches of sister taxa to taxa with variable or uniform associated pairs that have different female grouping character states*

Female social state	Total number of branches in cladogram	Number of basal branches of sisters with this state (unordered)	%	Number of basal branches of sisters with this state (ordered)	%
<2	33	6	18.2	4	12.1
2–5	34	5	14.7	5	14.7
>5	31	1	3.2	1	3.2
Solitary	41	5	12.2	5	12.2

Analysis is based on DELTRAN reconstructions using non-pair living, variable pairs, and uniform pairs either as unordered or as ordered character states. Although the choice for ordered states is plausible, it leads to the lumping together as homologous of rather different kinds of variable pairs among lemurs (the associated pairs in the cathemeral and diurnal species and dispersed pairs of *Lepilemur*). Hence, the unordered states option was also explored.
The solitary social state is added for comparison only (because associated pairs rarely arose from the solitary state).

the females do not widely advertise their oestrus or have their oestrous cycles highly synchronized (cf. Emlen & Oring, 1977; Dunbar, 2000).

There are not enough field data to assess whether the detailed prediction holds, but we can examine the validity of the broad precondition: lineages with pair-living species should show a stronger tendency towards exclusiveness of female ranges than those without pair-living species. Preliminary evaluation provides weak support for this hypothesis (Figure 4.2). In the galago clade, overlapping female ranges are reconstructed as ancestral, and only one species (*Galagoides zanzibaricus*) is found to live in variable pairs, whereas among the lemurs, female range exclusivity is ancestral, and two or more taxa (*Lepilemur* spp., *Cheirogaleus medius*) show dispersed pairs, one of them probably of the uniform kind. It is possible that pairs are even more common here. If social arrangements are considered ordered, variable pairs is the reconstructed ancestral state of this part of the lemur clade (see Figure 4.1). Although the variable pairs shown in this clade are associated ones (perhaps intermediate in *Phaner*), it is possible that variable pairs preceded infant carrying in this clade. Furthermore, in the *Tarsius* clade, female range exclusiveness is reconstructed as ancestral, and *T. spectrum* shows variable pairs (Gursky, 2000*a*). However, the lorises are also reconstructed as showing exclusive ranges, but there is no evidence for pairs among them.

What this analysis does not reveal, of course, is why females in some taxa have more or less exclusive ranges. Ecological explanations are traditionally sought for this, but social factors (e.g., Wolff & Peterson, 1998) may be relevant as well.

The evolution of associated pairs from gregarious ancestors

We can examine female group size in sister taxa to get a reasonable estimate of group size in the lineages that produced pairs. We inspect sister groups rather than reconstructed ancestral states because the species with pairs affect the reconstructed ancestral states but not those of the basal state of the sister groups (except to some extent in the DELTRAN option). The likelihood of any given ancestral state can then be expressed as the proportion of reconstructed states of the basal branch of the sister group from among all possible branches in the tree with the same character state. Table 4.4 gives various features for the ten known origins of uniform pair living in primates, as well as several of the five to seven origins of variable pairs. Variable and uniform associated pair-living taxa are least likely to have evolved from large female groups (Table 4.4), but with roughly equal frequency from social units with small mixed-sex groups (<2 females on average), or medium-sized, mixed-sex groups (2–5 females). This calculation ignores the resolution of lineages that have no pair-living species at all, but their inclusion would skew the expected states further toward large groups.

The sole example of a pair-living taxon whose sister lineage, the *Cebus–Saimiri* clade, lives in large groups is *Aotus*. It is possible that *Aotus* evolved a nocturnal lifestyle along with the change in social system, but it is difficult to imagine a species living in large groups becoming nocturnal. Hence, it seems more plausible that it has retained a pair-living lifestyle (shared with the callitrichids) and secondarily became nocturnal, whereas the sister lineage retained diurnality and secondarily went on to form larger groups. In conclusion, pair-living taxa have overwhelmingly arisen in lineages with small female group size.

The ecological causes for small female group size may well have been different in different lineages. We can discern three patterns, but explanations for them are speculative, given the limited testability of such historical scenarios.

First, many of the largest arboreal–terrestrial and medium-to-large arboreal primates live in bottom-up fission–fusion situations. These lineages contain taxa characterized by low predation risk in the classification of Nunn and van Schaik (2000). However, also due to their large size, and, where arboreal, their high locomotion costs, they face strong within-group feeding competition. Thus, females of these species tend towards foraging in small parties. Some of these species live in pairs, but others do not. Examples of the latter are spider monkeys, chimpanzees, and orangutans. In these species, female ranges overlap widely, female antagonism varies with food abundance, and association with males tends to be low, although they do live with a fixed set of males (chimpanzees, spider monkeys: Goodall, 1986; Symington, 1990), or preferentially associate and mate with only one local male (orangutans: Delgado & van Schaik, 2000). The absence of pair living in these species is most likely due to the combination of large female home range size and high range overlap. Males have more difficulty defending exclusive access to a female when females have large ranges, whereas high overlap makes it more feasible to pursue polygynous matings (cf. van Schaik & Dunbar, 1990; Dunbar, 2001).

Second, among the small-bodied Neotropical platyrrhines, small group size may, paradoxically, be attributable to unusually high predation risk. These small species, 1 kg or less, are exposed to a large array of mammalian and avian predators, and most, with the obvious exception of *Saimiri*, have responded to them by adopting a cryptic lifestyle (van Schaik & van Hooff, 1983; Janson & Goldsmith, 1995). They tend to concentrate in dense vegetation, such as vine tangles, patches of secondary forest, and understorey, and behave inconspicuously. This emphasis on avoidance of detection is incompatible with large group size. Obviously, the resulting very small group size should predispose them in the direction of pair living. This explanation is tentative, and difficult to test. It can be challenged on the grounds that some of the small callitrichids seem to be less than cryptic (Heymann, chapter 16). However, no obvious alternative interpretations exist.

Third, the high incidence of variable pair living among lemurs of various body sizes remains enigmatic. Their group sizes tend to be smaller than those of other primates (Kappeler & Heymann, 1996), but, unique among mammals, their groups have approximately equal adult sex ratios (Kappeler, 2000), despite clear trends towards male monopolization of matings (e.g., Overdorff, 1998; Ostner & Kappeler, 1999). Although several hypotheses have been proposed to explain this unusual group composition, none so far has been satisfactory (van Schaik & Kappeler, 1996). Indeed, it is possible that the groups in this lineage are small because variable pair living was the ancestral condition for all lemurs except *Daubentonia* (Figure 4.1; assuming that the female social arrangements are ordered states). Year-round male–female association in these species is expected because they all carry their infants and have lengthy life histories (van Schaik & Kappeler, 1997; see also The Origins and maintenance of uniform pair living, below). Van Schaik & Kappeler (1996) argued earlier that until recently, the majority were predominantly nocturnal, with some cathemeral activity, and thus lived in small groups which, because of intersexual bonding, were mainly pairs. However, this idea is difficult to test and strongly contested (e.g., Overdorff, 1998; Wright, 1999).

The exception: associated pairs from solitary ancestors

We should note here that the characterization of *Tarsius spectrum* is still somewhat preliminary. In *T. spectrum*, infants are frequently parked and carried in the mouth during the night (MacKinnon & MacKinnon, 1980). It is not known to what extent all *Tarsius* frequently park their infants; hence, it is not known whether the male–female association seen in *T. spectrum* is an inevitable consequence of frequent cache and carry (Gursky, 2000*a*), or whether it is a special, derived feature of the variable pair living of *T. spectrum*. *Phaner*, another taxon for which the available evidence once indicated the existence of associated pairs (van Schaik & Kappeler, 1997), is now best considered dispersed (Schülke & Kappeler, 2003).

If the infant-carrying, nocturnal *Avahi* is derived from a solitary ancestor, it made the same transition from solitary foraging to associated pairs. However, two alternative scenarios are more likely, and both involve a gregarious ancestral state. Thalmann (2001) assumed *Avahi* was previously diurnal and simply maintained its lifestyle subsequent to its transition to nocturnality, as did *Aotus*. In contrast, van Schaik & Kappeler (1996) assumed that most of the other lemurs were also nocturnal or cathemeral, and recently shifted toward diurnality, making *Avahi* simply one among many nocturnal, infant-carrying, prosimian primates (cf. Kappeler, 1998).

THE ORIGINS AND MAINTENANCE OF UNIFORM PAIR LIVING

Introductory remarks

In both dispersed and gregarious species, we argued that pairs originally arose in situations where the more typical social and mating system failed to materialize. Thus, pairs almost certainly started out variable and could only become uniform over time. What factors were responsible for the evolution of the situation in which deviations from pair living had become very rare?

Even if one of the partners would benefit from pair living, it is unlikely to happen consistently whenever the other player has viable alternatives. The transition from variable to uniform pair living requires a situation in which pair living is the Evolutionarily Stable Strategy (ESS), i.e., the stable and non-invadable outcome for both sexes, either because both players prefer this state or because one of them prefers it and the other cannot change it toward his or her preferred state (see Davies, 2000). Over time, the latter situation should produce changes in the behaviour or reproductive biology of the partner preferring the non-pair-living state, changes that maximize fitness under pair living and moving the optimum to pair living for both partners. We assume that, in equilibrium, uniform pairs are in this state; the acid test for uniform pairs is therefore whether removing the male leads to a clear loss of fitness, most commonly due to an increase in infant loss. Below, we will discuss the specific benefits to infant production or survival postulated by each particular hypothesis.

Hypotheses for the origin of uniform pairs must explain why the ancestral state was not favoured by natural selection. Among territorial, solitary, infant-caching females, this state was composed of variable pairs and solitary females and roaming males or dispersed multi-female units inside the range of a single male. Among infant-carrying females, the ancestral state was variable pairs and small, single-male/multi-female groups. The benefits that favoured the origin will often also be the ones that maintain it because these ancestral states are still possible as alternative states. However, changes in ecological conditions or in the (reproductive) biology of the species subsequent to the evolution of pair living may cause some of these original selective benefits to disappear over time or lead to the emergence of new selective benefits. It is difficult to identify examples of the former, but the most likely of the latter is direct male care.

We will first discuss the following possible selective agents for the maintenance of pair living: (i) male resource defence; (ii) infanticide avoidance; (iii) baby sitting; and (iv) mate guarding. After that, we will discuss the evolution of male carrying and provisioning.

Male defence of resources

Especially for gibbons, it has been argued that the main benefit to females of having a bonded male partner, i.e., of pair living, is that the male defends access to the female's food supply, and thus improves the female's nutritional status, and hence, indirectly, his own fitness (Raemaekers & Chivers, 1980; Brockelman & Srikosomatara, 1984). More recently, Müller (1999) suggested that male *Cheirogaleus medius*, by emerging earlier than females from hibernation defend the food supply for their mates and their offspring. This hypothesis does not predict uniform pairs, however, because there is no reason why males, given their normal preference for polygynous mating, need to associate with the female rather than defend the ranges of multiple females in areas of unusually high productivity (see van Schaik & Dunbar, 1990). Likewise, in the case of gibbons, such rich habitats could also be expected to contain small groups of females rather than individual females. Nonetheless, a clear prediction of this hypothesis is that males should defend the territory and its resources not just against other males, but also against other females.

This is not usually observed (e.g., Mitani, 1984). We will therefore not consider this hypothesis further.

Infanticide reduction and the transition to uniform pairs

Van Schaik and Dunbar (1990) hypothesized that associated pairs are found where females vulnerable to infanticide are antagonistic towards other females and thus have more or less exclusive ranges, but are associated with a male who provides protection against infanticidal males. This idea was later seen as a special case of the more general hypothesis (van Schaik & Kappeler, 1993, 1997) that the risk of infanticide by males favoured the evolution of year-round male–female association among all infant-carrying primate species. In these taxa, female solitary dispersion leads, by default, to variable associated pairs, and to uniform pairs if, in addition, the females are antagonistic (territorial). The argument therefore has two steps: (i) the evolution of year-round male–female association in infant-carrying primates; and (ii) the evolution of associated pairs where females in these species are mutually antagonistic.

Infanticide risk and counterstrategies in primates

Infanticide by males is adaptive when three conditions are met simultaneously: the male kills offspring he did not sire (at little cost to himself, e.g., injury or future access to females), the female returns to receptivity earlier than she would have otherwise, and the male has a high probability of siring that female's next infant. It is well known that the risk of infanticide by males among primates is among the highest in mammals (Hausfater & Hrdy, 1984). This vulnerability has a clearly defined biological basis: the slow development of infants (van Schaik, 2000*b*). The critical factor turns out to be the duration of lactation relative to gestation. The longer the lactation, relative to gestation, the less likely are post-partum matings, and the more likely time gains, and thus male fitness gains, emanating from loss of dependent offspring (van Schaik, 2000*c*). Unfortunately, data on the relative duration of lactation are not available for a critical set of primate species, so it is not certain whether all of the infant-cachers are below the infanticide-threshold value of relative lactation length and all infant carriers above it.

Because the loss of the currently dependent offspring curtails female fitness, especially in slow-breeding organisms, female counterstrategies are to be expected (Hrdy, 1979). Some of these are sexual in nature (van Noordwijk & van Schaik, 2000), but others should be social. Among the social counterstrategies, the role of permanent association with males is prominent. Not only will all males that have mated with the female refrain from attacking the infant (Hrdy 1979; Borries *et al.*, 1999; van Schaik, 2000*a*), but most importantly, the likely sire will defend the infant against potentially infanticidal, incoming males.

The role of this male is probably critical to the survival of the infant, and hence to the fitness of both mother and sire. First, experimental removal of this male, both in the wild and in captive situations, when followed by the introduction of one or more new males, reliably provokes infanticidal attacks in a large array of primate species, almost 40 at the last count (for review, see van Schaik, 2000*a*). Second, the male's critical role was demonstrated in a review based on directly observed cases of infanticide among wild primates: van Schaik (2000*a*, building on Struhsaker & Leland, 1987, and Sommer, 1994) found that prior to the infanticidal event the likely sire had been eliminated or incapacitated in 85% of all cases. Third, when examining variation in the estimated rate of infanticide across species and populations, the relative rate of replacement of the group's dominant male (relative to the average inter-birth interval) emerges as one of the main correlates (Janson & van Schaik, 2000). Various other patterns are also consistent with the infanticide-avoidance hypothesis. Thus, most behavioural changes between the three stages of the group's life cycle in Thomas' langurs (*Presbytis thomasi*) only make sense if group formation and dissolution are seen as reducing the risk of infanticide (Steenbeek, 2000). This even includes patterns in vigilance behaviour, resting heights, range use and overlap, and male morning calls.

If protection against infanticide by year-round association between males and females is indeed an adaptation against the risk of infanticide, then this phenomenon should be much more common when females are vulnerable to infanticide by males. Infant development is especially slow among those species in which females carry their infants (van Schaik & Kappeler, 1997). Indeed, virtually all primate species in which females carry their infants show male–female association, whereas, with one exception, this association is not shown in any of the species that do not carry their infants, i.e., cache them (van Schaik & Kappeler 1997).

Other factors show a less tight correlation with year-round male–female association among primates. However, a few intermediate cases are known; of particular interest is *Phaner*, about which more information is badly needed (see Schülke & Kappeler, 2003). Nonetheless, the hypothesis that year-round male–female association serves as an infanticide-reduction strategy is generally supported by its taxonomic distribution and by the common occurrence of males protecting infants.

Associated pairs where females are mutually antagonistic

If females live in groups with other females, intersexual association produces the well-known single-male or multi-male groups that are so common among primates. But if the number of infant-carrying females approaches one, then by default this produces variable pairs which, due to their gregariousness, will be noted as associated pairs. Obviously, this association must be maintained or else one would expect males to opt for dispersed, multi-female systems whenever defensibility is high enough. That this did not happen, with the exception of some of the largest fission–fusion species (*Ateles*, *Pongo*, *Pan troglodytes*), suggests that where females could afford to associate, they gained from maintaining the association with the single male. From an ecological perspective it can be argued that pairs will form when females can afford to live in very small groups, and that increasing group size is ecologically costly, but a group size of two adults and several associated maturing subadults is ecologically tolerable. However, in the richest habitats, i.e., those with the smallest home ranges, small groups of two or three females should be expected. Thus, it is fair to ask why they form *uniform* pairs, even during temporary food gluts and in the richest habitats.

Under the infanticide model, female antagonism could not only serve the ecological function of resource defence, but could also keep the number of females small. Smaller female group size reduces the risk of takeovers, which are likely to be accompanied by infanticide. In two species with small single-male groups, groups with a greater than average number of females were the exclusive target of takeover attempts by extra-group males (red howlers: Crockett & Janson, 2000; Thomas's langurs: Steenbeek & van Schaik, 2001). In these two species, mean female group sizes were small, at 2.5 and 3, respectively. Where it becomes even smaller, the size at which takeover risk is minimized is one (1) female because every group with as few as two would make a more attractive target.

Reducing the takeover risk by minimizing the number of female companions may be worth the inclusive fitness cost of forcing female relatives to leave (especially if this leaving would also lead to avoidance of inbreeding). In red howlers, resident females actively maintain small group size: as group size increases, the proportion of quickly expelled maturing females increases to near 100% (Pope, 2000). Targeted aggression, often resulting in the eviction of female relatives, may serve the same function in several lemur species (Vick & Pereira, 1989; Pereira, 1993). In Thomas' langurs, the relatively short duration of male tenure usually prevents the build-up of such large groups; most male turnover does not involve takeovers, but rather females leaving males unable to defend them as soon as they have no vulnerable infant (Steenbeek, 2000).

In the absence of variation in female group size, it is not easy to test this subsidiary hypothesis for pair-living species. However, data from gibbons suggest that habitats are normally saturated, and taking over already occupied territories should be the most common way in which maturing individuals establish themselves (see Brockelman *et al.*, 1998; *pace* Palombit, 1999). Hence, in normal populations there is a pool of would-be territory owners ready to strike (Cowlishaw, 1992), and making territories more attractive for takeover by having multiple females in them may reduce male tenure below the break-even point. Exceptions to this rule are expected where females can let female relatives reproduce alongside them without reducing their reproductive output, in which case the inclusive fitness benefits may outweigh the increased takeover risk.

Baby sitting

Among the infant-carrying primates, predation avoidance through male protection is an automatic by-product of male–female association. In the solitary infant-cachers, survival may be critically dependent on males guarding the nests of newborn infants. Fietz (1999, chapter 14) reports nest guarding by male *Cheirogaleus medius* and loss of the litter following disappearance of the male, both consistent with such a benefit. Unfortunately, sample size was small.

It is possible that there are other benefits to nest guarding (baby sitting), but not enough is known of prosimian biology to evaluate them. In a situation analogous to that of *Peromyscus californicus* (Ribble, chapter 5), it is possible that thermoregulatory benefits provided by the male may be important in some species. Since uniform pair living is rare among the infant-cachers, it is likely that the ecology of *Cheirogaleus* is unusual in some respect.

Male mate guarding

It can be argued that opportunities for male polygyny are so rare that the optimum male mating strategy is permanently to guard a single female (Palombit, 1999; Komers & Brotherton, 1997), i.e., it does not pay for the male to maintain assessment mechanisms for flexible decision making. This is remarkable for a mammal because at least dominant males are generally thought to prefer a polygynous mating system (e.g., Hawkes *et al.*, 1995), making it unlikely that variable pair living ever evolves to become uniform. Palombit (1999) suggested that in gibbons, guarding a single female is nonetheless the optimum choice for males, one in which males gained a direct benefit because only paired males would improve their chances for extra-pair matings. This argument assumes, of course, that the fitness benefits of gaining extra-pair matings outweigh the benefits of a polygynous strategy.

Note that all other hypotheses discussed here also expect males to guard mating access to the female with whom they form a pair because females may benefit from social monogamy but not necessarily from genetic monogamy. Scenarios can be envisioned in which mate guarding led to the origin of (variable) social monogamy, but if mate guarding is to explain the *maintenance* of uniform pair living, we must assume that females are either indifferent to or do not prefer living in pairs, i.e., do not derive strong benefits from males in terms of improved infant growth or survival (cf. Palombit, 1999). Dropping this assumption amounts to adopting one of the other hypotheses.

The mate-guarding hypothesis makes two predictions. The first is critical. It states that no effect of the loss of the male partner on infant fitness is expected under this hypothesis, whereas all other hypotheses expect solid benefits. In general, there is little evidence for or against this hypothesis, although in callitrichids, the role of males in increasing the unit's reproductive output is pronounced (e.g., Koenig, 1995; Heymann & Soini, 1999). In species vulnerable to infanticide, loss of the male leads to a clear increase in infant mortality (van Schaik, 2000*a*), but the lack of data for pair-living species makes it difficult to evaluate this idea. Because the hypothesis has been invoked especially for gibbons, it is important that systematic evidence be gathered to evaluate it for this taxon.

A second critical prediction is that there is no reason to expect any pair bonds at all because only the male is expected to benefit and no reciprocal interests exist. Yet, in all species clear pair bonds are present, as evidenced by reciprocal grooming and coordinated proximity (e.g., Pollock, 1979; Harcourt, 1991; Reichard, 1995*a*; Palombit, 1996; Curtis & Zaramody, 1999) and by separation-induced stress (Mendoza & Mason, 1986), even though males and females are not equally responsible for maintaining the relationship.

Thus, for now, this hypothesis lacks strong support. Nonetheless, Palombit (1999) decided in favour of the mate-guarding hypothesis for gibbons, based in large part on his observation that hylobatid males are mainly responsible for maintaining proximity. This was interpreted as reflecting the absence of substantial female benefit, and thus as arguing in favour of male mate guarding as the selective advantage keeping the male bonded to the female. This is a reasonable interpretation, and unless a viable alternative can be provided, one is inclined to consider the mate-guarding hypothesis live. Hence, we need to examine whether responsibility for maintaining a pair bond constitutes firm evidence in favour of the hypothesis.

Greater male responsibility for proximity maintenance may be due to differences in benefit, but also to (unacknowledged) differences in relative power. Relative responsibility for proximity (R), reflects relative benefits (B), corrected for relative power (P):

$$R_a / R_b = B_a / B_b \times P_a / P_b \quad (4.1)$$

where a and b are the two players. Power refers to the combined effect of dominance and leverage, i.e., any asymmetry in the relationship regardless of its origin. In monomorphic animals, power is mainly based on leverage (terminology follows Lewis, 2002), the ability to elicit behavioural asymmetries based on some influence other than force. Leverage can have a variety of sources, but what they have in common is that a can provide benefits to b that b cannot take by force. In pairs, the most obvious sources are a fertilizable egg (in an oestrous female), the ability to carry or provision young, the ability to defend a territory, possession of a territory, and the presence of alternative partners. Thus, imbalances in the local sex ratio may give the rarer sex significant leverage in a pair relationship, an example of the market effect (Noë *et al.*, 1991; Noë & Hammerstein, 1994).

It is often implicitly assumed that in pair-living organisms $P_a / P_b = 1$. However, there are many indications that the power ratio affects the value of R_a / R_b. In practice, this ratio is estimated by the Hinde index.[2] If a is the female, $R_a / R_b > 1$ corresponds to a positive value of the Hinde index. Thus, the positive values found in baboon friendships (*Papio anubis*: Palombit, 1999) reflect the relative rarity of male partners, former top-ranking males, and likely sires, each of which tends to have several female 'friends' (Palombit *et al.*, 1997). Likewise, the less negative value for the index in siamangs (*Hylobates syndactylus*) than in white-handed gibbons (*H. lar*: Palombit, 1996) can be attributed to the infant carrying of the siamang male, giving him greater leverage relative to the gibbon male. Negative values are expected in species with multi-male groups, and no special relationships or friendships. They are indeed observed in groups of the lemur *Eulemur fulvus rufus*, in which males outnumber females and do not contribute to infant rearing (Gerson, 2000); and in mountain gorillas (*Gorilla gorilla beringei*: Sicotte, 1994), where male responsibility for proximity increased when the females were sexually attractive. Reichard (1995*a*) obtained highly variable results in different groups of white-handed gibbons, which could be interpreted as being due to variation in the females' reproductive states in the different groups. Thus, the absolute values of the Hinde index only estimate the benefit ratio if power is held constant, and to measure changes in the actual benefit ratio we need to document changes over time or differences between otherwise similar situations. In conclusion, the value of the Hinde index cannot be used to draw inferences about the relative benefits drawn by each player without knowledge of each animal's leverage. Thus, Palombit's (1999) argument, while plausible, is not necessarily conclusive.

So, if males seem to prefer guarding a single female under all circumstances, i.e., ignore opportunities to incorporate multiple female territories into their own, this almost certainly implies that the presence or actions of the male produces fitness benefits for the female. Because such benefits also extend to the infant's sire, the optimum male strategy is then to be bonded to a single female, although this does not necessarily foreclose the male's option to pursue extra-pair matings. Nonetheless, in the discussion section, we will examine a modified version of the mate-guarding hypothesis to explain the maintenance of dispersed pairs.

Direct male care

It is widely held that the benefits of one form of direct male care (carrying and/or provisioning), are secondary, i.e., followed the evolution of uniform pair living (Dunbar, 1995) because they are not likely to be large enough to make pair living the ESS for the male (cf. Maynard Smith, 1977; Dunbar, 1995; Hawkes *et al.*, 1995). For the male, the benefits of pair living have become significant only after major changes in the life histories of females (twinning) and infants (rates of growth and development) (Ross & MacLarnon, 2000; Treves, 2000). Thus, these changes could not have evolved overnight, and could only have taken place once pair living was already permanent (and it clearly did not evolve in all lineages with uniform pair living). Likewise, the one-female, multi-male units found in a subset of callitrichids, and clearly linked to the need for male care of offspring (Goldizen, 1990; Koenig, 1995), are a sequel to the presence of intensive male care and changes in female life history made possible by this care.[3] Hence, while male care of infants cannot help explain the origins of uniform pair living, it does provide a perfectly valid explanation for its maintenance in the species with extensive male care (it is an exaptation *sensu* Gould & Vrba, 1982).

In all callitrichid genera, and in a few small cebids (*Aotus*, *Callicebus*), as well as (to a lesser extent) in the siamang, we see strong male involvement in the rearing of offspring, i.e., direct male care. A brief discussion of terminology is needed to clarify matters. To qualify as care, the male action must on average increase infant fitness (cf. male parental investment: Kleiman & Malcolm, 1981), and this is clearly not always the case (e.g., van Schaik & Paul, 1997). We use the term male care rather than the commonly used paternal care, because its use does not assume the caretaker is the actual sire. We follow Kleiman and Malcolm (1981) in defining direct care as activities targeted directly at the infant, such as carrying and provisioning, and indirect care as activities that make it possible for the female to allocate more resources to reproduction, such as defence of food and territory.

In the small platyrrhines, where the amount of allomaternal care determines reproductive success (Koenig, 1995), infants show much faster growth than in taxa without allomaternal care (Mitani & Watts, 1997; Ross & MacLarnon, 2000). Similar data for siamang are not available. These taxa, then, are expected to have truly uniform, single-female systems. Deviations toward multi-female states are not expected, either of the associated or the dispersed form. Infanticide by dominant females is an important means of preventing any of the other females in these groups (who function as helpers) from reproducing and thus competing for that help (Digby, 2000).

DISCUSSION

To date, we believe that only infanticide reduction among gregarious female primates and predation reduction through nest guarding (baby sitting) among solitary females are plausible pacemakers for the evolutionary transition to uniform pairs, and both are also additional plausible benefits for its maintenance. However, strong tests of the hypotheses discussed here are not yet possible, and there is some doubt if they ever will be. Tests of current function require fitness data on the alternative options, but the latter are not available, and even if we could create them experimentally, the animals might not respond appropriately to them due to limited reaction norms (cf. Sterck *et al.*, 1997). A related problem is that we can only test the current function version of adaptation, not the historical origin version. We did, however, run into an example where current function probably only arose secondarily. In extant callitrichids, male care is a major adaptive benefit to pair living. However, it is likely that among the ancestors of today's callitrichids, male–female association served to reduce risk of infanticide (e.g., Dunbar, 1995). Today, due to the faster female life history, they are no longer vulnerable to infanticide by males (van Schaik, 2000*a*). Thus, pairs in these species were 'not built by natural selection for their current roles' (Gould & Vrba, 1982, p. 6). Instead, male care is a co-opted exaptation, originally selected for by infanticide avoidance. This shift illustrates the difference between adaptation as current value and as reason for the historical origin of the trait, but also shows the difficulty of actually demonstrating the difference.

Similar limitations adhere to the evolutionary reconstructions. They rely on arguments that invoke plausibility (e.g., conditions producing variable pairs) or parsimony (reconstructed ancestral states), both of which are not always compelling.

In response to these weaknesses, we have relied strongly on consistency and relative plausibility: competing hypotheses must be consistent with all known facts and ideally explain several new facts about the species involved, and the fewer inconsistencies a hypothesis faces, the more faith one has in it.

Pairs in gregarious species: the infanticide hypothesis

The infanticide-reduction hypothesis has so far held up well against this scrutiny, but the need for much additional critical analysis remains. There is a clear link between the broad hypothesis of infanticide risk as the ultimate cause of permanent male–female association in primates and the specific one of the same factor as the ultimate cause of pair living in species with small female party size. Rejection of the broad hypothesis should automatically lead to the rejection of the specific one, but, clearly, acceptance of the broad hypothesis alone is not enough. Hence, we must examine specific cases of pair-living species in some detail, both by examining the evidence for the infanticide hypothesis and by examining alternative hypotheses. Van Schaik and Dunbar (1990) did much of this, and concluded that the evidence for three alternative hypotheses was not as strong as that for the infanticide hypothesis (although they assumed that female antagonism had an ecological basis, i.e., they did not link it to the reduction of takeover risk).

This hypothesis has been criticized, especially by those who believe infanticide is not a major potential source of mortality among the taxa involved (Palombit, 1999; Fuentes, 2000; Müller & Thalmann, 2000). Obviously, infanticide must occur in the species concerned, but its rate may be very low; indeed, the social counterstrategy may be so effective that infanticide has become very rare. Thus, the absence of observed infant mortality due to infanticide is not in itself strong evidence against the hypothesis (*pace* Palombit, 1999; Fuentes, 2000): infanticide risk would be high if we removed the social adaptation, i.e., removed associated males. However, if these species face a low takeover risk, regardless of the presence of a male, then the criticism is more potent. Palombit (1999) rejected the infanticide hypothesis based on the absence of floaters in his population of gibbons, arguing that there was no threat from the outside. However, this absence is clearly not a general condition among gibbons (Cowlishaw, 1992; Brockelman *et al.*, 1998), and is most likely due to a contagious disease affecting his population (M. Griffiths, personal communication).

Additional predictions can also be developed. Thus, as in other primate species with permanent male–female association (and without communal breeding; cf. van Schaik, 2000*c*), the disappearance of the group's male ('widowing') or the immigration of another male should be accompanied by predictable effects. First, as noted above, infanticide should become much more likely (U. Reichard, personal communication). Second, induced resorption or abortion should become more likely. Pregnancy termination has been recorded in white-handed gibbons by Palombit (1995) in exactly this context. Brockelman and Srikosomatara (1984) remarked on the long delay between pair formation and conception of the first offspring. Similar delays have been observed in other species (e.g., lions, Thomas' langurs: van Noordwijk & van Schaik, 2000; Steenbeek, 2000), and are consistent with a function to reduce the risk of infanticide. Third, we expect changes in female singing behaviour, and Cowlishaw (1992) noted a tendency in females to sing deceptive, duet-like songs when alone.

Other observations of pair-living primates are consistent with the infanticide hypothesis. Thus, patterns of male–infant interactions in between-group encounters in gibbons are also consistent with the infanticide-reduction hypothesis, assuming that paternities are confused through extra-pair matings (Sommer & Reichard, 2000). Moreover, extra-pair matings, observed in pair-living primates (Palombit, 1994; Reichard, 1995*b*; Fietz *et al.*, 2000) may have various functions (e.g., Keller & Reeve, 1995), but reduction of infanticide risk is also, in fact, one of them (van Noordwijk & van Schaik, 2000). For now, then, infanticide reduction seems to be the best supported function of year-round male–female association in infant-carrying primates and, by extension, to provide the most plausible explanation for uniform pairs in these primates.

Pairs in solitary foragers

We need much more observational evidence on male behaviour in all of the nocturnal infant-caching species. These species of solitary foragers may have come to live in variable pairs due to female range exclusivity, but it is unclear what kind of male contribution could turn variable pairs into uniform pairs. The great majority of the infant-caching species are probably not vulnerable to infanticide by males. Lactation is relatively short (van Schaik *et al.*, 1999). There is no evidence for infanticide by males in these species, either in captivity or (understandably) in the wild (see van Schaik, 2000*c*; but see Rasoloharijaona *et al.*, 2000). If protection of cached young against predators is indeed the major benefit, one would expect uniform dispersed pairs to be much more widespread among the nocturnal infant cachers with exclusive female ranges because whenever males can make some contribution to infant fitness, variable pairs should be common.

The data show that uniform pairs are not very common among infant-cachers, perhaps because lack of association during the active period makes extra-pair matings far more likely than among associated pairs (see Fietz *et al.*, 2000). Once more data become available, it will become possible to test the hypothesis that among the dispersed pairs, extra-pair matings will be more common than among associated pairs because mate guarding is much harder to maintain in dispersed pairs.

Alternative hypotheses?

While we believe that the two hypotheses discussed above are the most parsimonious, they may provide neither a correct nor a complete explanation of the evolution of pair living in all primates. Moreover, in other mammals with pairs, other selective advantages are also likely (see Brotherton & Komers, chapter 3; Kishimoto, chapter 10).

One other possible scenario among solitary female foragers is that the male's contribution is not to infant fitness directly, but to the female's feeding efficiency. Thus, where a female is traplining resources, and return times to feeding patches affect foraging efficiency, as for example with gum trees, she may prefer the situation where no other males roam through her territory exploiting her resources, except for the one male whose movements she can predict and influence. If the female does not derive strong benefits from polyandrous mating or from strong preferences for males with higher intrinsic variability, she may derive a net benefit from being in a pair. This scenario could apply not only to dispersed pairs, but also to associated pairs if she has priority of access (not uncommon among mammals: Kappeler, 1990).

The male benefits are less obvious, but a modified mate-guarding hypothesis may apply here. The male may benefit from permanently guarding the female and her territory, when he would otherwise be inferior to and at a disadvantage in comparison with other males. By being permanently territorial and by consistently destroying any signs of female reproductive state (e.g., overmarking her scent marks), males may make it very costly for other, stronger males to gain information about the guarded female's reproductive state. Because it is to the female's benefit not to have additional males ranging inside her territory, she does not resist her male's attempts to mask signals of her reproductive state. In the absence of information, dominant males may find it difficult to know when to challenge territory owners. While this need not lead to uniform pair living in all cases, it may produce a system in which most males are socially monogamous (and probably genetically monogamous as well).

When to expect direct male care

In many pair-living primate species, males show virtually no involvement in rearing offspring (Table 4.2). Hence, the question that needs answering is why in some taxa with uniform pair living direct male care of infants did evolve, whereas in others it did not. This is not, as suspected earlier (Kleiman, 1977; Wright, 1990), simply a question of body size and an allometric relationship with size of neonates, so that small-bodied species produce relatively large young that require much allomaternal care. High levels of male care are observed in callitrichids, *Aotus* and *Callicebus*, but not in the similar-sized lemurs or *Pithecia*. Likewise, siamang, roughly twice the size of the other gibbons, shows more male care than any of the other gibbons. Neither is there a simple correlation with seasonality: strictly seasonal breeding might be expected to lead to reduced mate guarding and opportunities for extra-pair matings, and thus more care for infants. However, the highly seasonal lemurs show remarkably little male care.

At present, this problem remains unsolved, indeed largely unaddressed, and we offer only two from among several possible suggestions. First, the extent of male carrying depends on the compatibility between investment in mate guarding and infant care. Where the two activities are incompatible, infant care is almost certain to lose out (cf. Hawkes *et al.*, 1995). Thus, where conflicts with extra-group males (neighbours or floaters) are common and often escalated, males that carry infants may lose condition (cf. Sanchez *et al.*, 1999) and thus be at greater risk of being ousted by the other males. At the same time, they also place the infants more directly at risk during the aggressive interactions. Increased conflict with extra-group males may have ecological (e.g., group density) or demographic reasons (e.g., strongly male-biased adult sex ratios in the population).

Second, if investment in opportunities for additional, extra-pair matings is warranted by fitness returns, this almost inevitably competes with infant care. This idea may overlap with the first one because extra-pair matings are most common when groups meet (e.g., Reichard 1995*b*, chapter 13), and frequent between-group encounters may favour reduced male care for infants.

Note that we did not list the probability of paternity, even though males are expected to invest more in their own offspring. However, it could be argued that the threat of extra-pair matings reduces the benefits of pair living for males to the point that they should prefer to pursue polygynous matings under all conditions. On the face of it, this is a strong argument. If pair living is a superior strategy for the male because of improved infant survival for the primary female, extra-pair matings should reduce certainty of the paternity of the infant whose survival the male's behaviour is supposed to improve, making pair living less beneficial for the male. Nonetheless, empirical data do not provide very strong support for it (e.g., van Schaik & Paul, 1997). The most plausible reason for this is that males in this system have very few other options once they are paired, given demographically-based female leverage, and will therefore have a very high desertion threshold (cf. Dunbar, 1995; Gowaty, 1997).

CONCLUSIONS

We have investigated the evolutionary origins and potential adaptive significance of pair living. Pair living as a social system (in the sense of social units with a single breeding female) is common among primates, and has evolved independently several times. Our analyses suggest that the two major kinds of pairs, associated and dispersed, evolved from ancestors with different social systems. Historically, variable pair living almost certainly preceded the uniform variety. Variable pairs must have started when the original social system favouring polygynous mating systems failed: in very small, mixed-sex groups of infant-carrying species and, less commonly, in species where females are territorial solitary foragers who cached their infants.

Our review of the major selective benefits for the maintenance of pairs suggested that male care almost certainly evolved secondarily. However, the prominence of male care for infants is very patchy and not well understood. Of the original benefits for associated pairs, infanticide avoidance remains a strong candidate hypothesis: the ancestral state already involved mixed-sex groups, and in populations with extremely small group size, infanticide risk is minimized by female antagonism, which leads to pairs. Protection of infants against predation may provide the major selective advantage among the dispersed pairs of infant-caching species. However, other hypotheses are possible and need to be examined. For that, we especially need more data on the demographic consequences of social change.

Acknowledgements
We thank Ulrich Reichard and Christophe Boesch for the invitation to the very productive workshop, and Peter Brotherton, Claudia Fichtel, Joanna Fietz, Eckhard Heymann, Bart Kempenaers, Signe Preuschoft, Ulrich Reichard, Beverly Strassmann, Nicola Uhde, Maria van Noordwijk, and other participants, as well as A. P. Møller and anonymous reviewers for valuable discussion and comments.

Notes

1 The social system can be divided into social organization and social structure. Social structure refers to the patterning of social relationships, but because social pair bonds can clearly be recognized in all primate pairs, this should not affect the distinction between associated and dispersed pairs.

2 The Hinde index (Hinde, 1983) is the proportion of all approaches due to one partner minus the proportion of all departures that are due to the same partner. A positive value for one partner indicates that this partner is more responsible for maintaining the bond than the other one.

3 These single-breeding-female units receiving help from multiple helpers, sometimes multiple adult males, are clearly derived from straightforward social monogamy. This is why, in the introduction, we included these systems under the broader heading of SBFU.

References

Atsalis, S. (2000). Spatial distribution and population composition of the brown mouse lemur (*Microcebus rufus*) in Ranomafana National Park, Madagascar, and its implications for social organization. *American Journal of Primatology*, **51**, 61–78.

Barlow, G. W. (1988). Monogamy in relation to resources. In *The Ecology of Social Behavior*, ed. C. N. Slobodchikoff, pp. 55–79. San Diego: Academic Press.

Bearder, S. K. (1987). Lorises, bushbabies, and tarsiers: diverse societies in solitary foragers. In *Primate Societies*, ed. B. B. Smuts, D. L. Cheney, R. M. Seyfarth, R. W. Wrangham & T. T. Struhsaker, pp. 11–24. Chicago: University of Chicago Press.

Borries, C., Launhardt, K., Epplen, C., Epplen, J. T. & Winkler, P. (1999). Males as infant protectors in Hanuman langurs (*Presbytis entellus*) living in multimale groups: defence pattern, paternity, and sexual behaviour. *Behavioral Ecology and Sociobiology*, **46**, 350–6.

Brockelman, W. Y. & Srikosomatara, S. (1984). Maintenance and evolution of social structure in gibbons. In *The Lesser Apes: Evolutionary and Behavioural Biology*, ed. H. Preuschoft, D. J. Chivers, W. Y. Brockelman & C. N. Creel, pp. 298–323. Edinburgh: Edinburgh University Press.

Brockelman, W. Y., Reichard, U., Treesucon, U. & Raemaekers, J. J. (1998). Dispersal, pair formation and social structure in gibbons (*Hylobates lar*). *Behavioral Ecology and Sociobiology*, **42**, 329–47.

Charles-Dominique, P. (1977). *Ecology and Behaviour of Nocturnal Primates*. New York: Columbia University Press.

Charles-Dominique, P. & Hladik, M. (1971). Le lepilemur du sud de Madagascar: écologie, alimentation, et vie sociale. *La Terre et La Vie*, **25**, 3–66.

Cords, M. (1987). Forest guenons and patas monkeys: male–male competition in one-male groups. In *Primate Societies*, ed. B. B. Smuts, D. L. Cheney, R. M. Seyfarth, R. W. Wrangham & T. T. Struhsaker, pp. 98–111. Chicago: University of Chicago Press.

Cowlishaw, G. (1992). Song function in gibbons. *Behaviour*, **121**, 131–53.

Crockett, C. M. & Janson, C. H. (2000). Infanticide in red howlers: female group size, male membership, and a possible link to folivory. In *Infanticide by Males and its Implications*, ed. C. P. van Schaik & C. H. Janson, pp. 75–98. Cambridge: Cambridge University Press.

Curtis, D. & Zaramody, A. (1999). Social structure and seasonal variation in the behaviour of *Eulemur mongoz*. *Folia Primatologica*, **70**, 79–96.

Davies, N. B. (2000). Multi-male breeding groups in birds: ecological causes and social conflict. In *Primate Males*, ed. P. M. Kappeler, pp. 11–20. Cambridge: Cambridge University Press.

Delgado, R. & van Schaik, C. P. (2000). The behavioral ecology and conservation of the orangutan (*Pongo pygmaeus*): a tale of two islands. *Evolutionary Anthropology*, **9**, 201–18.

Digby, L. (2000). Infanticide by female mammals: implications for the evolution of social systems. In *Infanticide by Males and its Implications*, ed. C. P. van Schaik & C. H. Janson, pp. 423–46. Cambridge: Cambridge University Press.

Dunbar, R. I. M. (1995). The mating system of callitrichid primates: I. Conditions for the coevolution of pair bonding and twinning. *Animal Behaviour*, **50**, 1057–70.

(2000). Male mating strategies: a modeling approach. In *Primate Males*, ed. P. M. Kappeler, pp. 259–68. Cambridge: Cambridge University Press.

(2001). The economics of male mating strategies. In *Economics in Nature: Social Dilemmas, Mate Choice and Biological Markets*, ed. R. Noë, J. A. R. A. M. van Hooff & P. Hammerstein, pp. 245–69. Cambridge: Cambridge University Press.

Emlen, S. T. & Oring, L. W. (1977). Ecology, sexual selection, and the evolution of mating systems. *Science*, **197**, 215–23.

Fietz, J. (1999). Monogamy as a rule rather than exception in nocturnal lemurs: the case of the fat-tailed dwarf lemur, *Cheirogaleus medius*. *Ethology*, **105**, 259–72.

Fietz, J., Zischler, H., Schwiegk, C., Tomiuk, J., Dausman, K. H. & Ganzhorn, J. U. (2000). High rates of extra-pair young in the pair-living fat-tailed dwarf lemur, *Cheirogaleus medius*. *Behavioral Ecology and Sociobiology*, **49**, 8–17.

Fleagle, J. G. (1999). *Primate Adaptation and Evolution*, 2nd Edition. New York: Academic Press.

Fuentes, A. (2000). Hylobatid communities: changing views on pair bonding and social organization in hominoids. *Yearbook of Physical Anthropology*, **43**, 33–60.

Futuyma, D. J. (1998). *Evolutionary Biology*, 3rd Edition. Sunderland, Massachusetts: Sinauer Associates.

Gerson, J. S. (2000). Social Relationships in Wild Red-fronted Brown Lemurs (*Eulemur fulvus rufus*). Ph.D. thesis, Duke University, Durham, North Carolina.

Goldizen, A. W. (1990). A comparative perspective on the evolution of tamarin and marmoset social systems. *International Journal of Primatology*, **11**, 63–80.

Goodall, J. (1986). *The Chimpanzees of Gombe*. Cambridge, Massachusetts: Harvard University Press.

Gould, S. J. & Vrba, E. S. (1982). Exaptation – a missing term in the science of form. *Paleobiology*, **8**, 4–15.

Gowaty, P. A. (1997). Sexual dialectics, sexual selection, and variation in reproductive behavior. In *Feminism and Evolutionary Biology*, ed. P. A. Gowaty, pp. 351–84. New York: Chapman & Hall.

Gursky, S. (2000*a*). Sociality in the spectral tarsier, *Tarsius spectrum*. *American Journal of Primatology*, **51**, 89–101.

(2000*b*). Allocare in a nocturnal primate: data on the spectral tarsier, *Tarsius spectrum*. *Folia Primatologica*, **71**, 39–54.

Harcourt, C. S. (1991). Diet and behaviour of a nocturnal lemur, *Avahi laniger*, in the wild. *Journal of Zoology, London*, **223**, 667–74.

Harcourt, C. S. & Nash, L. T. (1986). Social organization of galagos in Kenyan coastal forest: I. *Galago zanzibaricus*. *American Journal of Primatology*, **10**, 339–55.

Hausfater, G. & Hrdy, S. B. (ed.) (1984). *Infanticide: Comparative and Evolutionary Perspectives*. New York: Aldine de Gruyter.

Hawkes, K., Rogers, A. R. & Charnov, E. L. (1995). The male's dilemma: increased offspring production is more paternity to steal. *Evolutionary Ecology*, **9**, 662–77.

Heymann, E. W. & Soini, P. (1999). Offspring number in pygmy marmosets, *Cebuella pygmaea*, in relation to group size and the number of adult males. *Behavioral Ecology and Sociobiology*, **46**, 400–4.

Hinde, R. A. (1983). *Primate Social Relationships*. Oxford: Blackwell Scientific Publications.

Hrdy, S. B. (1979). Infanticide among animals: a review, classification, and examination of the implications for the reproductive strategies of females. *Ethology and Sociobiology*, **1**, 13–40.

(1986). Empathy, polyandry, and the myth of the coy female. In *Feminist Approaches to Science*, ed. R. Bleier, pp. 119–46. New York: Pergamon Press.

Janson, C. H. & Goldsmith, M. (1995). Predicting group size in primates: foraging costs and predation risk. *Behavioral Ecology*, **6**, 326–36.

Janson, C. H. & van Schaik, C. P. (2000). The behavioral ecology of infanticide by males. In *Infanticide by Males and its Implications*, ed. C. P. van Schaik & C. H. Janson, pp. 469–94. Cambridge: Cambridge University Press.

Kappeler, P. M. (1990). Female dominance in *Lemur catta*: more than just female feeding priority? *Folia Primatologica*, **55**, 92–5.

(1997*a*). Determinants of primate social organization: comparative evidence and new insights from Malagasy lemurs. *Biological Reviews*, **72**, 111–51.

(1997*b*). Intrasexual selection in *Mirza coquereli*: evidence for scramble competition polygyny in a solitary primate. *Behavioral Ecology and Sociobiology*, **41**, 115–28.

(1998). Nests, tree holes, and the evolution of primate life histories. *American Journal of Primatology*, **46**, 7–33.

(1999). Convergence and nonconvergence in primate social systems. In *Primate Communities*, ed. J. Fleagle, C. Janson & K. Reed, pp. 158–70. Cambridge: Cambridge University Press.

(2000). Causes and consequences of unusual sex ratios among lemurs. In *Primate Males*, ed. P. M. Kappeler, pp. 55–63. Cambridge: Cambridge University Press.

Kappeler, P. M. & Heymann, E. W. (1996). Nonconvergence in the evolution of primate life history and socio-ecology. *Biological Journal of the Linnean Society*, **59**, 297–326.

Kappeler, P. M. & van Schaik, C. P. (2002). Evolution of primate social systems. *International Journal of Primatology*, **23**, 707–40.

Keller, L. & Reeve, H. K. (1995). Why do females mate with multiple males? The sexually selected sperm hypothesis. *Advances in the Study of Behavior*, **24**, 291–315.

Kleiman, D. G. (1977). Monogamy in mammals. *Quarterly Review of Biology*, **52**, 39–69.

(1981). Correlations among life history characteristics of mammalian species exhibiting two extreme forms of monogamy. In *Natural Selection and Social Behavior*, ed. R. D. Alexander & D. W. Tinkle, pp. 332–44. New York: Chiron Press.

Kleiman, D. G. & Malcolm, J. R. (1981). The evolution of male parental investment in mammals. In *Parental Care in Mammals*, ed. D. J. Gubernick & P. H. Klopfer, pp. 347–87. New York: Plenum Press.

Koenig, A. (1995). Group size, composition, and reproductive success in wild common marmosets (*Callithrix jacchus*). *American Journal of Primatology*, **35**, 311–17.

Komers, P. E. & Brotherton, P. N. M. (1997). Female space use is the best predictor of monogamy in mammals. *Proceedings of the Royal Society of London, Series B*, **264**, 1261–70.

Lehman, S. M., Prince, W. & Mayor, M. (2001). Variations in group size in white-faced sakis (*Pithecia pithecia*): evidence for monogamy or seasonal congregations? *Neotropical Primates*, **9**, 96–101.

Lewis, R. J. (2002). Beyond dominance: the importance of leverage. *Quarterly Review of Biology*, **77**, 149–64.

MacKinnon, J. & MacKinnon, K. (1980). The behavior of wild spectral tarsiers. *International Journal of Primatology*, **1**, 361–79.

Maddison, W. P. & Maddison, D. R. (1992). *MacClade: Analysis of Phylogeny and Character Evolution, version 3*. Sunderland, Massachusetts: Sinauer Associates.

Maynard Smith, J. (1977). Parental investment: a prospective analysis. *Animal Behaviour*, **25**, 1–9.

Mendoza, S. P. & Mason, W. A. (1986). Contrasting responses to intruders and to involuntary separation by monogamous and polygynous New World monkeys. *Physiology & Behavior*, **38**, 795–801.

Mitani, J. C. (1984). The behavioral regulation of monogamy in gibbons (*Hylobates muelleri*). *Behavioral Ecology and Sociobiology*, **15**, 225–9.

Mitani, J. C. & Watts, D. (1997). The evolution of non-maternal caretaking among anthropoid primates: do helpers help? *Behavioral Ecology and Sociobiology*, **40**, 213–20.

Müller, A. E. (1999). Social organization of the fat-tailed dwarf lemur (*Cheirogaleus medius*) in northwestern Madagascar. In *New Directions in Lemur Studies*, ed. B. Rakotosamimanana, H. Rasamimanana, J. U. Ganzhorn & S. M. Goodman, pp. 139–57. New York: Kluwer Academic.

Müller, A. E. & Thalmann, U. (2000). Origin and evolution of primate social organisation: a reconstruction. *Biological Reviews*, **75**, 405–35.

Newton, P. N. & Dunbar, R. I. M. (1994). Colobine monkey society. In *Colobine Monkeys: Their Ecology, Behaviour and Evolution*, ed. A. G. Davies & J. F. Oates, pp. 311–46. Cambridge: Cambridge University Press.

Noë, R. & Hammerstein, P. (1994). Biological markets: supply and demand determine the effect of partner choice in cooperation, mutualism and mating. *Behavioral Ecology and Sociobiology*, **35**, 1–11.

Noë, R., van Schaik, C. & van Hooff, J. (1991). The market effect: an explanation for pay-off asymmetries among collaborating animals. *Ethology*, **87**, 97–118.

Nunn, C. L. & van Schaik, C. P. (2000). Social evolution in primates: the relative roles of ecology and intersexual conflict. In *Infanticide by Males and its Implications*, ed. C. P. van Schaik & C. H. Janson, pp. 388–419. Cambridge: Cambridge University Press.

Ostner, J. & Kappeler, P. M. (1999). Central males instead of multiple pairs in redfronted lemurs, *Eulemur fulvus rufus* (Primates, Lemuridae)? *Animal Behaviour*, **58**, 1069–78.

Overdorff, D. J. (1998). Are *Eulemur* species pair-bonded? Social organization and mating strategies in *Eulemur fulvus rufus* from 1988–1995 in Southeast Madagascar. *American Journal of Physical Anthropology*, **105**, 153–66.

Palombit, R. A. (1994). Extra-pair copulations in a monogamous ape. *Animal Behaviour*, **47**, 721–3.

(1995). Longitudinal patterns of reproduction in wild female siamang (*Hylobates syndactylus*) and white-handed gibbons (*Hylobates lar*). *International Journal of Primatology*, **16**, 739–60.

(1996). Pair bonds in monogamous apes: a comparison of the siamang, *Hylobates syndactylus*, and the white-handed gibbon, *Hylobates lar*. *Behaviour*, **133**, 321–56.

(1999). Infanticide and the evolution of pair bonds in nonhuman primates. *Evolutionary Anthropology*, **7**, 117–29.

Palombit, R. A., Seyfarth, R. M. & Cheney, D. L. (1997). The adaptive value of "friendships" to female baboons: experimental and observational evidence. *Animal Behaviour*, **54**, 599–614.

Pereira, M. E. (1993). Agonistic interaction, dominance relation, and ontogenetic trajectories in ringtailed lemurs. In *Juvenile Primates: Life History, Development, and Behavior*, ed. M. E. Pereira & L. A. Fairbanks, pp. 285–305. New York: Oxford University Press.

Pollock, J. I. (1979). Female dominance in *Indri indri*. *Folia Primatologica*, **31**, 143–64.

Pope, T. (2000). The evolution of male philopatry in Neotropical monkeys. In *Primate Males*, ed. P. M. Kappeler, pp. 219–35. Cambridge: Cambridge University Press.

Purvis, A. (1995). A composite estimate of primate phylogeny. *Philosospical Transactions of the Royal Society, Series B*, **348**, 405–21.

Purvis, A. & Webster, A. (1999). Phylogenetically independent comparisons and primate phylogeny. In *Comparative Primate Socioecology*, ed. P. C. Lee, pp. 44–70. Cambridge: Cambridge University Press.

Radespiel, U. (2000). Sociality in the gray mouse lemur (*Microcebus murinus*) in Northwestern Madagascar. *American Journal of Primatology*, **51**, 21–40.

Raemaekers, J. J. & Chivers, D. J. (1980). Socio-ecology of Malayan forest primates. In *Malayan Forest Primates: 10 Years' Study in Tropical Rain Forest*, ed. D. J. Chivers, pp. 279–316. New York: Plenum Press.

Rasoloharijaona, S., Rakotosamimanana, B. & Zimmermann, E. (2000). Infanticide by a male Milne-Edwards' sportive lemur (*Lepilemur edwardsi*) in Ampijoroa, NW-Madagascar. *International Journal of Primatology*, **21**, 41–5.

Reichard, U. (1995*a*). Sozial- und Fortpflanzungsverhalten von Weisshandgibbons (*Hylobates lar*): eine Freilandstudie im Thailändischen Khao Yai Regenwald. Ph.D. thesis, University of Göttingen.

(1995*b*). Extra-pair copulations in a monogamous gibbon (*Hylobates lar*). *Ethology*, **100**, 99–112.

Ross, C. & Jones, K. E. (1999). Socioecology and the evolution of primate reproductive rates. In *Comparative Primate Socioecology*, ed. P. C. Lee, pp. 73–110. Cambridge: Cambridge University Press.

Ross, C. & MacLarnon, A. (2000). The evolution of non-maternal care in anthropoid primates: a test of the hypotheses. *Folia Primatologica*, **71**, 93–113.

Sanchez, S., Pelaez, F., Gil-Bürman, C. & Kaumanns, W. (1999). Costs of infant-carrying in the cotton-top tamarin (*Saguinus oedipus*). *American Journal of Primatology*, **48**, 99–111.

Schülke, O. & Kappeler, P.M. (2003). So near and yet so far: stable territorial pairs but low cohesion between pair-partners in a nocturnal lemur, *Phaner furcifer*. *Animal Behaviour* (in press).

Sicotte, P. (1994). Effect of male competition on male–female relationships in bi-male groups of mountain gorillas. *Ethology*, **97**, 47–64.

Sommer, V. (1994). Infanticide among the langurs of Jodhpur: testing the sexual selection hypothesis with a long-term record. In *Infanticide and Parental Care*, ed. S. Parmigiani & F. S. vom Saal, pp. 155–98. London: Harwood Academic Publishers.

Sommer, V. & Reichard, U. (2000). Rethinking monogamy: the gibbon case. In *Primate Males*, ed. P. M. Kappeler, pp. 159–68. Cambridge: Cambridge University Press.

Steenbeek, R. (2000). Infanticide by males and female choice in Thomas's langurs. In *Infanticide by Males and its Implications*, ed. C. P. van Schaik & C. H. Janson, pp. 153–77. Cambridge: Cambridge University Press.

Steenbeek, R. & van Schaik, C. P. (2001). Competition and group size in Thomas's langurs (*Presbytis thomasi*): the folivore paradox revisited. *Behavioral Ecology and Sociobiology*, **49**, 100–10.

Sterck, E. H. M., Watts, D. P. & van Schaik, C. P. (1997). The evolution of female social relationships in nonhuman primates. *Behavioral Ecology and Sociobiology*, **41**, 291–309.

Sterling, E. (1993). Patterns of range use and social organization in aye-ayes (*Daubentonia madagascariensis*) on Nosy Mangabe. In *Lemur Social Systems and their Ecological Basis*, ed. P. M. Kappeler & J. U. Ganzhorn, pp. 1–10. New York: Plenum Press.

Struhsaker, T. T. & Leland, L. (1987). Colobines: infanticide by adult males. In *Primate Societies*, ed. B. B. Smuts, D. L. Cheney, R. M. Seyfarth, R. W. Wrangham & T. T. Struhsaker, pp. 83–97. Chicago: University of Chicago Press.

Symington, M. M. (1990). Fission–fusion social organization in *Ateles* and *Pan*. *International Journal of Primatology*, **11**, 47–61.

Thalmann, U. (2001). Food resource characteristics in two nocturnal lemurs with different social behavior: *Avahi occidentalis* and *Lepilemur edwardsi*. *International Journal of Primatology*, **22**, 287–324.

Treves, A. (2000). Prevention of infanticide: the perspective of infant primates. In *Infanticide by Males and its Implications*, ed. C. P. van Schaik & C. H. Janson, pp. 223–38. Cambridge: Cambridge University Press.

van Noordwijk, M. A. & van Schaik, C. P. (2000). Reproductive patterns in eutherian mammals: adaptations against infanticide? In *Infanticide by Males and its Implications*, ed. C. P. van Schaik & C. H. Janson, pp. 322–60. Cambridge: Cambridge University Press.

van Schaik, C. P. (2000*a*). Infanticide by male primates: the sexual selection hypothesis revisited. In *Infanticide by Males and its Implications*, ed. C. P. van Schaik & C. H. Janson, pp. 27–60. Cambridge: Cambridge University Press.

(2000*b*). Vulnerability to infanticide: patterns among mammals. In *Infanticide by Males and its Implications*, ed. C. P. van Schaik & C. H. Janson, pp. 61–71. Cambridge: Cambridge University Press.

(2000*c*). Social counterstrategies against male infanticide in primates and other mammals. In *Primate Males*, ed. P. M. Kappeler, pp. 34–52. Cambridge: Cambridge University Press.

van Schaik, C. P. & Dunbar, R. I. M. (1990). The evolution of monogamy in large primates: a new hypothesis and some crucial tests. *Behaviour*, **115**, 30–62.

van Schaik, C. P. & Kappeler, P. M. (1993). Life history, activity period and lemur social systems. In *Lemur Social Systems and their Ecological Basis*, ed. P. M. & J. U. Ganzhorn, pp. 241–60. New York: Plenum Press.

(1996). The social systems of gregarious lemurs: lack of convergence due to evolutionary disequilibrium? *Ethology*, **102**, 915–41.

(1997). Infanticide risk and the evolution of male–female association in primates. *Proceedings of the Royal Society of London, Series B*, **264**, 1687–94.

van Schaik, C. P. & Paul, A. (1997). Male care in primates: does it ever reflect paternity? *Evolutionary Anthropology*, **5**, 152–6.

van Schaik, C. P. & van Hooff, J. A. R. A. M. (1983). On the ultimate causes of primate social systems. *Behaviour*, **85**, 91–117.

van Schaik, C. P., van Noordwijk, M. A. & Nunn, C. L. (1999). Sex and social evolution in primates. In *Comparative Primate Socioecology*, ed. P. C. Lee, pp. 204–40. Cambridge: Cambridge University Press.

Vick, L. G. & Pereira, M. E. (1989). Episodic targeting aggression and the histories of *Lemur* social groups. *Behavioral Ecology and Sociobiology*, **25**, 3–12.

Warren, R. & Crompton, R. (1997). A comparative study of the ranging behaviour, activity rhythms and sociality of *Lepilemur edwardsi* (Primates, Lepilemuridae) and *Avahi occidentalis* (Primates, Indriidae) at Ampijoroa, Madagascar. *Journal of Zoology, London*, **243**, 397–415.

Wolff, J. O. & Peterson, J. A. (1998). An offspring-defense hypothesis for territoriality in female mammals. *Ethology, Ecology & Evolution*, **10**, 227–39.

Wright, P. C. (1990). Patterns of paternal care in primates. *International Journal of Primatology*, **11**, 89–101.

(1999). Lemur traits and Madagascar ecology: coping with an island environment. *Yearbook of Physical Anthropology*, **42**, 31–72.

CHAPTER 5

The evolution of social and reproductive monogamy in *Peromyscus*: evidence from *Peromyscus californicus* (the California mouse)

David O. Ribble

INTRODUCTION

The genus *Peromyscus* (deer mice) is an attractive group in which to study the evolution of social and mating behaviours. This genus includes over 50 species (Carleton, 1989) that are widely distributed across North and Central America from coast to coast and from the northern subarctic to Panama (Kirkland & Layne, 1989). The diversity in body sizes among *Peromyscus* ranges from 13 to 77 g (Millar, 1989) and exceeds that of most other genera. Phylogenetic relationships among species of *Peromyscus* are relatively well understood (Stangl & Baker, 1984), although the systematics of *Peromyscus* is an active area of study (e.g., Rogers & Engstrom, 1992; Bradley *et al.*, 2000). Most relevant to this chapter, populations and species of *Peromyscus* exhibit a variety of social behaviours and mating systems (Wolff, 1989), with social monogamy, and particularly reproductive monogamy, being relatively rare. Since monogamy is rare among *Peromyscus*, those *Peromyscus* species that exhibit monogamous behaviours may reveal important factors in the evolution of the genus.

One of the best studied monogamous species within the genus is *P. californicus* (California mouse). Association patterns, biparental care, and mating exclusivity indicate that this species is socially and reproductively monogamous, and I will begin by reviewing these elements. Furthermore, recent field experiments demonstrate that male care is critical for offspring survival and is the salient feature of monogamy in this species. I will then review the ecology of female and male home range use and spatial organization and paternal care in other *Peromyscus* species. Finally, within a phylogenetic framework, I will examine the evolution of monogamy and paternal care in *Peromyscus* by mapping male and female spacing patterns as well as male paternal behaviour. This comparative look at monogamy in the genus can provide clues to the maintenance and evolution of monogamy in *P. californicus*. My objective in this chapter is to explore the evolution of the reproductive strategies of *P. californicus* in the larger context of what is known about other *Peromyscus* species in order to gain a better understanding of the evolution of monogamous mating systems.

THE MONOGAMOUS MATING SYSTEM OF *PEROMYSCUS CALIFORNICUS*, THE CALIFORNIA MOUSE

Peromyscus californicus is one of the larger species of *Peromyscus* (*ca.* 40 g); it is distributed in California south of the San Francisco Bay down to northern Baja California along the coastal ranges and into the western foothills of the Sierra Nevada (Merritt, 1978). Within its range, *P. californicus* is associated with dense chaparral habitats in the south and broad-leaved forests in the north (Merritt, 1974). The breeding season typically begins with the onset of winter rains in November and extends until the dry summer months (Ribble, 1991). The average number of litters per female per breeding season is 2.35 (2SE = 0.38; Ribble, 1992*b*). Water availability, rather than photoperiod or food resources, regulates breeding activity in males (Nelson *et al.*, 1995), which is consistent with the species' relatively poor physiological capacities for maintaining internal water balance (MacMillen, 1964). Breeding males live on average 342.2 days (2SE = 97.2) and breeding females 280.9 days (2SE = 124.0), but it is not unusual for breeding males and females to live for more than one year (Ribble, 1992*b*).

Peromyscus californicus males tend to have larger ranges than females, but unlike most *Peromyscus* (Wolff, 1989), these males have very little intrasexual overlap, resulting in mated pairs having largely overlapping home ranges that are statistically distinguishable from those of adjacent mated pairs (Ribble & Salvioni, 1990). These mated pairs remain together as long as both members of the pair are alive, with individuals switching to a new mate only after their first mate dies (Ribble, 1991). The amount of time fathers spend in the nest at night, presumably caring for offspring, is comparable to the amount of time spent by lactating mothers. Paternal care has been documented extensively in the laboratory (Gubernick & Alberts, 1987, 1989), and persists even when cages are enlarged or males are presented with other females (Gubernick & Addington, 1994). In natural populations, mated pairs mate exclusively with each other. All offspring from 28 families over a two-year period resulted from exclusive matings between single male and female pairs (Ribble, 1991). Extra-pair fertilizations were not detected using DNA fingerprinting, similar to the Malagasay giant rat (Sommer, chapter 7) but unlike the case for the fat-tailed lemur (Fietz, chapter 14). Thus, based on association patterns, biparental care, and mating exclusivity, *P. californicus* is monogamous, both socially and reproductively.

Survival of offspring to weaning age is high relative to other *Peromyscus* species (Ribble, 1992*a*). Litter size at weaning (mean = 1.73, 2SE = 0.22) in the field is close to the range of litter sizes at birth reported for female *P. californicus* (range 1.8–2.5). Parity (number of births) appears to have no effect on litter size, but interbirth interval does increase with parity. Interbirth intervals involving mate switches are significantly longer than intervals for pairs that remain together. Lifetime reproductive success (LRS; number of offspring weaned during lifetime) was similar between males (mean = 4.4, 2SE = 1.68) and females (mean = 4.7, 2SE = 1.41) during a three-year study (Ribble, 1992*a*), but the standardized variance in LRS for males was twice that of females. The number of days that individuals were mated was positively correlated with LRS for both sexes. Maximum weight was also correlated with female LRS. Time to first litter was negatively correlated with LRS in males, implying that stochastic demographic features do affect male LRS.

Unlike the socially monogamous fat-tailed lemur, in which offspring remain in their family group for one or more breeding seasons (Fietz, chapter 14), *P. californicus* offspring leave their natal home range prior to the birth of the next litter (Ribble, 1992*b*). Once offspring leave their natal home ranges, natal dispersal patterns are sex-dependent, with females being more dispersive and males more philopatric (Ribble, 1992*b*). Female-biased dispersal is unusual for mammals, but it is more common among socially monogamous birds (Greenwood, 1980, 1983). In *P. californicus*, females that disperse tend to be from natal litters with significantly more females than from natal litters of those that remained philopatric, implying that females disperse due to competition. Male-biased philopatry is probably due to the monogamous mating system of this species (Ribble, 1992*b*).

Monogamy in *P. californicus* does not appear to be caused by female dispersion (Ribble & Salvioni, 1990; Ribble, 1991). Both male and female home range sizes are inversely correlated with population density, but even at high densities some males had territories large enough to encompass multiple females, yet they did not do so (Ribble, 1991). Mated males also failed to respond to unmated females in adjacent territories.

There is experimental evidence to indicate that male care in *P. californicus* enhances offspring survival, particularly under cold environmental conditions or when the parents must work for food. In the laboratory, under warm, ambient temperatures and with food provided *ad libitum*, *P. californicus* females can successfully rear offspring without any paternal care (Dudley, 1974*a*, *b*; Gubernick *et al.*, 1993). But Gubernick *et al.* (1993) and Cantoni and Brown (1997) have shown that the father's presence increases offspring survival in cold ambient temperatures and when parents must work for their food.

In the field under natural conditions, Gubernick and Teferi (2000) have experimentally demonstrated the critical importance of male care for offspring survival in the same populations that I studied. They removed 11 mated males within three days of the birth of their mated female's first litter, and compared the number of young that emerged to 14 females with their mated male present. There was no difference in the number of young born to the father-present pairs (mean = 1.9 ± 0.4 [2SE]) compared to the father-removed pairs (2.1 ± 0.4), but the number of young that emerged was significantly greater in the father-present pairs (1.5 ± 0.2 vs. 0.6 ± 0.4). Six of the females that had their partners removed went on to successfully reproduce with a new male partner, and their reproductive success was significantly greater with their new partner than their efforts without a male present (Gubernick & Teferi, 2000).

The social organization, mating system, and biparental care of *P. californicus* is strikingly similar to many monogamous birds that exhibit low extra-pair fertilizations (EPFs) and large contributions of paternal care by males (e.g., Black, 2001; Haggerty *et al.*, 2001; Quillfeldt *et al.*, 2001). For birds, it has been suggested that in species with low EPF rates, males should contribute to offspring care (Birkhead & Møller, 1996). Among mammals, female gestation and lactation typically emancipate males from care of the young (Kleiman, 1977; Barlow, 1988), and males usually maximize reproductive success by securing additional matings rather than investing in their offspring (Trivers, 1972). Paternal care, then, is an essential feature of male reproductive strategies in *P. californicus*, which is unusual compared with other *Peromyscus* species.

ECOLOGY OF SPATIAL ORGANIZATION IN *PEROMYSCUS*

Traditionally, most studies of *Peromyscus* have focused on the widespread *P. maniculatus* and *P. leucopus*. Wolff (1989) reviewed and summarized social behaviour of *Peromyscus*. Since Wolff's review, using modern techniques of radio telemetry and molecular biology, numerous studies of *P. maniculatus*, *P. leucopus*, and other species of *Peromyscus* have furthered our understanding of spatial organization and mating systems in *Peromyscus*. These advances are important because previously home ranges and spatial organization were largely determined by live-trapping individuals. We have demonstrated that for *Peromyscus*, live-trapping, compared to radio telemetry, significantly underestimates home range size, particularly at low densities (Ribble *et al.*, 2002). Thus, more and better information on home range use and spatial organization in *Peromyscus* has become available since Wolff's review.

Most studies of *Peromyscus* have indicated that females occupy home ranges that are mutually exclusive from adjacent females; hence females are solitary both spatially and socially (Table 5.1). In general, *Peromyscus* do not select home ranges that contain a *specific* food resource since they tend to be omnivorous (Kaufman & Kaufman, 1989). Females typically choose home ranges that contain a variety of resources, and compared with males, females tend to be more selective in their home ranges (Bowers & Smith, 1979). On the other hand, female selection of habitats may be limited by the availability of suitable nesting sites, and females may select habitats on that basis (Scheibe & O'Farrell, 1995). There are, however, documented cases of communal or group nesting by female *P. maniculatus* and *P. leucopus* (Howard, 1949; Hansen, 1957; Millar & Derrickson, 1992; Wolff, 1994). These cases appear to be due to the inclusion of female offspring in the nests of their mothers, and do not result in any noticeable decreases in the reproductive success of the reproductive females (Wolff, 1994). Furthermore, non-offspring nursing has been reported in *P. leucopus* (Jacquot & Vessey, 1994).

For females that are solitary and territorial, it is generally accepted that they defend their home ranges from other females in order to defend resources that are critical during the energetically demanding periods of gestation and lactation (Ostfeld, 1990). Females tend to demonstrate more aggressive territorial behaviours than males, particularly at higher densities (Wolff, 1989). Female home range size is typically inversely correlated with population density (Metzgar, 1971; Madison, 1977; Ribble & Salvioni, 1990; Ribble & Stanley, 1998), but not always (Wolff, 1985). Experimental studies of food addition usually indicate that addition of food results in smaller female home ranges (reviewed in Wolff, 1989). Wolff (1993) and Wolff and Peterson (1998) have suggested that female small mammals, including *Peromyscus*, may be territorial to protect young from infanticide, primarily by adjacent females (pup-defence hypothesis). Unfortunately, there are few experimental data that discriminate between the food-defence and pup-defence hypotheses of territoriality in *Peromyscus* or mammals in general (Wolff, 1993). Whatever the reason for mutually exclusive use of space by females, the spatial pattern of females is thought to select for spacing patterns among male *Peromyscus*.

Male home ranges are usually larger than female home ranges (Ribble & Stanley, 1998; but see Madison, 1977), and male spacing patterns are more variable than female spacing patterns across *Peromyscus* species (Table 5.1). Male spacing patterns vary from monogamy (one male overlaps one primary female with little intrasexual overlap), to roving (one male overlaps several females with extensive intrasexual overlap between males), to polygyny (one male overlaps several females with little intrasexual overlap between males) (Table 5.1). Male spacing in populations of *P. leucopus* and *P. maniculatus* has been shown to vary across subspecies and populations in different habitats. For example, male montane *P. maniculatus nubiterrae* tend to

Table 5.1. *Documented spacing characteristics and paternal behaviour of* Peromyscus *species.* Onychomys *is included as an outgroup for comparative purposes*

Taxon	Female spacing	Male spacing	Paternal care			References
			Laboratory	Field	Best evidence	
Onychomys	Solitary	Roving	Y	N	N	Horner & Taylor, 1968; Frank & Heske, 1992; Stapp, 1999
Peromyscus crinitus	Solitary	ND	N	ND	N	Eisenberg, 1963
P. boylii	Solitary	Roving	ND	N	N	Ribble & Stanley, 1998; Kalcounis-Rueppell, 2000
P. eremicus	Solitary	Roving	N	ND	N	Hatton & Meyer, 1973; Lewis, 1972; Eisenberg, 1968
P. californicus	Solitary	Monogamous	Y	Y	Y	This chapter
P. melanocarpus	ND	ND	Y	ND	Y	Rickart, 1977; Rickart & Robertson, 1985
P. mexicanus	Solitary	ND	Y	ND	Y	Rickart, 1977; Duquette & Millar, 1995
P. truei	Solitary	Roving	ND	N	N	Hall & Morrison, 1997; Ribble & Stanley, 1998
P. leucopus	Solitary & gregarious	Monogamous Roving Polygynous	Y	Y[a]	Y[a]	Wolff, 1989; Wolf & Cicirello, 1989, 1991; Schug *et al.*, 1992; Xia & Millar, 1988, 1989
P. polionotus	Solitary	Monogamous	Y	Y	Y	Blair, 1951; Smith, 1966; Foltz, 1981
P. maniculatus	Solitary & gregarious	Monogamous Roving Polygynous	Y	Y[a]	Y[a]	Horner, 1947; Howard, 1949; Xia & Millar, 1986; Wolff, 1989; Wolff & Cicirello, 1989, 1991; Ribble & Millar, 1996

[a] Presence of paternal care in some populations but not others.
ND, no data.

be socially monogamous, while male *P. m. bairdii* tend to be polygynous or roving (Wolff & Cicirello, 1991). Wolff and Cicirello speculated that, because of the cooler breeding season of montane *P. m. nubiterrae*, there may be selection for males to invest in paternal care. Variability among different populations or subspecies may also be due to the density and dispersion of females. For example, Wolff and Cicirello (1990) have shown in *P. leucopus* that when females are at lower densities and widely dispersed, males adopt a non-territorial roving strategy. At higher female densities, males defend smaller home ranges that contain two to four females. Thus, the density of females largely determines the spatial organization and home range use of males. Experimental food addition usually does not influence male spacing patterns (Wolff, 1989), although we have found in *P. boylii* that food addition results in greater reductions in male than in female home range size (Ribble, unpublished data).

In *Peromyscus* with solitary females and roving males, genetic evidence indicates that litters can be sired by multiple males (Birdsall & Nash, 1973; Xia & Millar, 1991; Ribble & Millar, 1996). Based on the reproductive patterns observed, the roving male spacing pattern is often referred to as a promiscuous mating system (e.g., Heske & Ostfeld, 1990). Rarely, however, do studies of promiscuous spacing patterns have genetic evidence on offspring paternity.

Male parental behaviour is poorly understood in most natural populations of *Peromyscus* due to the difficulty of studying this behaviour in nocturnal, secretive individuals. There have been many studies of *Peromyscus* in the laboratory demonstrating that males will care for offspring if the females will allow them (e.g., Horner, 1947; Eisenberg, 1963; Table 5.1). Some species, for example *P. leucopus*, will exhibit paternal care in the laboratory (Horner, 1947; Hartung & Dewsbury, 1979), but in larger enclosures or in the field, paternal care is not observed (Xia & Millar, 1988). Many of the species in Table 5.1 have been observed exhibiting paternal behaviour in the laboratory, but only three, *P. californicus*, *P. polionotus*, and *P. m. nubiterrae* have unequivocally been demonstrated to be paternal in the field. The evidence for paternal care includes long periods of occupation in a nest that contains offspring (Ribble & Salvioni, 1990) or sampling of nests that contain both male and females (Foltz, 1981). Two of these species, *P. polionotus* and *P. californicus*, have also been shown to exhibit mating exclusivity, that is to say the socially monogamous male is also the genetic partner (Foltz, 1981; Ribble, 1991).

The discrepancy between laboratory and field observations of paternal care at least indicates that males have the ability to care for offspring, but either female aggression or ecological situations prevent males from being paternal. For example, Schug *et al.* (1992) have shown that the genetic father in *P. leucopus* was found to be associated with pups in nest boxes only after their weaning, but not before. Also, Wolff and Cicirello (1991) observed *P. leucopus* fathers present in nest boxes with pups in 32% of litters. Thus, in some species of *Peromyscus* paternal care appears to be variable.

To conclude, female *Peromyscus* tend to have smaller, solitary home ranges and male associations can vary from monogamous to roving to polygynous. The density of females appears to be a determinant of whether males adopt a roving strategy rather than defending their home range, or defending a home range in a polygynous social organization. Paternal behaviour has been documented in many laboratory situations, but little is known about paternal behaviour in natural populations. Monogamous tendencies have been reported for *Peromyscus* (e.g., Hartung & Dewsbury, 1979; Dewsbury, 1981), but with the exception of *P. californicus* and *P. polionotus*, these are not very well understood.

COMPARATIVE VIEW OF MONOGAMY IN *PEROMYSCUS*

The most complete phylogeny to date of *Peromyscus* is that of Stangl and Baker (1984), and is based on karyotypic data. This phylogeny was also used by Langtimm and Dewsbury (1991) to examine variation in copulatory behaviour of *Peromyscus*. I used this phylogeny and included as the outgroup a commonly recognized one, *Onychomys* (grasshopper mice) (Carleton, 1989; Langtimm & Dewsbury, 1991). Character states from Table 5.1 were mapped on the phylogeny of *Peromyscus* using MacClade, Version 4.0 (Maddison & Maddison, 2000). I made no assumptions about the evolutionary sequence in which characters changed. Ambiguities in character tracings were resolved using the DELTRAN procedure, which delays changes away from the root of the phylogeny (Maddison & Maddison, 2000). Information that was not available (Table 5.1) was not scored. Female spacing patterns were scored as solitary (little or no overlap between home ranges) or gregarious (largely overlapping home ranges, usually accompanied with nest-sharing), based on spatial overlap during the breeding season. Species with both solitary and gregarious female spacing were scored as gregarious. Male spacing patterns were scored as monogamous, roving, polygynous, or variable if populations exhibited multiple patterns. No species has been documented as being solely polygynous; those species with polygyny have also been documented as being monogamous and roving. Paternal care was scored based on the best available evidence. If a species has exhibited male care in the laboratory but not in the field, then they were considered non-paternal. If a species has exhibited paternal behaviour in the laboratory, has other life history traits consistent with paternal care (e.g., Dewsbury, 1981), and there has been no conflicting information from the field, then they were considered paternal (Table 5.1).

Based on the phylogenetic patterns of female and male spacing patterns, it appears that the ancestral social organization of *Peromyscus* is one in which females are distributed in a solitary fashion and males rove across larger home ranges (Figure 5.1). The only cases of polygynous spacing by males are also in the species in which females have been documented as being gregarious, i.e., *P. maniculatus* and *P. leucopus*. The only two species of *Peromyscus* with well-documented, and exclusively male monogamous spacing patterns are

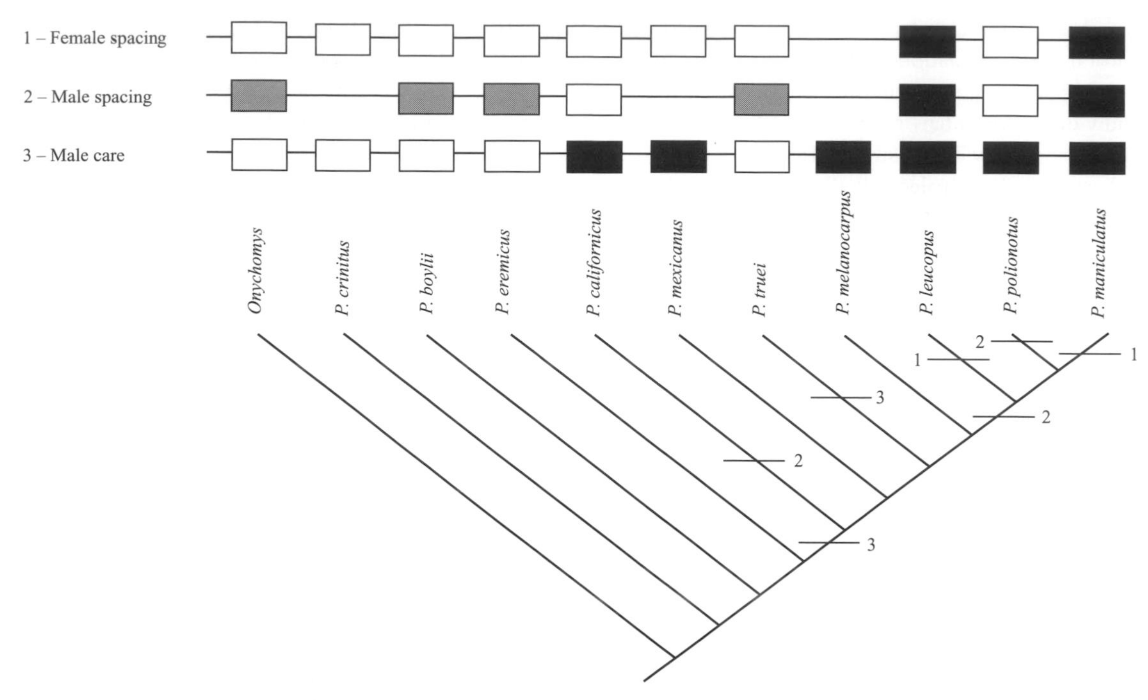

Figure 5.1. Spacing and paternal behaviours of *Peromyscus* overlaid on a cladogram modified from Stangl and Baker (1984). Character states (described in Table 5.1) are: **female spacing**: white – solitary, black – solitary and gregarious; **male spacing**: grey – roving, white – monogamous, black – monogamous, roving, and polygynous; **male care**: white – not present, black – present. Characters with no data are indicated without boxes. Branches in which the character state changes are indicated with horizontal bars and labelled with the character number.

P. californicus and *P. polionotus*. Based on the phylogenetic relationships, it appears that this feature is the result of homoplasy or convergent evolution. Based on limited information, and primarily from the laboratory, male care is potentially a relatively common feature of males. Thus, the presence of monogamous spacing and mating exclusivity in *P. californicus* and *P. polionotus* may be due to the paternal investment route suggested by Brotherton and Komers (chapter 3).

The only clear cases of male care in natural populations that appear to be fixed are, again, from *P. californicus* and *P. polionotus*. If male care has evolved twice in each of these lineages, then are there similarities in the ecologies and life histories of these two species? *P. polionotus* is one of the smallest *Peromyscus* species at an average weight of 14 g and *P. californicus* is one of the larger at 37 g (data from Millar, 1989). The average litter size of *P. polionotus* is 3.7, while that of *P. californicus* is around 2. Across *Peromyscus*, litter size tends to be inversely correlated with body size (Rickart, 1977), the smallest species producing larger litters. Part of this correlation may be because the larger species tend to be tropical and more *K*-selected species than the smaller species, which exist in more variable temperate environments and are thus more *r*-selected (Rickart, 1977). However, *P. californicus* is not a tropical species and it has a very small litter size. Body size is positively correlated with both individual neonate weight and entire litter mass in *Peromyscus* (Millar, 1989). If neonate weight or litter weight are adjusted by adult weight, an interesting pattern appears relative to litter size (Figure 5.2). Litter size is not correlated with relative neonate weight, but it is positively correlated with relative litter mass. And thus, in the *Peromyscus* species, *P. californicus* has one of the smallest litter sizes, and one of the smallest relative neonate and litter weights (Figure 5.2).

Any investment by the male in parental care will decrease his chance to secure additional matings (Trivers, 1972; Maynard Smith, 1977; Kurland & Gaulin, 1984), so why should male *P. californicus* invest in his offspring and mate exclusively? Other species of *Peromyscus* (*P. boylii*, *P. truei*, and *P. maniculatus*) that are syntopic

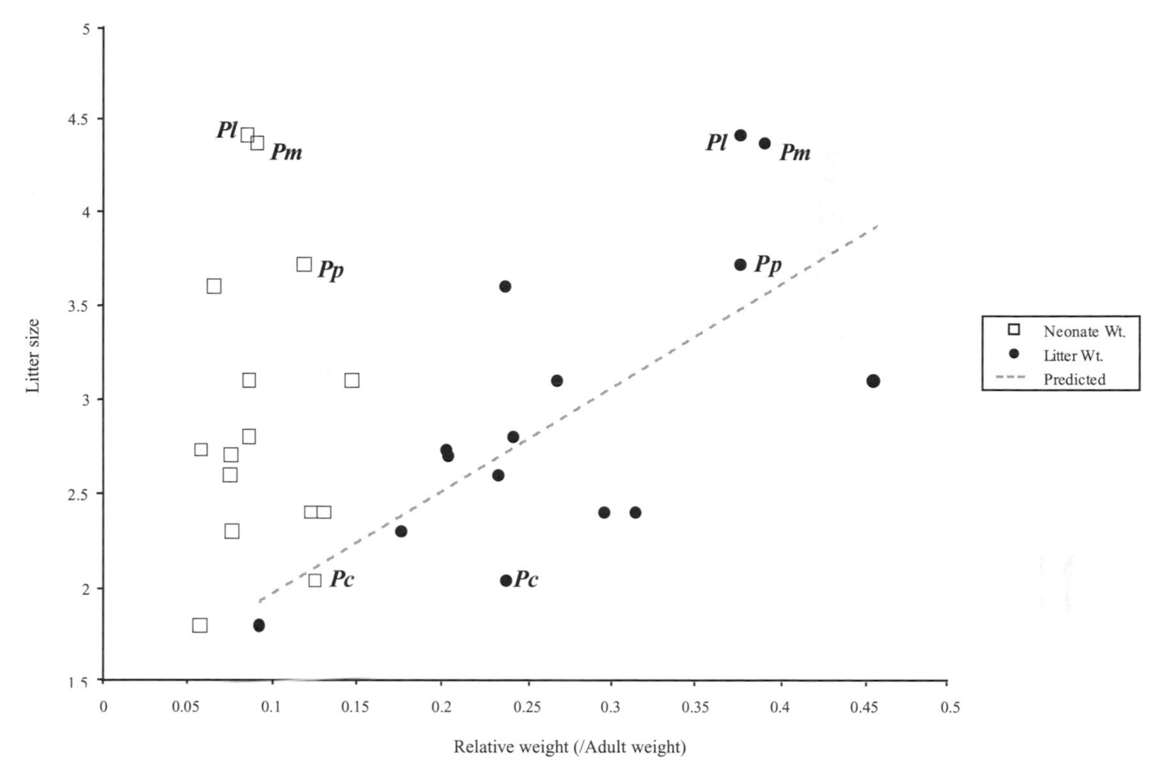

Figure 5.2. Relationships between relative weight (wt (g)/ adult weight (g)) of individual neonate mass and entire litter mass to litter size for *Peromyscus* species. Data were taken from Millar (1989). The predicted line was calculated from a regression analysis of all data. Select species discussed in the text are: ***Pl***, *P. leucopus*; ***Pm***, *P. maniculatus*; ***Pp***, *P. polionotus*; ***Pc***, *P. californicus*.

with *P. californicus* do not exhibit paternal care or monogamy (Ribble & Millar, 1996; Ribble & Stanley, 1998; Kalcounis-Rueppel, 2000). Furthermore, female *P. californicus* can raise at least some offspring without the male (Gubernick & Teferi, 2000). From the male's perspective, one can ask how many litters must a non-paternal male produce to equal the males that do engage in paternal care activities. Ribble (1992*a*) reported that average lifetime reproductive success for males was 4.5 weaned offspring. Assuming the average weaning success without paternal care is 0.6 offspring, then non-paternal males would have to mate and produce 7.5 litters to equal the average LRS for the average parental males (4.5/0.6). Since females produce an average of 2.5 litters in their lifetime, a non-parental male would have to mate with at least three females if each female produced 2.5 litters, to match the reproductive success of males that engage in paternal behaviours. The average lifespan of males in the field is almost one year (342 days), in which the breeding season is restricted from November to June (Ribble, 1992*a*). Thus, it would appear that a male's best strategy is to mate and pair with only one female. Sommer (chapter 7) argued that for the Malagasy giant rat (*Hypogeomys antimena*), the male's contribution to offspring survival must outweigh the costs of lost mating opportunities due to the impact of predation. There is no evidence to suggest that male *P. californicus* provide protection from predators, but rather it is direct paternal care that increases offspring survival (Ribble, 1990; Gubernick & Teferi, 2000).

But why is male care critical for offspring survival in *P. californicus* compared with other *Peromyscus*? I would suggest that male reproductive success is maximized by investing in care of the offspring because of the small litter size and relative mass of the litter in this species. In *Mus musculus*, offspring from smaller litters are energetically less efficient at converting milk to body weight (König *et al.*, 1988) and mothers spend significantly more time caring for smaller litters (König & Markl, 1987). In both cases, the authors attributed these effects to greater heat loss of smaller litters simply due to the number of bodies huddled in the nest. For *P. californicus*, it may be that with such small litters, fathers can contribute significantly to the growth of their offspring simply by huddling over them, providing warmth while the female is away from the nest foraging. *P. californicus* offspring are ectothermic up to 15 days postpartum (Gubernick & Alberts, 1987), and any warmth provided by parents would allow the offspring to invest their energy in growth. If litter size were larger, as is the case with most other *Peromyscus* (see review in Millar, 1989), the contribution of the male would be diminished by the thermal advantages of a larger litter and his reproductive success would then likely be maximized by securing other matings. Thus, I am suggesting that male *P. californicus* can contribute more to the growth of their offspring than most mammals due to the relatively (in terms of other *Peromyscus* species) small litter size of this species.

The thermal disadvantage of small litter size may be further exacerbated due to the timing of the breeding season. *P. californicus* begins breeding with the onset of winter rainfall in November and continues until the dry summer months (Ribble, 1992*a*). Consequently, many offspring are born during the coldest months of the year and the father's contributions to keeping the offspring warm may be critical during these months. Other *Peromyscus* species that coexist with *P. californicus* (*P. boylii*, *P. truei*, and *P. maniculatus*) do not initiate breeding until the warmer spring months (unpublished observation). Most *Peromyscus*, regardless of litter or body size, produce two or three litters per breeding season (Millar, 1989), and the earlier breeding by *P. californicus* is probably related to its longer interbirth intervals (Ribble, 1992*a*).

Females and their offspring are also likely to benefit indirectly from the presence of the male (Wittenberger & Tilson, 1980). Offspring are weaned at a heavier weight in the presence of the male in the laboratory (Dudley, 1974*a*), and fathers could provide protection from predators and conspecifics (Ribble & Salvioni, 1990). In laboratory experiments in which individuals must forage for food, the male's presence resulted in four times more offspring in a 74-day period compared with females without male help (Cantoni & Brown, 1997). In natural populations, survival from birth to emergence has been estimated at 30% in other *Peromyscus* species without paternal care (Millar & Innes, 1983; Ribble, unpublished data), which is similar to the survival rates observed by Gubernick and Teferi (2000) in *P. californicus* where the father was not present. The high survival rate of offspring in *P. californicus* is no doubt due to biparental care. Furthermore, females who switch mates have longer interbirth intervals, possibly decreasing female lifetime reproductive success (Ribble, 1992*a*). In addition to the direct benefits of male care, males could also contribute in indirect ways to offspring survival by providing protection from infanticide (Agrell *et al.*, 1998). In the experiments by Gubernick and Teferi (2000), all 11 females that had their mates removed remated with a new male. In all 11 cases, the new male took up residence *after* the female had ceased lactating, suggesting that the new male was not responsible for loss of any offspring. Thus it appears that the evolution of male care in *P. californicus* is not due to the benefits of protection against infanticide (Gubernick & Teferi, 2000).

In contrast to *P. californicus*, *P. polionotus* has a litter size and litter mass similar to other species (e.g., *P. maniculatus* and *P. leucopus*, Figure 5.2) that do not exhibit monogamy. This species is confined to the southern USA, which is arguably warmer. This species does, however, build extensive burrows, and it has been suggested by Smith (1966) that it takes both sexes to maintain the burrow. Thus, it is likely that monogamy has evolved for different reasons among the genus *Peromyscus*.

SUMMARY

Komers and Brotherton (1997) examined male care and monogamy in mammals, concluding that the ancestral species of *Peromyscus* may have had a tendency towards monogamy and paternal behaviour. Their analyses depended primarily on secondary literature that has been contradicted by more recent primary sources. For example, most secondary literature sources describe the genus *Onychomys* as exhibiting a monogamous social organization, based on live-trapping studies. Recent field studies using radio telemetry demonstrate otherwise (Frank & Heske, 1992; Stapp, 1999). There are also problems, as indicated above, with interpreting male parental behaviours based solely on laboratory studies.

I have demonstrated that the likely ancestral social organization of *Peromyscus* is one of solitary females, with males adopting a roving strategy of home range use that can result in a promiscuous reproductive mating system. And there are at least two species of *Peromyscus* (*P. californicus* and *P. polionotus*) with monogamous social and reproductive mating systems that appear to have evolved independently, and probably for different adaptive reasons.

Data from *P. californicus* suggest that the larger body size, smaller litter size, and relative litter mass may influence male reproductive strategies so that they mate exclusively and invest heavily in their offspring. Body size has long been recognized as important for various mammalian life history traits (see reviews of Clutton-Brock & Harvey, 1983; Sauer & Slade, 1987), but perhaps the importance of body size for mating systems has not been adequately appreciated in mammals. A notable exception is Jarman's (1974) analysis of mating systems among African antelope. He concluded that the interaction between body size and feeding ecology has influenced the evolution of mating systems between antelope species. Body size has also been recognized as an important trait in bird mating systems (Amadon, 1959; Wiley, 1974).

If the relationships between body size, litter size, and food resources are some of the principal factors accounting for monogamy in *P. californicus*, then other large-bodied, small litter size *Peromyscus* should provide important tests of this hypothesis. Based on reproductive tactics and behaviour observed in the laboratory, Rickart (1977) has suggested that *P. melanocarpus* (mean adult weight = 59 g; mean litter size = 1.8) and *P. mexicanus* (mean adult weight = 53.4 g; mean litter size = 2.1) may be monogamous. Unfortunately, little is known about the mating systems of other large-bodied *Peromyscus* in natural populations. There is obviously much to be learned about the evolution of monogamy from studies of *Peromyscus*. I hope this review will stimulate more work on the social ecology of lesser known species of *Peromyscus*.

References

Agrell, J., Wolff, J. O. & Ylonen, H. (1998). Counter-strategies to infanticide in mammals, costs and consequences. *Oikos*, **83**, 507–17.

Amadon, D. (1959). The significance of sexual differences in size among birds. *Proceedings of the American Philosophical Society*, **103**, 531–6.

Barlow, G. W. (1988). Monogamy in relation to resources. In *The Ecology of Social Behaviour*, ed. C. N. Slobodchikoff, pp. 55–79. San Diego, California: Academic Press.

Birdsall, D. A. & Nash, D. (1973). Occurrence of successful multiple insemination of females in natural populations of deer mice (*Peromyscus maniculatus*). *Evolution*, **27**, 106–10.

Birkhead, T. R. & Møller, A. P. (1996). Monogamy and sperm competition in birds. In *Partnerships in Birds: The Study of Monogamy*, ed. J. M. Black, pp. 323–43. Oxford: Oxford University Press.

Black, J. M. (2001). Fitness consequences of long-term pair bonds in barnacle geese: monogamy in the extreme. *Behavioural Ecology*, **12**, 640–5.

Blair, W. F. (1951). Population structure, social behavior, and environmental relations in a natural population of the beachmouse. *Contributions from the Laboratory of Vertebrate Biology, University of Michigan*, **48**, 1–47.

Bowers, M. A. & Smith, H. D. (1979). Differential habitat utilization by sexes of the deermouse, *Peromyscus maniculatus*. *Ecology*, **60**, 869–75.

Bradley, R. D., Tiemann-Boege, I., Kilpatrick, C. W. & Schmidly, D. J. (2000). Taxonomic status of *Peromyscus boylii sacarensis*, inferences from DNA sequences of the mitochondrial cytochrome-B gene. *Journal of Mammalogy*, **81**, 875–84.

Cantoni, D. & Brown, R. E. (1997). Paternal investment and reproductive success in the California mouse, *Peromyscus californicus*. *Animal Behaviour*, **54**, 377–86.

Carleton, M. D. (1989). Systematics and evolution. In *Advances in the Study of* Peromyscus *(Rodentia)*, ed. G. L. J. Kirkland & J. N. Layne, pp. 7–141. Lubbock, Texas: Texas Tech University Press.

Clutton-Brock, T. H. & Harvey, P. H. (1983). The functional significance of variation in body size among mammals. In *Advances in the Study of Mammalian Behavior*, ed. J. F. Eisenberg & D. G. Kleiman, pp. 632–63. *Special Publication of the American Society of Mammalogists*, No. 7. Shippensburg, Pennsylvania.

Dewsbury, D. A. (1981). An exercise in the prediction of monogamy in the field from laboratory data on 42 species of Muroid rodents. *Biologist*, **63**, 138–62.

Dudley, D. (1974*a*). Contributions of paternal care to the growth and development of the young in *Peromyscus californicus*. *Behavioral Biology*, **11**, 155–66.

(1974*b*). Paternal behavior in the California mouse, *Peromyscus californicus*. *Behavioral Biology*, **11**, 247–52.

Duquette, L. S. & Millar, J. S. (1995). The effect of supplemental food on life-history traits and demography of a tropical mouse *Peromyscus mexicanus*. *Journal of Animal Ecology*, **64**, 348–60.

Eisenberg, J. F. (1963). The intraspecific behavior of some cricetine rodents of the genus *Peromyscus*. *American Midland Naturalist*, **69**, 240–6.

(1968). Behavior patterns. *Special Publications, American Society of Mammalogy*, **2**, 451–95.

Foltz, D. W. (1981). Genetic evidence for long-term monogamy in a small rodent, *Peromyscus polionotus*. *American Naturalist*, **117**, 665–75.

Frank, D. H. & Heske, E. J. (1992). Seasonal changes in space use patterns in the southern grasshopper mouse *Onychomys torridus torridus*. *Journal of Mammalogy*, **73**, 292–8.

Greenwood, P. J. (1980). Mating systems, philopatry, and dispersal in birds and mammals. *Animal Behaviour*, **28**, 1140–62.

(1983). Mating systems and the evolutionary consequences of dispersal. In *The Ecology of Animal Movement*, ed. I. R. Swingland & P. J. Greenwood, pp. 116–31. Oxford: Clarendon Press.

Gubernick, D. J. & Addington, R. L. (1994). The stability of female social and mating preferences in the monogamous California mouse, *Peromyscus californicus*. *Animal Behaviour*, **47**, 559–67.

Gubernick, D. J. & Alberts, J. R. (1987). The biparental care system of the California Mouse, *Peromyscus californicus*. *Journal of Comparative Psychology*, **101**, 169–77.

(1989). Postpartum maintenance of paternal behaviour in the biparental California mouse, *Peromyscus californicus*. *Animal Behaviour*, **37**, 656–64.

Gubernick, D. J. & Teferi, T. (2000). Adaptive significance of male parental care in a monogamous mammal. *Proceedings of the Royal Society of London, Series B*, **267**, 147–50.

Gubernick, D. J., Wright, S. L. & Brown R. E. (1993). The significance of father's presence for offspring survival in the monogamous California mouse, *Peromyscus californicus*. *Animal Behaviour*, **46**, 539–46.

Haggerty, T. M., Morton, E. S. & Fleischer, R. C. (2001). Genetic monogamy in Carolina Wrens (*Throthorus ludovicianus*). *The Auk*, **118**, 215–19.

Hall, L. S. & Morrison, M. L. (1997). Den and relocation site characteristics and home ranges of *Peromyscus truei* in the white mountains of California. *Great Basin Naturalist*, **57**, 124–30.

Hansen, R. M. (1957). Communal litters of *Peromyscus maniculatus*. *Journal of Mammalogy*, **38**, 525.

Hartung, T. G. & Dewsbury, D. A. (1979). Paternal behavior in six species of muroid rodents. *Behavioral and Neural Biology*, **26**, 466–78.

Hatton, D. C. & Meyer, M. E. (1973). Paternal behavior in cactus mice (*Peromyscus crinitus*). *Bulletin of the Psychonometry Society*, **2**, 330.

Heske, E. J. & Ostfeld, R. S. (1990). Sexual dimorphism in size, relative size of testes, and mating systems in North American voles. *Journal of Mammalogy*, **71**, 510–19.

Horner, B. E. (1947). Paternal care of young mice of the genus *Peromyscus*. *Journal of Mammalogy*, **28**, 31–6.

Horner, B. E. & Taylor, J. M. (1968). Growth and reproduction in the southern grasshopper mouse. *Journal of Mammalogy*, **49**, 644–60.

Howard, W. E. (1949). Dispersal, amount of inbreeding, and longevity in a local population of prairie deermice on the George Reserve, Michigan. *Contributions from the Laboratory of Vertebrate Biology, University of Michigan*, **43**, 1–50.

Jacquot, J. J. & Vessey, S. H. (1994). Non-offspring nursing in the white-footed mouse, *Peromyscus leucopus*. *Animal Behaviour*, **48**, 1238–40.

Jarman, J. P. (1974). The social organisation of antelope in relation to their ecology. *Behaviour*, **48**, 215–67.

Kalcounis-Rueppell, M. C. (2000). Breeding Systems, Habitat Overlap, and Activity Patterns of Monogamous and Promiscuous Mating in *Peromyscus californicus* and *P. boylii*. Ph.D. thesis, University of Western Ontario, London, Ontario.

Kaufman, D. W. & Kaufman, G. A. (1989). Population biology. In *Advances in the Study of* Peromyscus *(Rodentia)*, ed. G. L. J. Kirkland & J. N. Layne, pp. 233–70. Lubbock, Texas: Texas Tech University Press.

Kirkland, G. L. J. & Layne, J. N. (ed.) (1989). *Advances in the Study of* Peromyscus *(Rodentia)*. Lubbock, Texas: Texas Tech University Press.

Kleiman, D. G. (1977). Monogamy in mammals. *Quarterly Review of Biology*, **52**, 39–69.

Komers, P. E. & Brotherton, P. N. M. (1997). Female space use is the best predictor of monogamy in mammals. *Proceedings of the Royal Society of London, Series B*, **264**, 1261–70.

König, B. & Markl, H. (1987). Maternal care in house mice: I. The weaning strategy as a means for parental manipulation of offspring quality. *Behavioral Ecology and Sociobiology*, **20**, 1–9.

König, B., Riester, J. & Markl, H. (1988). Maternal care in house mice (*Mus musculus*): II. The energy cost of lactation as a function of litter size. *Journal of Zoology, London*, **216**, 195–210.

Kurland, J. A. & Gaulin, S. J. C. (1984). The evolution of male paternal investment, effects of genetic relatedness and feeding ecology on the allocation of reproductive effort. In *Primate Paternalism*, ed. D. W. Taub, pp. 259–308. New York: Van Nostrand Reinhold.

Langtimm, C. A. & Dewsbury, D. A. (1991). Phylogeny and evolution of rodent copulatory behaviour. *Animal Behaviour*, **41**, 217–25.

Lewis, A. W. (1972). Seasonal population changes in the cactus mouse, *Peromyscus eremicus*. *The Southwestern Naturalist*, **17**, 85–93.

MacMillen, R. E. (1964). Population ecology, water relations, and social behavior of a southern California semidesert rodent fauna. *University of California Publications in Zoology*, **71**, 1–59.

Maddison, W. P. & Maddison, D. R. (2000). *MacClade 4. Analysis of Phylogeny and Character Evolution*. Sunderland, Massachusetts: Sinauer Associates.

Madison, D. M. (1977). Movements and habitat use among interacting *Peromyscus leucopus* as revealed by radiotelemetry. *Canadian Field Naturalist*, **91**, 273–81.

Maynard Smith, J. (1977). Parental investment, a prospective analysis. *Animal Behaviour*, **25**, 1–9.

Merritt, J. F. (1974). Factors influencing the local distribution of *Peromyscus californicus* in northern California. *Journal of Mammalogy*, **55**, 102–14.

(1978). *Peromyscus californicus. Mammalian Species*, **85**, 1–6.

Metzgar, L. H. (1971). Behavioral population regulation in the woodmouse, *Peromyscus leucopus*. *American Midland Naturalist*, **86**, 434–48.

Millar, J. S. (1989). Reproduction and development. In *Advances in the Study of* Peromyscus *(Rodentia)*, ed. G. L. J. Kirkland & J. N. Layne, pp. 169–231. Lubbock, Texas: Texas Tech University Press.

Millar, J. S. & Derrickson, E. M. (1992). Group nesting in *Peromyscus maniculatus*. *Journal of Mammalogy*, **73**, 403–7.

Millar, J. S. & Innes, D. G. L. (1983). Demographic and life cycle characteristics of montane deer mice. *Canadian Journal of Zoology*, **61**, 574–85.

Nelson, R. J., Gubernick, D. J. & Blom, J. M. C. (1995). Influence of photoperiod, green food, and water availability on reproduction in male California mice (*Peromyscus californicus*). *Physiology and Behavior*, **57**, 1175–80.

Ostfeld, R. S. (1990). The ecology of territoriality in small mammals. *Trends in Ecology and Evolution*, **5**, 411–15.

Quillfeldt, P., Schmoll, T., Peter, H.-U., Epplen, J. T. & Lubjuhn, Y. (2001). Genetic monogamy in Wilson's Storm-Petrel. *The Auk*, **118**, 242–8.

Ribble, D. O. (1990). Population and Social Dynamics of the California mouse (*Peromyscus californicus*). Ph.D. thesis, University of California, Berkeley.

(1991). The monogamous mating system of *Peromyscus californicus* as revealed by DNA fingerprinting. *Behavioral Ecology and Sociobiology*, **29**, 161–6.

(1992*a*). Dispersal in a monogamous rodent, *Peromyscus californicus*. *Ecology*, **73**, 859–66.

(1992*b*). Lifetime reproductive success and its correlates in the monogamous rodent, *Peromyscus californicus*. *Journal of Animal Ecology*, **61**, 457–68.

Ribble, D. O. & Millar, J. S. (1996). The mating system of northern populations of *Peromyscus maniculatus* as revealed by radiotelemetry and DNA fingerprinting. *Ecoscience*, **3**, 423–8.

Ribble, D. O. & Salvioni, M. (1990). Social organization and nest co-occupancy in *Peromyscus californicus*, a monogamous rodent. *Behavioral Ecology and Sociobiology*, **26**, 9–15.

Ribble, D. O. & Stanley, S. (1998). Home ranges and social organization of syntopic *Peromyscus boylii* and *P. truei*. *Journal of Mammalogy*, **79**, 932–41.

Ribble, D. O., Wurtz, A. E., McConnell, E. K., Buegge, J. J. & Welch, K. C. J. (2002). A comparison of home ranges of two species of *Peromyscus* using trapping and radiotelemetry data. *Journal of Mammalogy*, **83**, 260–6.

Rickart, E. A. (1977). Reproduction, growth and development in two species of cloud forest *Peromyscus* from southern Mexico. *The University of Kansas Natural History Museum Occasional Papers*, **67**, 1–22.

Rickart, E. A. & Robertson, P. B. (1985). Peromyscus melanocarpus. *Mammalian Species*, **241**, 1–3.

Rogers, D. S. & Engstrom, M. D. (1992). Evolutionary implications of allozymic variation in tropical *Peromyscus* of the *mexicanus* species group. *Journal of Mammalogy*, **73**, 55–69.

Sauer, J. R. & Slade, N. A. (1987). Size-based demography of vertebrates. *Annual Review of Ecology and Systematics*, **18**, 71–90.

Scheibe, J. S. & O'Farrell, J. O. (1995). Habitat dynamics in *Peromyscus truei*, eclectic females, density dependence, or reproductive constraints? *Journal of Mammalogy*, **76**, 368–75.

Schug, M. D., Vessey, S. H. & Underwood, E. M. (1992). Paternal behavior in a natural population of white-footed mice (*Peromyscus leucopus*). *American Midland Naturalist*, **127**, 373–80.

Smith, M. H. (1966). The Evolutionary Significance of Certain Behavioral, Physiological, and Morphological Adaptations of the Old-field Mouse, *Peromyscus polionotus*. Ph.D. thesis, University of Florida.

Stangl, F. B., Jr & Baker, R. J. (1984). Evolutionary relationships in *Peromyscus*, congruence in chromosomal, genic, and classical data sets. *Journal of Mammalogy*, **65**, 643–54.

Stapp, P. (1999). Size and habitat characteristics of home ranges of northern grasshopper mice (*Onychomys leucogaster*). *The Southwestern Naturalist*, **44**, 101–5.

Trivers, R. L. (1972). Parental investment and sexual selection. In *Sexual Selection and the Descent of Man*, ed. B. Campbell, pp. 136–79. Chicago: Aldine-Atherton.

Wiley, R. H. (1974). Evolution of social organization and life-history patterns among grouse. *Quarterly Review of Biology*, **49**, 201–27.

Wittenberger, J. F. & Tilson, R. L. (1980). The evolution of monogamy, hypotheses and evidence. *Annual Review of Ecology and Systematics*, **11**, 197–232.

Wolff, J. O. (1985). The effects of density, food, and interspecific interference on home range size in *Peromyscus leucopus* and *Peromyscus maniculatus*. *Canadian Journal of Zoology*, **63**, 2657–62.

(1989). Social behavior. In *Advances in the Study of* Peromyscus *(Rodentia)*, ed. G. L. J. Kirkland & J. N. Layne, pp. 271–91. Lubbock, Texas: Texas Tech University Press.

(1993). Why are female small mammals territorial? *Oikos*, **68**, 364–70.

(1994). Reproductive success of solitary and communally nesting white-footed mice and deer mice. *Behavioral Ecology*, **5**, 206–9.

Wolff, J. O. & Cicirello, D. M. (1989). Field evidence for sexual selection and resource competition infanticide in white-footed mice. *Animal Behaviour*, **38**, 637–42.

(1990). Mobility versus territoriality, alternative reproductive strategies in white-footed mice. *Animal Behaviour*, **39**, 1222–24.

(1991). Comparative paternal and infanticidal behavior of sympatric white-footed mice (*Peromyscus leucopus noveboracensis*) and deermice (*P. maniculatus nubiterrae*). *Behavioral Ecology*, **2**, 38–45.

Wolff, J. O. & Peterson, J. A. (1998). An offspring-defense hypothesis for territoriality in female mammals. *Ethology, Ecology and Evolution*, **10**, 227–39.

Xia, X. & Millar, J. S. (1986). Sex-related dispersion of breeding deer mice in the Kananaskis Valley, Alberta. *Canadian Journal of Zoology*, **64**, 933–6.

(1988). Paternal behavior by *Peromyscus leucopus* in enclosures. *Canadian Journal of Zoology*, **66**, 1184–7.

(1989). Dispersion of adult male *Peromyscus leucopus* in relation to female reproductive status. *Canadian Journal of Zoology*, **67**, 1047–52.

(1991). Genetic evidence of promiscuity in *Peromyscus leucopus*. *Behavioral Ecology and Sociobiology*, **28**, 171–8.

PART II

Reproductive strategies of socially monogamous males and females

CHAPTER 6

Social functions of copulation in the socially monogamous razorbill (*Alca torda*)

Richard H. Wagner

INTRODUCTION

The fundamental purpose of copulation is sperm transfer. Additionally, however, animals in a wide range of taxa copulate outside the breeding period (reviewed in Bagemihl, 1999), suggesting that copulation is used for multiple purposes. In birds, there have long been reports of pairs copulating outside the female fertile period (Birkhead & Møller, 1992), and a social role for such behaviour, such as the appraisal and acquisition of mates, has been proposed (Colwell & Oring, 1989; Tortosa & Redondo, 1992; Heg *et al.*, 1994). Because copulation requires physical contact it has been viewed as a form of tactile communication (Dewsbury, 1988). In monogamous white storks (*Ciconia ciconia*), which copulate at high rates but rarely perform extra-pair copulations (EPCs), it has been suggested that females use male copulation performance to appraise the balance, size, and body condition of prospective mates (Tortosa & Redondo, 1992). This hypothesis rests on the prediction that copulation rate or other aspects of its performance by either sex reflects individual quality. Supporting this prediction is the finding in tree sparrows (*Passer montanus*) that within-pair copulation frequency covaried with female reproductive performance (Heeb, 2001). Another hypothesis is that copulation is used as a display by males or pairs to signal their possession of the territory (Strahl, 1988; Negro & Grande, 2001).

In monogamous species that share parental duties and form long-term pair bonds, mate choice can have major effects on fitness (Black, 1996). Monogamous pairs of black-legged kittiwakes (*Rissa tridactyla*) usually breed together in multiple years, but divorce is common after breeding failure, suggesting that mates switch in order to acquire better or more compatible partners (Coulson & Thomas, 1983). In general, animals that form long-term bonds should be selected to exploit any available cues that reveal the quality and compatibility of prospective mates, and such cues may be provided by copulation performance.

Despite its potential to explain certain poorly understood aspects of sexual behaviour, interest in these hypothetical social uses of copulation has been relatively neglected in birds. One reason for this oversight is that the exciting topic of sperm competition has overwhelmed the study of other aspects of copulation behaviour in birds. For over a decade, sperm competition has been a dominant topic in behavioural ecology (e.g., Birkhead & Møller, 1998; Møller, chapter 2). With the advent of molecular techniques such as DNA fingerprinting, remarkable and unexpected patterns of parentage have been revealed that have dramatically altered the way biologists think about mating systems, and monogamy in particular. Ideas involving sperm competition have explained, for example, the marked variation in copulation frequency among species, ranging from one to 600 per clutch: in species in which males are constrained from mate guarding, males pursue the paternity assurance tactic of copulating frequently with their mates so that their sperm will outnumber those of possible extra-pair males (Birkhead *et al.*, 1987). The study of sperm competition and paternity has shown much higher than expected variance in male fertilization success, opening the floodgates to studies of sexual selection in socially monogamous species (Petrie & Kempenaers, 1998). The focus on sperm competition in the study of bird copulation behaviour has been invaluable. Nevertheless, in many bird species, males and females perform certain sexual behaviours which suggest that copulation also plays a social role independent of sperm competition. Throughout this chapter I refer to 'social' in the general sense of a sexual behaviour involving interactions between individuals that do not directly lead to fertilization, such as using copulation for mate appraisal and pair bonding. The aim of this chapter is to synthesize results from my field study of razorbills (*Alca torda*), which illustrate the need to look beyond sperm competition to understand fully copulation behaviour in socially monogamous birds.

SITE, TIME, AND SPECIES

Razorbills are socially monogamous, long-lived, colonial seabirds which breed at temporal and boreal latitudes on the coasts of the North Atlantic (Cramp, 1985). Approximately 90% of mates re-pair each year while both partners are alive (Wagner, 1991*a*), resulting in the maintenance of a large number of partnerships over multiple years. They spend most of the year swimming and foraging underwater for fish, coming ashore only to breed. The species is sexually monomorphic and both sexes share about equally in incubation, brooding, and chick feeding (Wagner, 1992*a*), although the male invariably escorts the fledgling to sea, where he continues to provision it for approximately six weeks until it is independent (Harris & Birkhead, 1985). Biparental care is essential because the egg must be incubated continuously for 34 days, necessitating cooperation by both parents. Furthermore, both sexes are required to brood and feed the nestling and both defend the nesting site (Harris & Birkhead, 1985).

I studied razorbills on Skomer Island, Wales, UK (51° 40′ N, 05° 15′ W) in 1988 and 1989. Using a telescope, I observed razorbill behaviour at two mating arenas outside the nesting colony from a hide at a distance of 25 m. The two arenas comprise horizontal ledges, outcropping from a steep rocky slope. In both years, from early April to late May, I observed the arenas every morning during copulation activity for a total of 225 hours. I dictated the following information into a cassette tape recorder: the time of arrival of each marked bird; whether upon arrival the bird's mate was present, absent, or arrived simultaneously. I recorded all mountings (i.e., when a male stepped onto the back of a female, apparently attempting copulation), which were classified in four ways: within-pair copulation; unsuccessful within-pair copulation attempt; extra-pair copulation; unsuccessful extra-pair copulation attempt. Successful copulations were those in which the male made cloacal contact with the female, and presumably achieved insemination. Unsuccessful within-pair copulation attempts were those in which the female did not cooperate by lifting her tail or allowing the male to balance on her back. I recorded whether successful within-pair copulations were terminated by the male or female. Females terminated copulations by standing upright, causing the male to lose his balance, and males terminated copulations by dismounting while the female remained receptive in the prone copulation posture.

Most razorbills in the study colony bred under boulders in order to protect their eggs and chicks from predators (Wagner, 1992*b*). A consequence of this was that the boulders limited the ability of breeders to observe and interact with their neighbours. However, both males and females congregated on the open ledges outside the nesting area at densities up to 50 times that of the boulder colony. Males and females breeding in a given boulder colony consistently attended a mating arena within view of their nesting territory. Thus, neighbours breeding at relatively low densities encountered each other at much higher frequencies and closer distances in the open mating arenas while also apparently attempting to maintain surveillance over their nesting territory. Individuals tended to occupy approximately the same spots in the arena on a daily basis and even frequented the same places in consecutive years (Wagner, 1992*b*). These conditions provided opportunities for breeding neighbours to appraise each other as competitors and future mates over the course of multiple years.

Near the study colony, about 18 pairs of razorbills bred on open ledges, but unlike boulder breeders, these individuals rarely or never visited the arenas. A likely explanation for this difference is that ledge breeders were able to interact socially at or near their breeding sites and therefore were not forced to leave their nests undefended in order to socialize. In this regard, ledge-breeding razorbills resemble common guillemots [murres] (*Uria aalge*), which also breed on ledges, although at much higher densities (Birkhead, 1977), and also do not attend mating arenas. These facts suggest that EPCs and other kinds of encounters with non-mate conspecifics are an important part of the breeding strategies of razorbills because when such social interactions are limited, both males and females aggregate elsewhere.

Razorbills in the boulder colony performed 74% of 529 within-pair copulations and the remainder near their nests ($N = 20$ marked pairs that were watched simultaneously by two observers in the arena and colony). Pairs performed a mean of 26 (SD $= 9.8$, range

9–47) copulations comprising *c.* 80 cloacal contacts in the month preceding egg laying. Most (82%) extra-pair copulations were also performed in the arenas. Eighty-seven per cent of 53 males attempted EPCs and 96% of 49 females received EPC attempts in the arenas (Wagner, 1992*b*). The frequency of these behaviours, combined with the clear visibility of large numbers of birds, enabled me to record 6000 mating events.

FEMALE CONTROL OF COPULATION

Although the ability of females to control copulation enables them to use it for social purposes such as mate appraisal, female control is often difficult to evaluate. Earlier reports often stressed a lack of female control and described observations of apparently forced extra-pair copulation attempts by males aggressively pursuing females and sometimes achieving cloacal contact (Birkhead *et al.*, 1985; Emlen & Wrege, 1986; Morton, 1987). However, an alternative interpretation is that females in certain species incite sexual chases in order to assure insemination from dominant males (Cox & LeBoeuf, 1977), or resist insemination as a ploy to test males (Westneat *et al.*, 1990), as has been suggested in several studies (Montgomerie & Thornhill, 1989; Wagner, 1991*b*; Hoi & Hoi-Leitner, 1997; Neudorf *et al.*, 1997). Female control of copulation has received increased attention (Eberhard, 1996; Gowaty, 1996), although its mechanisms have not been studied in detail in many avian species. Direct evidence of active pursuit of EPCs by females has been derived from reports of female northern fulmars (*Fulmaris glacialis*: Hatch, 1987; Hunter *et al.*, 1992), black-capped chickadees (*Parus atricapillis*: Smith, 1988), blue tits (*Parus caeruleus*: Kempenaers *et al.*, 1995), and least flycatchers (*Epidonax minimus*: Tarof & Ratcliffe, 2000) visiting males in their territories and accepting EPCs. In tree swallows (*Tachycineta bicolor*), Lifjeld and Robertson (1992) provided experimental evidence of female control by demonstrating that females which were faithful to their mates remained so after the males were removed. Furthermore, faithful females copulated with a replacement male later than did females that had engaged in EPCs prior to the removal of their mates. An indirect form of evidence of female pursuit of EPCs is a positive relationship between mate guarding and extra-pair paternity in blue tits (Kempenaers *et al.*, 1995) and purple martins (*Progne subis*: Wagner *et al.*, 1996), which suggests that males guard their mates not only to prevent access *of* other males, but also *to* other males. However, the pursuit of EPCs by females is not evidence that they cannot be forced. In fact, in some species such as the common guillemot, females were involved in apparently forced and unforced EPCs (Birkhead *et al.*, 1985; Hatchwell, 1988). In other birds, especially ducks, males possess an intromittent sexual organ and forced copulation is well known (McKinney *et al.*, 1983). Even in species in which males lack such an organ, it may be possible for males to use aggressive behaviour to coerce females into accepting EPCs (Westneat *et al.*, 1990).

If females use copulation behaviour to appraise males, they probably do so not just at one but at different stages of their encounters with males. Females may simply observe males copulating, they may allow males to mount but not inseminate, or they may permit insemination. In order for females to obtain information about males during mountings (Dewsbury, 1988) without incurring unwanted inseminations, females must be able to control whether or not insemination occurs. Females could not, for example, allow males to mount them in order to test or 'sample' the male without the ability to avoid unwanted inseminations. The degree to which females control copulation varies among species and should substantially determine their ability to use copulation for mate appraisal. In razorbills, there were several lines of evidence indicating that females exercise substantial control over all levels of copulation, both extra-pair and within-pair. The following account is drawn from Wagner (1991*c*).

The frequent visitations by female razorbills to the mating arenas indicate that females in this species actively pursue EPCs. Unlike in the breeding colony, where both sexes defend the nesting site, the mating arenas contain no resources other than mates, and females often visited and remained in the arenas when their mates were absent. I witnessed instances of a female accepting an EPC in the mating arena while her mate was guarding the breeding site, as well as of a female accepting an EPC in the breeding colony while her mate attended the mating arena (Wagner, 1992*b*). And while the obvious willingness of females to pursue EPCs does not exclude the possibility that males can also force EPCs (Westneat *et al.*, 1990), direct observations suggest that EPCs are not forced in this species. The sexes are similar in size (Wagner, 1999), making it difficult for males to force or coerce females, which possess similarly powerful bills that constitute dangerous weapons. Both sexes have long, stiff tails, and females use theirs to guard their cloacas. Razorbills are awkward on land and males often struggle to balance on the backs of even cooperative females. Based on over 500 observations of EPC attempts in the mating arenas, males appeared unable to force females to lift their tails.

More objective evidence of female control involves the common occurrence of repeated, failed EPC attempts with the same female. Often females arrived in the arenas, were mounted and aggressively pursued by one or several males, and then would sometimes fly away. However, females would sometimes remain stationary, in effect allowing a male to make several repeated attempts within one minute, but which females always successfully resisted. In one year, 23 out of 49 females allowed repeated EPC attempts. Females did not remain in the arena during repeated EPC attempts to be with their mates because during 95% of repeated attempts the female's mate was absent. An additional, related line of evidence of female control is that males ceased attempting EPCs from persistently unwilling females. After resisting EPC attempts upon their arrival, females were subsequently often ignored by numerous other males. These observations suggest that, rather than attempting to coerce females, males were testing the female's willingness to accept an EPC.

Another line of evidence of female control derives from observations of females visiting the arenas and receiving extra-pair mountings after they had laid the egg. As the season progressed and an increasing proportion of females became fertile, there was a marked acceleration of male persistence in attempting EPCs (Figure 6.1*a*). Whereas, early in the season, prior to egg laying, EPC attempts were relatively infrequent and fighting among males was uncommon (only one fight per male per 10 hours), fighting increased tenfold by the median egg-laying date (i.e., the peak of female fertility: Figure 6.1*b*). During and soon after this peak, females arrived, often appeared to incite male–male competition to mount them, and then departed. Subsequent analyses revealed that females often visited the arenas after egg laying, when fertilization was no longer likely. (A replacement egg is occasionally laid if the original egg is lost; however, this did not occur among the females in my sample.) None of the 55 EPC attempts with post-laying females resulted in cloacal contact compared with 22 of 107 EPC attempts with pre-laying females.

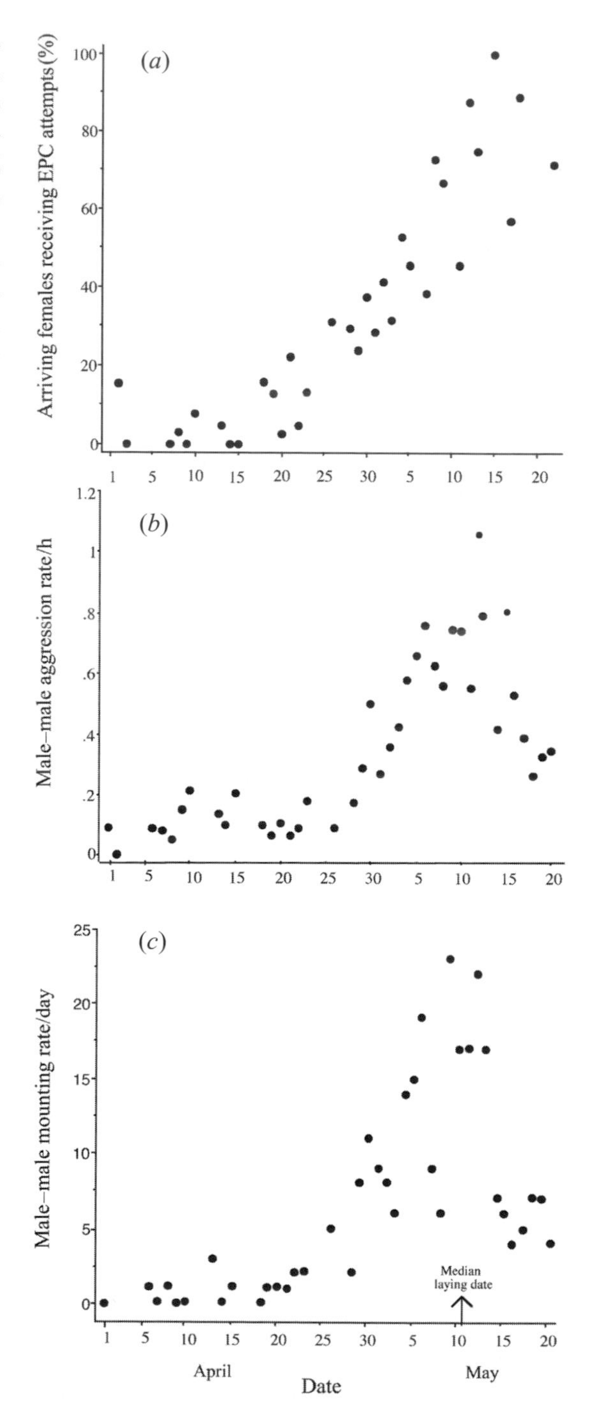

Figure 6.1. The relationship between date and (*a*) percentage of females receiving EPC attempts upon arrival to the arenas; (*b*) male–male aggression rate (fights per male/h); and (*c*) male–male mounting rate (number of male–male mountings per day). (*a*) and (*b*) from Wagner, 1992*e*; (*c*) from Wagner, 1996*a*.

This suggests that females controlled copulation and always resisted cloacal contact after laying because there was no advantage to being inseminated then, whereas pre-laying females may have been attempting to obtain extra-pair fertilizations.

COPULATION AND PAIR FORMATION

Razorbills provide three kinds of evidence that copulation is involved in mate appraisal and pair bonding: (i) an observational account of pair formation; (ii) extra-pair mountings after laying; and (iii) extra-pair copulations at the end of the breeding cycle. If copulation is used for pair bonding it should be performed during courtship, even when fertilization is not possible. Pair formation is difficult to observe because it requires monitoring marked individuals over extended periods in various locations; however, I did record one detailed account of a pair's first encounters and subsequent pairing in which copulation played a central role. The following is a summary of that account (Wagner, 1991*a*).

An unpaired male M had been seen in the arena attempting to copulate with paired and unpaired females. Unpaired female F was four years old, the youngest age at which razorbills are known to breed, while five is the modal age of first breeding (Lloyd & Perrins, 1977). F visited the mating arena 16 times in 12 days and was mounted by six different males, accepting a copulation from one of them. Several days later F encountered M for the first time and rejected his attempt to mount, but remained in place. Two minutes later, F permitted a 40-second copulation from M comprising three cloacal contacts. In the next 16 minutes, F refused three more copulation attempts by M. Overall, F visited the arena on 10 days and received 15 copulation attempts from several different males, of which she rejected 12. Eight of the 15 mountings were performed by M.

Several days after their first encounter in the mating arena, M and F were spotted prospecting nesting sites in the colony, where they performed one copulation that was terminated by F, who soon after departed to sea for the year. M remained in the colony but failed to procure a mate. However, early in the following year, M and F consistently attended the arena together as a pair. Prior to egg laying, M attempted 27 copulations of which F accepted nine. In that time, F received six copulation attempts from several different males and accepted none. The pair bred in the colony near where they had prospected the previous year, and produced an egg, which hatched. This account implies that: (i) copulation is used for mate sampling by females and possibly by males as well; (ii) copulation rejection by females may be involved in pair bonding and sampling; and (iii) copulation and other social encounters performed in one year may contribute to pair bonding in the next year.

NON-FERTILIZABLE EXTRA-PAIR COPULATIONS

EPCs after egg laying

Razorbills performed extra-pair mountings or copulations during two phases of the breeding cycle when fertilization did not occur: after laying and at the end of the season (Wagner, 1991*b*). Above, I suggested that the failure of males to inseminate females after they had laid is evidence of female control of extra-pair copulation. The fact that post-laying females visit the arenas at all, when doing so nearly always results in extra-pair mountings, also suggests that females were actively seeking such mountings. During the ten days after laying, a minimum of 20 (59%) of 34 females visited the arenas. Eighteen (90%) of the 20 females received between one and 11 EPC attempts. One female received ten EPC attempts from five males in three minutes. Overall, males made 30% of their EPC attempts with post-laying females. Notably, the number of EPC attempts per day per female was 300% higher in the ten days after than before laying. This difference is probably explained by the absence of females' mates after laying; before laying, arriving females received EPC attempts three times more often when their mates were absent (44% of 106 arrivals) than when they were present (15% of 166).

The timing of post-laying arena visits is intriguing because they occurred when females were assured of not encountering their mates. In the pre-laying phase, visiting females usually found their mates already in the arenas or the pair arrived together. A typical occurrence was for a female to spend the night in the breeding colony incubating the egg while her mate roosted at sea. (Mates were never observed roosting together in the colony and no razorbill was ever seen roosting on land except when incubating or brooding.) After the male relieved the female on the nest, the female departed to sea to forage, first stopping at the mating arena to rendezvous with other males without any possibility of interference from her mate. Prior to egg laying, EPC attempts always failed in the presence of the female's mate, either because the female resisted cloacal contact or her mate disrupted the mounting. Thus, before laying, females were usually limited in their ability to interact with other males, whereas after laying, females were able to receive mountings from multiple males without detection by their mates.

An interesting question raised by visitations to the mating arenas by post-laying females is whether females were actively inciting male–male aggression in order to evaluate competing males. During the laying period, 49% of 71 unescorted female arrivals were immediately followed by fights among males. At this time, the sex ratio was strongly male-biased and an arriving female typically faced numerous males that were aggressively competing for EPCs. In one well-documented example, a female that visited the arena six days after egg laying was immediately mounted upon arrival by male A, who dismounted and attacked male B. The female was then mounted by male C, who she rejected, and then by D, who was disrupted by E. Male B then remounted and was rejected by the female; B then attacked C. The female then walked across the entire arena, which resulted in four more fights, during which the female departed. Overall, in under two minutes, the female received four EPC attempts from three males and incited seven fights involving nine males. In several other detailed examples, unescorted females also arrived in the arena, received mountings from multiple males, and generated male–male aggression, all within a brief time span. Although it is evident that females triggered male fighting, it is difficult to determine whether they did so deliberately, as some of these accounts suggest, or whether fighting is a by-product of females seeking extra-pair mountings.

EPCs after fledging

The notion that females exploit the absences of their mates to encounter other males is supported by observations of female behaviour at the end of the breeding cycle. At the end of the breeding cycle, after the male departs to sea with the fledgling, the female remains in the colony, possibly to signal ownership of the breeding site to individuals prospecting for a site for the following year (Wanless & Harris, 1986). My observations confirmed that females defend the site, but in addition, they frequently consort and copulate with extra-pair males at this time. Consortships comprise males and females visiting prospective nesting sites for the following year, allopreening, and sitting side by side in bodily contact, none of which are performed by two members of the same sex. At least 20 (57%) of 35 females that produced a fledgling consorted with other males at the end of a breeding season. Eighteen of the 20 females consorted with one male each while the other two consorted sequentially with two males. Nine of the 20 females received EPCs and seven of the nine accepted between one and three copulations. Females permitted a higher percentage of EPC attempts to result in cloacal contact at the end of the season (10/21 = 48%) than in the pre-laying period (19/91 = 21%). A divorced male who had consorted with a female at the end of one season bred with that female the following year, suggesting that post-breeding consortships and EPCs may be involved in the establishment of future pair bonds.

Non-fertilizable EPCs differ in the post-laying and post-breeding periods in ways that suggest that they have different functions. At the end of the breeding cycle, females usually consorted and copulated with one extra-pair male consistently over a period of days. In contrast, post-laying females tended to encounter multiple males for very brief periods. While it is plausible that post-breeding EPCs and consortships are used to ensure the development of pair bonds, the brief and rapid post-laying mountings might be for a different purpose, such as the appraisal of males for extra-pair fertilizations in the following year. A more parsimonious explanation is that females simply gather whatever information about males is available, and different kinds of information exist during different phases of the breeding cycle.

It is fascinating that when females sought mountings in the post-laying period they never permitted cloacal contact but they frequently did so at the end of the season. If females risk venereal disease from promiscuous copulations (Hamilton, 1990), then perhaps post-laying females visit the arenas to gain information about males by permitting mountings but avoiding the risk of infection by refusing cloacal contact. This interpretation raises the question of why females permit cloacal contact during EPCs at the end of the season. A possible explanation is that since diseases may be transmitted through semen (Hillgarth, 1990) and semen production tends to cease outside the reproductive period (Sturkie, 1986), the risk of infection is higher during the laying period. The avoidance of cloacal contact after egg laying argues against the suggestion of Lombardo *et al.* (1999) that microbes transmitted during copulation may be beneficial. The fact that females commonly allowed cloacal contact at the end of the season suggests the possibility that full copulations are more useful for mate appraisal than mountings without cloacal contact.

Why do razorbills invest in mate assessment at the end of the breeding season instead of waiting until the beginning of the following year? It may in part simply be that they exploit any available opportunities for mate assessment, and that the combination of longevity and breeding site faithfulness allows them to use information over a number of years. The fact that males are away at sea with the fledgling enables females to interact with prospective mates in the absence of their current mates. Even if the males that females consort with at the end of the season are not superior to their own mates, it may nevertheless be adaptive for females to build an insurance pair bond in the event that their own mates either do not return or divorce them. By already having established a bond with another male during the previous year, a female may reduce competition in the search for a mate at the beginning of the next breeding season.

MATE GUARDING BY FEMALES

Mate guarding is usually defined in terms of males protecting their paternity by preventing other males from gaining access to their fertile mates (Birkhead & Møller, 1992). If a male razorbill attempts EPCs in the presence of another male's mate, the male mate will often disrupt the attempt (although usually the female ends the mounting even before her mate can interfere: Wagner, 1991*c*). A male that fails to assure his paternity may pay the cost of investing an entire breeding season in rearing the offspring of another male while producing no offspring of his own. It is therefore not surprising that the literature contains abundant examples of males closely guarding their mates during the females' fertile period (Birkhead & Møller, 1992). On the other hand, from the female perspective there is no risk of losing parentage if their mates fertilize other females. Given the absence of this obvious risk for females, mate guarding by females is much less discussed (but see Petrie, 1992; Eens & Pinxten, 1995). Nevertheless, female razorbills were observed breaking up their mates' EPC attempts. To disrupt the mounting, females attacked and drove away the extra-pair female and/or pushed or pecked their mates. One female attacked her mate after she had disrupted his EPC attempt. I witnessed nine females disrupting 18 of their mates' EPC attempts. During several extra-pair mountings that the pair female did not disrupt, the pair female approached but was unable to disrupt the mounting because it was ended, either by the extra-pair female or by interference from another male. Females disrupted their mates' EPC attempts exclusively, never disrupting any of 1591 copulation attempts made by other males.

In addition to disrupting their mates' EPC attempts, females also appeared to prevent their mates from gaining access to extra-pair females before a mounting could occur. When one male approached an arriving female his mate stepped between them and another male mounted the female. Another female prevented her mate from mounting an arriving female by stabbing him with her bill as he approached her. Even though their mates attended the arenas on 67% of the same days, males made only 6% of 245 EPC attempts when their mates were present, probably as a result of female interference. This discrepancy is produced largely because males increase their arena attendance during the last several days before laying (see Figure 1 in Wagner, 1992*c*), while the female remains at sea to feed and build nutritional reserves for the egg.

Why do females disrupt their mates' EPCs? If razorbills use copulation for mate appraisal and pair bonding it is plausible that females risk losing their mates to an extra-pair female in a subsequent breeding attempt. An alternative explanation is that females attempt to monopolize their mates' sperm supply in order to ensure fertilization. This is unlikely given that females reject 45% of their mates' mountings and terminate 82% of copulations (see the following section), while the male continues further insemination attempts. High frequencies of rejection were also observed in the aftermath of female disruption of their mates' EPC attempts; following the disruption, six of ten females refused copulations from their mates, and of the four that were accepted, all were terminated by the female.

Another alternative explanation is that as in some polygynous species, such as red-winged blackbirds (*Agelaius phoeniceus*), females compete over male parental contributions (Yasukawa & Searcy, 1982). However, in 71 breeding attempts by biparental razorbills, males were never observed contributing to another nest (Wagner, 1992*a*). A third possibility is that females lower the risk of venereal infection by reducing their mates' promiscuous behaviour. This seems plausible in light of the finding that females always avoided cloacal contact during post-laying extra-pair mountings. On the other hand, 50% of females did allow cloacal contact from one to seven males prior to laying, suggesting that the risk of infection is less costly than the unknown benefits of pre-laying EPCs.

FEMALE CONTROL OF WITHIN-PAIR COPULATION AND MATE REJECTION

While a number of workers have studied the ability of females to control extra-pair copulation, there has been much less interest in the ability of females to control copulation with their own mates. I examined this aspect in detail in razorbills and found that females exercise considerable control, allowing them potentially to use within-pair as well as extra-pair copulation for social purposes. The control of within-pair copulation by female razorbills was evident by their frequent rejections of their mates (see Wagner, 1996*a*).

When I began quantifying razorbill copulations, I noticed that males attempted to mount their mates, which either assumed the prone posture, allowing the male to balance on her back, or remained upright, preventing mounting. However, I often also witnessed an intermediate situation in which a female allowed her mate to balance almost long enough to inseminate, but an instant before cloacal contact could be made would alter her posture or lower her tail, ending the mounting. Initially, I was not interested in failed copulation attempts because I began by asking questions regarding sperm competition, such as whether males attempted to copulate frequently to assure paternity. However, it soon became obvious that females rejected their mates so often that why they did so was itself an interesting new question. At the first level of rejection, females simply refuse to allow the male to mount them. In many bird species, one or both sexes solicit copulation using clear, ritualized behaviours. However, in razorbills, neither sex performs clear solicitation. To attempt to initiate a copulation, a male would sometimes sidle up to his mate and gently allopreen her head, while seeming to attempt to nudge the female to sit. The female often responded by either assuming a prone posture and allowing the male to mount, or remaining upright or turning away, apparently rejecting a copulation solicitation. Because of the subtlety of these interactions, I did not quantify them. However, they are perplexing and a future worker might gain insights by examining such behaviours closely using video.

At the second level of mate rejection, which was easy to quantify, females allowed mounting but prevented insemination. All ($N = 37$) females rejected at least some of their mate's copulation attempts ($N = 1531$), and the mean percentage of attempts rejected was a surprisingly high 45%, with rejection frequency varying markedly among females, from 9% to 70% of male attempts. At the third level of rejection, females allowed insemination but terminated the copulation by standing upright or only slightly increasing the angle of their breast relative to the ground, while the male was attempting additional cloacal contacts. Females terminated a mean of 82% of within-pair copulations, again with a wide range among females of 42–100%. The mean duration of copulations terminated by females was 33 seconds versus 61 seconds for those ended by males. Because the number of cloacal contacts increased with copulation duration ($r_s = 0.94$, $N = 881$, $P < 0.0001$), females prevented their mates from achieving numerous inseminations by prematurely terminating copulations. These observations illustrate that females not only control EPC, but they also control within-pair copulation to a strikingly precise degree. The prevalence of female rejection suggests a conflict between mates that can be examined in terms of the competing hypotheses of sperm competition and mate appraisal (Wagner, 1996*a*).

Sperm competition might explain female rejection behaviour if females attempt to influence which male achieves fertilization by rejecting and accepting relative numbers of copulations from their mates versus extra-pair males. This prediction is based on the assumption that when multiple males inseminate the same females, the male which transfers the largest amount of sperm has a higher chance of achieving fertilization (Birkhead & Møller, 1992). The sperm competition hypothesis predicts that females that accept EPCs will reject their mates more frequently than do females that avoid EPCs. However, the opposite result was found (Wagner, 1992*d*): females that accepted EPCs rejected their mates significantly less (31%) than females that avoided EPCs (49%).

The inability of sperm competition to explain variation in female rejection behaviour suggests that females may reject their mates because of conflicts involving the pair bond rather than fertilization. The 'testing of the bond' hypothesis (Wagner, 1996*a*) proposes that females frequently reject their mates in order to test their commitment to contribute parental care after laying. Male care is essential because the female alone cannot incubate for 34 days, and the nestling requires brooding and feeding from both parents (Harris & Birkhead, 1985). A female risks losing her large investment in the egg, which is 14% of body mass (Lloyd, 1979) if her mate switches to another female or continues attending the arena to attempt EPCs rather than share in incubation. A male's willingness to remain with his mate and attempt copulation despite repeated rejection may inform a female about the male's commitment to perform parental duties after laying. This reasoning follows Zahavi's (1977) observation that mates often test their pair bond by imposing costs on one another. The 'testing of the bond' hypothesis predicts that females react to their mates' EPC attempts by rejecting their copulation attempts. This prediction is supported by a significant correlation between the number of EPC attempts made by males and the percentage of copulation attempts rejected by their mates (Figure 6.2; $r_s = 0.44$, $N = 27$, $P = 0.026$) (Wagner, 1996*a*).

Whereas females may test their mate's commitment to provide parental care, males should test the female for paternity assurance. Females which accept EPCs might risk either a reduction in male parental effort (Møller, 1988) or divorce unless they provide their mates with high confidence of paternity by accepting a large percentage of copulation attempts. This risk might explain the surprising finding that females which accepted EPCs were significantly more receptive to their mates than were females which avoided EPCs, i.e., females that lowered their mates' confidence of paternity by their receptivity to EPCs may be less able to test their mates by rejecting them frequently without risking retaliation.

WHY DO MALES MOUNT MALES?

A final use of copulation for social purposes by razorbills is the mounting of males by other males (Wagner, 1996*b*). Male–male mountings have been recorded for a number of bird species (Meanly, 1955; Armstrong, 1965; Mills, 1994; reviewed in Bagemihl, 1999). Some observers have suggested that male–male mountings may be used to establish or maintain dominance relationships (Fujioka & Yamagishi, 1981; Jamieson & Craig, 1985; Lombardo *et al.*, 1994). Other observers have regarded such mountings as a case of mistaken identity, especially in sexually monomorphic species such as common guillemots (Birkhead *et al.*, 1985) and Laysan albatrosses (*Diomedea immutabilis*: Fisher, 1971), in which males are

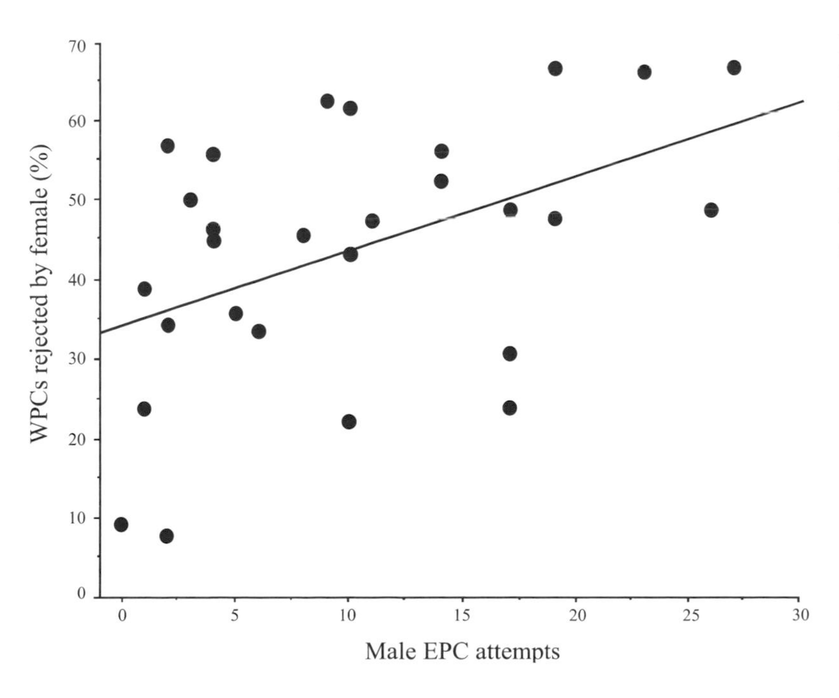

Figure 6.2. The relationship between the percentage of copulation attempts females accepted from their mates and the number of extra-pair copulations their mates attempted ($r_s = 0.44$, $N = 27$, $P = 0.026$). (Adapted from Wagner, 1996*a*.)

copulation rejection into the vexing and much studied relationship between confidence of paternity and male parental investment (Wright, 1998; Møller, chapter 2). Females are expected to assure their mates' paternity in exchange for male parental care (Trivers, 1972), yet female rejection should reduce male paternity assurance. The reason for high frequencies of rejection by female razorbills is unclear; however, sperm competition is unlikely to be involved, suggesting that a social explanation exists. The 'testing of the bond' hypothesis is consistent with the findings, but we now need to determine whether and how rejection may test the male's commitment to sharing parental duties with his mate. The wide variation among species in female within-pair copulation rejection (Wagner, 1996*a*), along with variation among females of the same species, implies that conflicts between mates exist in many contexts. Given the broad interest in the 'battle of the sexes' (Gowaty, 1996), female rejection behaviour should be exploited by researchers as an easily measured variable that can be studied to address intersexual conflict.

Ostensibly, razorbills are an unusual bird species, making it fair to ask how applicable conclusions drawn from their behaviour are to other species. Razorbills are the only monogamous species in which female disruption of their mates' EPCs has been reported. They are also the first in which females have been observed actively pursuing non-fertilizable EPCs. The clarity of these observations is attributable to the mating arenas. Female–female aggression is common in birds but is difficult to interpret when it occurs in a territory containing resources such as nests and food. However, in the arena it is clear that females are defending their pair bond. Female razorbills may be typical of females in other colonial species in seeking extra-pair mountings after egg laying; however, in other species such behaviour may be disguised by the nesting colony. For example, upon returning to their breeding sites, post-laying female common guillemots are frequently mounted by neighbouring males (Hatchwell, 1988). Like female razorbills, female guillemots also resist such mountings; but it would be difficult to determine whether female guillemots seek the mountings since they receive them at their nests. Female razorbills resisted the EPC attempts in the mating arenas, yet their visits to the arenas suggest that they actively sought them. Thus, the seemingly unusual behaviours performed by razorbills may be common in other species but have gone unnoticed. Razorbills, except for their attendance at the mating arenas, may in fact behave quite similarly to many other monogamous, colonial species. And the arenas are simply by-products of the social constraints imposed on them by their boulder colonies.

The existence of mating arenas for a monogamous species suggests a general point about social behaviour, namely that observing and interacting with neighbours is sufficiently important that when it is constrained by the geography of the breeding colony, birds will congregate elsewhere. This perspective reverses conventional thinking: rather than view social behaviour as an adaptation to high-density breeding, it may be a cause of it (Wagner, 1993, 1997).

The prevalence of deliberate male–male mountings is another social aspect of copulation that may be much more common among bird species than realized. It would be easy to assume that such mountings in a sexual monomorphic species are cases of mistaken identity. However, their occurrence in the mating arena allows more clear observations and their consequent interpretation than when such behaviour occurs in crowded colonies, where aggression may break out over defence of various resources. The evidence that males mount males as a fighting tactic adds another dimension to copulation behaviour in a species which clearly demonstrates that there is more to copulation than the transfer of sperm.

References

Armstrong, E. A. (1965). *Bird Display and Behaviour*. London: Lindsay Drummund Ltd.

Bagemihl, B. (1999). *Biological Exuberance*. New York: St. Martin's Press.

Birkhead, T. R. (1977). The effect of habitat and density on breeding success in the Common Guillemot *Uria aalge*. *Journal of Animal Ecology*, **46**, 751–64.

Birkhead, T. R. & Møller, A. P. (1992). *Sperm Competition in Birds: Evolutionary Causes and Consequences*. London: Academic Press.

(1998). Sperm competition, sexual selection and different routes to fitness. In *Sperm Competition and Sexual Selection*, ed. T. R. Birkhead & A. P. Møller, pp. 757–81. New York: Academic Press.

Birkhead, T. R., Johnson, S. D. & Nettleship, D. N. (1985). Extra-pair matings and mate guarding in the common murre *Uria aalge*. *Animal Behavior*, **33**, 608–19.

Birkhead, T. R., Atkin, L. & Møller, A. P. (1987). Copulation behaviour of birds. *Behaviour*, **101**, 101–38.

Birkhead, T. R., Hatchwell, B. J., Linder, R., Blomqvist, D., Pellat, E. J., Griffiths, R. & Lifjeld, J. (2001). Extra-pair paternity in the common murre. *Condor*, **103**, 158–62.

Black, J. M. (1996). Pair bonds and partnerships. In *Partnerships in Birds: The Study of Monogamy*, ed. J. M. Black, pp. 3–20. Oxford: Oxford University Press.

Colwell, M. A. & Oring, L. W. (1989). Extra-pair mating in the spotted sandpiper: a female mate acquisition tactic. *Animal Behavior*, **38**, 675–84.

Coulson, J. C. & Thomas, C. S. (1983). Mate choice in the kittiwake gull. In *Mate Choice*, ed. P. Bateson, pp. 361–76. Cambridge: Cambridge University Press.

Cox, C. R. & LeBoeuf, B. J. (1977). Female incitation of male competition: a mechanism of sexual selection. *American Naturalist*, **111**, 317–35.

Cramp, S. (ed.) (1985). *The Birds of the Western Palearctic*, Volume 4. Oxford: Oxford University Press.

Dewsbury, D. A. (1988). Copulatory behavior as courtship communication. *Ethology*, **79**, 218–34.

Eberhard, W. G. (1996). *Female Control: Sexual Selection by Cryptic Female Choice*. Princeton: Princeton University Press.

Eens, M. & Pinxten, R. (1995). Intersexual conflicts over copulation in the European starling: evidence for the female mate guarding hypothesis. *Behavioural Ecology and Sociobiology*, **36**, 71–81.

Emlen, S. T. & Wrege, P. H. (1986). Forced copulations and intraspecific brood parasitism: two costs of social living in the white-fronted bee-eater. *Ethology*, **71**, 2–29.

Ens, B., Choudhury, S. & Black, J. M. (1996). Mate fidelity and divorce in monogamous birds. In *Partnerships in Birds: The Study of Monogamy*, ed. J. M. Black, pp. 344–401. Oxford: Oxford University Press.

Fisher, H. E. (1971). The Laysan albatross: its incubation, hatching and associated behaviors. *Living Bird*, **10**, 19–78.

Fujioka, M. & Yamagishi, S. (1981). Extramarital and pair copulations in the Cattle Egret. *The Auk*, **98**, 134–44.

Gowaty, P. A. (1996). Battles of the sexes and origins of monogamy. In *Partnerships in Birds: The Study of Monogamy*, ed. J. M. Black, pp. 53–9. Oxford: Oxford University Press.

Hamilton, W. D. (1990). Mate choice near or far. *American Zoologist*, **30**, 341–52.

Harris, M. P. & Birkhead, T. R. (1985). Breeding ecology of the Atlantic Alcidae. In *The Atlantic Alcidae*, ed. D. N. Nettleship & T. R. Birkhead, pp. 155–204. London: Academic Press.

Hatch, S. A. (1987). Copulation and mate guarding in the northern fulmar. *The Auk*, **104**, 450–61.

Hatchwell, B. J. (1988). Intraspecific variation in extra-pair copulation and mate defence in Common Guillemots *Uria aalge*. *Behaviour*, **107**, 157–85.

Heeb, P. (2001). Pair copulation frequency correlates with female reproductive performance in tree sparrow *Passer montanus*. *Journal of Avian Biology*, **32**, 120–6.

Heg, D., Ens, B. J., Burke, T., Jenkins, L. & Kruijt, J. P. (1994). Why does the typically monogamous oystercatcher (*Haematopus ostralegus*) engage in extra-pair copulations? *Behaviour*, **126**, 247–89.

Hillgarth, N. (1990). Sexual Selection and Parasites in Pheasants. D. Phil. thesis, University of Oxford, Oxford.

Hoi, H. & Hoi-Leitner, M. (1997). An alternative route to coloniality in the bearded tit: females pursue extra-pair fertilizations. *Behavioral Ecology*, **8**, 113–19.

Hunter, F. M., Burke, T. & Watts, S. E. (1992). Frequent copulation as a method of paternity assurance in the northern fulmar. *Animal Behavior*, **44**, 149–56.

Hunter, F. M., Harcourt, R., Wright, M. & Davis, L. S. (2000). Strategic allocation of ejaculates by male Adelie penguins. *Proceedings of the Royal Society of London, Series B*, **267**, 1541–5.

Insley, S. J., Parades, R. & Jones, I. L. (2003). Sex differences in razorbill *Alca torda* parent offspring vocal recognition. *Journal of Experimental Biology*, **206**, 25–31.

Jamieson, I. G. & Craig, J. L. (1985). Male–male and female–female courtship and copulation behaviour in a communally breeding bird. *Animal Behavior*, **35**, 1251–3.

Kempenaers, B., Verheyen, B. & Dhont, A. A. (1995). Mate-guarding and copulation behaviour in monogamous and polygynous blue tits: do males follow the best-of-a-bad-job? *Behavioral Ecology and Sociobiology*, **36**, 33–42.

Lifjeld, J. T. & Robertson, R. J. (1992). Female control of extra-pair fertilizations in tree swallows. *Behavioral Ecology and Sociobiology*, **31**, 89–96.

Lloyd, C. S. (1979). Factors affecting breeding of Razorbills *Alca torda* on Skokholm. *Ibis*, **121**, 165–76.

Lloyd, C. S., Perrins, C. M. (1977). Survival and age of first breeding in the razorbill *Alca torda*. *Bird-Banding*, **48**, 239–52.

Lombardo, M., Bosman, R. M., Faro, C. A., Houttman, S. G. & Kluisza, T. S. (1994). Homosexual copulations by male tree swallows. *Wilson Bulletin*, **106**, 555–7.

Lombardo, M. P., Thorpe, P. A. & Power, H. W. (1999). The beneficial sexually transmitted microbe hypothesis of avian copulation. *Behavioral Ecology*, **10**, 333–7.

McKinney, F., Derrickson, S. R. & Minneau, P. (1983). Forced copulation in waterfowl. *Behaviour*, **86**, 250–94.

Meanly, B. (1955). A nesting study of the little blue heron in eastern Arkansas. *Wilson Bulletin*, **67**, 84–99.

Mills, J. A. (1994). Extra-pair copulations in the red-billed gull: females with high quality, attentive males resist. *Behaviour*, **128**, 42–64.

Møller, A. P. (1988). Paternity and paternal care in the Swallow *Hirundo rustica*. *Animal Behaviour*, **36**, 996–1005.

Montgomerie, R. & Thornhill, R. (1989). Fertility advertisement in birds: a means of inciting male–male competition? *Ethology*, **81**, 209–20.

Morton, E. S. (1987). Variation in mate guarding intensity by male purple martins. *Behaviour*, **101**, 211–24.

Negro, J. J. & Grande, J. M. (2001). Territorial signalling: a new hypothesis to explain frequent copulation in raptorial birds. *Animal Behaviour*, **62**, 803–9.

Neudorf, D. L., Stutchbury, B. J. M. & Piper, W. H. (1997). Covert extra-territorial behavior of female hooded warblers. *Behavioral Ecology*, **8**, 595–600.

Petrie, M. (1992). Copulation frequency in birds: why do females copulate more than once with the same male? *Animal Behavior*, **44**, 790–2.

Petrie, M. & Kempenaers, B. (1998). Why does the proportion of extra-pair paternity vary within and between species? *Trends in Ecology and Evolution*, **13**, 52–7.

Smith, S. M. (1988). Extra-pair copulations in Black-Capped Chickadees: the role of the female. *Behaviour*, **107**, 15–23.

Strahl, S. D. (1988). The social organization and behaviour of the *Hoatzin Opisthocomus hoazin* in central Venezuela. *Ibis*, **130**, 483–502.

Sturkie, P. D. (1986). *Avian Physiology*, 4th Edition. New York: Springer-Verlag.

Tarof, S. A. & Ratcliffe, L. M. (2000). Pair formation and copulation behavior in least flycatcher clusters. *Condor*, **102**, 832–7.

Tortosa, F. S. & Redondo, T. (1992). Frequent copulations despite low sperm competition in white storks (*Ciconia ciconia*). *Behaviour*, **121**, 288–313.

Trivers, R. L. (1972). Parental investment and sexual selection, In *Sexual Selection and the Descent of Man 1871–1971*, ed. B. G. Campbell, pp. 136–79. Chicago: Aldine Press.

Wagner, R. H. (1991*a*). Pair-bond formation in the razorbill. *Wilson Bulletin*, **103**, 682–5.

(1991*b*). Evidence that female razorbills control extra-pair copulations. *Behaviour*, **118**, 157–69.

(1991*c*). The use of extra-pair copulations for mate appraisal by razorbills, *Alca torda*. *Behavioral Ecology*, **2**, 198–203.

(1992*a*). Confidence of paternity and parental effort in razorbills. *The Auk*, **109**, 556–62.

(1992*b*). Behavioural and habitat-related aspects of sperm competition in razorbills. *Behaviour*, **123**, 1–26.

(1992*c*). Mate-guarding by monogamous female razorbills. *Animal Behaviour*, **44**, 533–8.

(1992*d*). The pursuit of extra-pair copulations by monogamous female razorbills: how do females benefit? *Behavioural Ecology and Sociobiology*, **29**, 455–64.

(1992*e*). Extra-pair copulations in a lek: the secondary mating system of monogamous razorbills. *Behavioural Ecology and Sociobiology*, **31**, 63–71.

(1993). The pursuit of extra-pair copulations by female birds: a new hypothesis of colony formation. *Journal of Theoretical Biology*, **163**, 333–46.

(1996*a*). Why do female birds reject copulations from their mates? *Ethology*, **102**, 465–80.

(1996*b*). Male–male mountings by a sexually monomorphic bird: mistaken identity or fighting tactic? *Journal of Avian Biology*, **27**, 209–14.

(1997). Hidden leks: sexual selection and the clustering of avian territories. In *Avian Reproductive Tactics: Female and Male Perspectives*, ed. P. G. Parker & N. Burley. *Ornithological Monographs*, Volume 49, pp. 123–45. Washington, DC: American Ornithologists' Union.

(1999). Sexual size dimorphism and assortative mating in razorbills. *The Auk*, **116**, 542–4.

Wagner, R. H., Schug, M. D. & Morton, E. S. (1996). Condition-dependent control of paternity by female purple martins: implications for coloniality. *Behavorial Ecology and Sociobiology*, **38**, 379–89.

Wanless, S. & Harris, M. P. (1986). Time spent at the colony by male and female Guillemots *Uria aalge* and Razorbills *Alca torda*. *Bird Study*, **33**, 168–76.

Westneat, D. F., Sherman, P. W. & Morton, M. L. (1990). The ecology and evolution of extra-pair copulations in birds. In *Current Ornithology*, Volume 7, ed. D. M. Power, pp. 331–69. New York: Plenum Press.

Wright, J. (1998). Paternity and paternal care. In *Sperm Competition and Sexual Selection*, ed. T. R. Birkhead & A. P. Møller, pp. 117–45. New York: Academic Press.

Yasukawa, K. & Searcy, W. A. (1982). Aggression in female Red-winged Blackbirds: a strategy to ensure male parental investment. *Behavioral Ecology and Sociobiology*, **11**, 13–17.

Zahavi, A. (1977). The testing of a bond. *Animal Behaviour*, **25**, 246–7.

CHAPTER 7

Social and reproductive monogamy in rodents: the case of the Malagasy giant jumping rat (*Hypogeomys antimena*)

Simone Sommer

INTRODUCTION

Why males should mate exclusively with one partner and after mating abstain from searching for additional females is difficult to understand in mammalian species, and several hypotheses have been proposed to explain the evolution of monogamy (Clutton-Brock, 1989). Widely accepted explanations for the evolution of monogamy in mammals include female dispersion, with female ranges being too large or too dispersed to allow males to defend more than one female, or the need for biparental care (e.g., Kleiman, 1977; Wittenberger & Tilson, 1980; Kleiman & Malcolm, 1981; Clutton-Brock, 1989). However, a recent phylogenetic analysis indicates that monogamy evolved significantly more often in the absence of paternal care than in its presence (Komers & Brotherton, 1997). Females were not widely dispersed and obligate monogamy without paternal care was found in a small antelope (Kirk's dik-dik, *Madoqua kirki*: Brotherton & Rhodes, 1996; Komers, 1996; Brotherton & Komers, chapter 3).

Biparental care appears to improve offspring survival in most monogamous fish (Barlow, 1984), birds (Lack, 1968; Wittenberger & Tilson, 1980), and mammals (Kleiman, 1977; Clutton-Brock, 1989). For example, substantial improvement in offspring survival was found in the monogamous, biparental California mouse (*Peromyscus californicus*: Gubernick & Teferi, 2000) and in the obligate monogamous Djungarian hamster (*Phodopus campbelli*: Wynne-Edwards, 1987). It is generally assumed that the fitness benefits gained by pair-living males should outweigh the costs of lost mating opportunities (Trivers, 1972; Kleiman & Malcolm, 1981). Widely accepted hypotheses for biparental care are the **'resource defence hypothesis'** (Kleiman, 1977; Rutberg, 1983), the **'infanticide avoidance hypothesis'** (Hausfater & Hrdy, 1984; Labov *et al.*, 1985; van Schaik & Dunbar, 1990; van Schaik & Kappeler, 1993), and the **'predation avoidance hypothesis'** (Dunbar & Dunbar, 1980; Dickman, 1992). The **'resource hypothesis'** predicts that limited food resources, and therefore high energetic and physiological costs of reproduction, may explain male assistance in rearing young. This male care comprises either food provisioning or defending home ranges of an appropriate size that will meet increased demands during reproduction (Kleiman, 1977). Depending on whether all-purpose territories are defended throughout the year or a larger area is required to ensure vital resources during reproduction, home range sizes should be stable or increase after the birth of offspring. Male resource defence has been assumed to be important in several monogamous primates, particularly gibbons (*Hylobates* spp.: Rutberg, 1983; Leighton, 1987).

The evolution of social monogamy can also be explained by the **'infanticide avoidance hypothesis'**. In species where new male residents commit infanticide, they apparently benefit by eliminating parental investment in offspring they have not sired and reducing the time before females enter their next fertile period, hence increasing their own reproductive rate. The father, however, incurs appreciable loss of fitness if the infants he sired are lost due to infanticide, and he is expected to protect his progeny against infanticidal conspecifics (Hausfater & Hrdy, 1984; Labov *et al.*, 1985; van Schaik & Kappeler, 1993). The **'infanticide avoidance hypothesis'** was specifically proposed to account for the evolution of monogamy in primates (van Schaik & Dunbar, 1990) and rodents (Cockburn, 1988). However, almost all records of infanticide in wild populations have been from polygynous species, and it is often females, not males, who commit it (Labov *et al.*, 1985).

Males can also reduce the risk of predation (**'predation avoidance hypothesis'**). Male anti-predator behaviour includes active defence (golden jackal, *Canis aureus*), baby sitting (aardwolf, *Proteles cristatus*), carrying (marmosets, *Callithrix* spp.), remaining between alarm sources and offspring (gibbons, *Hylobates* spp.), and acting as sentinels (dwarf mongoose, *Helogale parvula*) (reviewed in Kleiman & Malcolm, 1981; Woodroffe & Vincent, 1994). For example, reduction of offspring predation is the favoured explanation for the occurrence of monogamy in the klipspringer (*Oreotragus oreotragus*: Dunbar & Dunbar, 1980).

If protection of offspring against infanticidal conspecifics or predators accounts for the evolution of monogamy, males would be expected to associate closely with those offspring. In species where offspring are carried (e.g., Titi monkey, *Callicebus moloch*: Anzenberger, 1992) or become active very soon after birth, male range size might not be curtailed. However, in species where offspring begin exploring the parental habitat with initially limited movement capabilities (Malagasy giant jumping rat, *Hypogeomys antimena*: this study), close association between male and offspring limits daily ranges and reduction of the range size is therefore expected. It has also been shown that when predators are present, the prey's daily home range is reduced (prairie voles, *Microtus ochrogaster*: Desy *et al.*, 1990).

When offspring are targeted by a predator (or at least are more vulnerable to predation), males should increase their efforts to protect them. However, this might also bring males into more frequent contact with predators. In species where juveniles and adults are vulnerable to the same predator species, males would therefore increase their own predation risk by protecting their offspring. Darwinian fitness can be enhanced either by increasing the fitness of current offspring or by investing in the fitness of future offspring (for a review of the sexual conflict over parenting, see Westneat & Sargent, 1996). Given their own risk of being eaten, it should only be profitable for a male to invest in current offspring if the expected life reproductive success is higher than risking the offspring's life but increasing the probability of future, successfully reared offspring (e.g., in species with a limited number of chances for reproduction). In general, this means that parents must decide whether to protect their residual reproductive value (decreasing their own predation risk) or to invest in the welfare of current offspring, thereby increasing their own risk of being eaten. When a trade-off has to be made between reproduction and predator avoidance, the degree of risk taking during the current reproductive event should be related to the probability of future reproductive opportunities. In the marine fish *Gobius niger*, young individuals refrained from reproducing when exposed to a predator, while older individuals readily spawned in the same situation (for a review on predation risk as a cost of reproduction, see Magnhagen, 1991).

The following predictions arise from these theoretical considerations. First, if it is profitable for males to invest in the welfare of current offspring instead of pursuing additional mates, behavioural traits to protect current young and to improve their survival would be expected. If males are vulnerable to the same predator species as their offspring, they must reduce predation risk by increased vigilance and alarm calling. A predator's chances of success are quite slim if the prey detects the predator before it is close enough to attack (Caro, 1986). Therefore, it can be assumed that a potentially successful way of avoiding predation is to detect predators in time (van Schaik & van Noordwijk, 1989). Increased detection of predators by vigilance on elevated spots was reported in dwarf mongooses (*Helogale parvula*: Kleiman & Malcolm, 1981; Woodroffe & Vincent, 1994). Another method for determining if predators are entering the territory might be patrolling the range borders. Adult male Capuchin monkeys (both *Cebus albifrons* and *C. apella*) were more vigilant than adult females, spent less time feeding and foraging, and were at the periphery of the group's range more often than females. The increased vigilance of adult males was reflected in their superior performance in the detection of predators (van Schaik & van Noordwijk, 1989). Alarm calls might not function to warn specific individuals, but rather also to alert other animals, which could cause stalking predators to give up. Since the ability of an animal to detect predators intruding on its territory depends on how frequently it patrols, the prediction would be that males patrol more during periods of high risk and stay within close enough proximity to their offspring so that they can warn them of danger.

Second, when behavioural traits to protect current young are favoured, males may encounter predators more frequently, which in turn may lead to sex-biased predation. There is significant documentation of male-biased predation in sexually dimorphic, polygynous species. It is considered to be the cost associated with male ornaments and with intensive mate-acquisition behaviour, in turn leading to an increased probability of encountering predators (e.g., Fitzgibbon, 1990; Clutton-Brock, 1991; Magnhagen, 1991; Owen-Smith, 1993). However, in monogamous mammals that are sexually monomorphic in size or appearance, this should not be the case, and so far has not been documented. Evidence for male-biased predation connected to protection of young would, for the evolution of monogamy in sexually monomorphic mammals, support the role of predation risk to offspring (Sommer, 2000).

Which behavioural traits should then be expected in females? A common aspect of female parental care is resource acquisition (Kleiman & Malcolm, 1981). If females are also necessary for the protection of their offspring and both parents are equally vigilant in efforts to detect predators, then no sex-biased predation should occur in monogamous mammals. However, if females cannot increase their protection of an already exploring offspring (perhaps because of the requirements of a subsequent, still lactating, younger litter), and the burden of predator detection and protection of offspring can be shifted to males, it would pay females to preserve their residual reproductive value for the future, which again may lead to male-biased predation. The increased time and energy made available to females if males participate in parental tasks can then be invested in those future offspring (Sommer, 2000). This scenario is also in the male's interest if the pair mate for life. The ability to assign parentage makes it possible to determine the genetic payoff for different observed behavioural strategies (reviewed in Hughes, 1998).

Ecological constraints on the evolution of monogamy, sex-specific reproductive strategies under the risk of predation, and genetic contributions were investigated in the Malagasy giant jumping rat (*Hypogeomys antimena*). The largest extant endemic rodent of Madagascar always lives in social monogamy. A male and a female stay together until one mate dies. In this chapter, I combine field and genetic studies to investigate current hypotheses concerning the ecological basis for the evolution of monogamy and male care, and the extent of genetic contribution of the social father in this obligate pair-bonded rodent. For this purpose, sex-specific home range sizes and behavioural responses with respect to resource requirements and mortality risk of young were analysed. The inheritance of alleles of the major histocompatibility complex (MHC) were used to investigate family relationships. The welfare of offspring after replacement of one parent was noted.

SITE, TIME, AND SPECIES

The Malagasy giant jumping rat (*Hypogeomys antimena*) is one of the key species of the highly threatened, dry deciduous forests on the western coast of Madagascar. The modern distribution of *H. antimena* has been reduced recently to a fragmented area of a suitable habitat of 200 km^2 within a geographic range that measures less than 20×40 km and lies north of the town of Morondava near the western coast of Madagascar (Sommer & Hommen, 2000; Ganzhorn *et al.*, 2001; Sommer *et al.*, 2002*a*, *b*). *H. antimena* is strictly nocturnal; pairs and their offspring spend the day in underground burrows. Burrows can be evenly or randomly distributed and are needed for raising offspring and for protection against predation and heat during the day. Digging of new burrows is rare. After the death of resident animals, burrows are occupied by new immigrants (Sommer, 2000). Both sexes are territorial. Animals from neighbouring burrows defend exclusive territories throughout the year irrespective of food abundance or reproductive state. Only burrow mates have overlapping, similar-sized home ranges. The mean territory size varies between 3.1 and 3.5 ha. Territory borders are marked with urine, faeces, and scent gland deposits (Cook *et al.*, 1991; Sommer, 1996, 1997). If a neighbouring territory is unused (because the respective pair was predated), pairs might temporarily increase their home range to claim the area. However, polygyny was never observed. The species forages on the forest floor for fallen fruit, seeds, and leaves. It is also known to dig for roots and tubers and to strip the bark of saplings. Reproduction takes place during the rainy season (December–March). Females can give birth twice during this period, each time to a single offspring. Offspring spend the first four to six weeks of their lives in the burrow and begin to leave it regularly during the next four weeks. Capture/recapture and radio-tracking data yielded figures of 50% and 57% mortality, respectively, for offspring until maturity (Sommer, 2000, 2001, 2003). Male offspring leave the parental burrow and territory when they are approximately a year old (before the next breeding period) and can reproduce immediately. However, female offspring show delayed dispersal and stay with their parents for one more reproductive season. Females are probably not sexually mature before the age of two years. In the remaining geographical range of *H. antimena* only two top predator species feed on terrestrial vertebrates over 1 kg: the fossa, *Cryptoprocta ferox* (Viverridae), is Madagascar's largest extant mammalian carnivore (Albignac, 1973). *Acrantophis dumerili* (Boidae), one of three boa species on Madagascar, averages just under 2 m in length, though it can reach up to 3 m (Glaw & Vences, 1994).

Field studies investigating the biology, population ecology, demography, and social behaviour of *H. antimena* have been carried out since 1992 in the 12 500 ha forestry concession of the Centre de Formation Professionnelle Forestière de Morondava in the Kirindy Forest/CFPF (20° 03′ S, 44° 39′ E) at the research station of the German Primate Center (DPZ, Göttingen, Germany). A detailed description of the area and the field methods is given in Ganzhorn and Sorg (1996) and Sommer (2000), respectively. Within a 100-ha study area, as many existing burrow systems (about 30) as possible were identified, regularly monitored, and classified as active or inactive. All animals were live-trapped in front of their burrow systems at least once per year. Captured animals were anaesthetised with an intramuscular injection of ketamin hydrochloride, sexed and weighed, and the usual body measurements (body, tail, ear, hind foot, testis, head length, and head width) were recorded. Between 1992 and 2000, 179 individuals of *H. antimena* were individually marked and genetically characterized.

Forty-one different animals from ten neighbouring burrows were fitted with radio collars. The sex ratio of radio-collared individuals was balanced. Additionally,

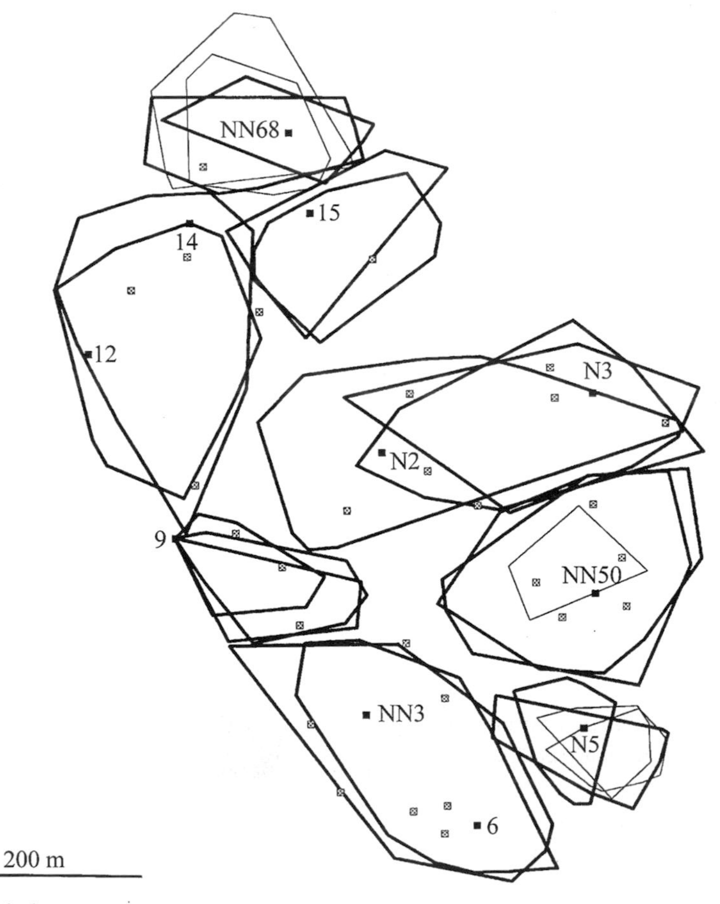

Figure 7.1. Distribution of burrows (squares) and territories in the study site. The main burrows are identified either by a number or a number–letter combination. Only family members sharing the same burrow have overlapping ranges.

newborn or immigrant animals were marked with a passive, integrated transponder (Trovan, Germany). After disinfecting, 2–3 mm of the tip of the tail were cut and preserved in 70% ethanol. The wound closed within a few minutes and was completely healed within five days. Animals were then released in front of their burrow holes. Intensive radio tracking (373 h, 3900 locations) throughout their entire nocturnal activity periods during different times of the year provided information on the ecology, behavioural, and life history characteristics of the species. Patrolling was measured by distance moved per time unit. Home ranges of radio-collared individuals were analysed using the method of minimum convex polygon (Mohr, 1947). Possible predation of radio-collared rats was checked daily during the field seasons and once a week for the rest of the year by verifying that all animals spent the day in their respective burrows. Otherwise, the radio signal was followed until the radio collar was found in the forest. The species responsible for the predation could be unequivocally identified because all radio transmitters were still working after predation and could be located in the forest. The radio collars of animals attacked by *C. ferox* showed clear signs of the predator's teeth and the deformation caused by the characteristic neck-bite. In addition, remains of the rodent's skull, colon, and gall bladder, as well as *Cryptoprocta* faeces, could often be found near the attack site. In cases of predation by *A. dumerili*, a boa could either be followed by radio tracking a *Hypogeomys*' collar for a couple of nights, or the regurgitated radio collar could be found at the resting site of the snake (e.g., unused *Hypogeomys* burrows or old, hollow stems of fallen trees).

The observed relationships of 48 families were all substantiated by genetic analyses (Sommer & Tichy, 1999; Sommer, unpublished data). The segregation of alleles of two loci (DQA and DRB) of the MHC were investigated after polymorphous chain reaction (PCR) amplification using the methods of direct sequencing and single-strand conformation polymorphism (SSCP). Primer sequences and methodological details are given in Sommer and Tichy (1999) and Sommer *et al.* (2002). SSCP is one of the most sensitive methods for quickly detecting nucleotide substitutions (Girman, 1996; Law *et al.*, 1996). Single base-pair changes should be detected in 99% of 100–300 bp fragments (Lessa & Applebaum, 1993). Homozygous and heterozygous animals can be distinguished. SSCP analysis relies on the fact that the mobility of a single-stranded DNA molecule in a non-denaturing gel is not only determined by its size, but also by its nucleotide sequence, which governs its three-dimensional structure (Orita *et al.*, 1989). At least three examples of all alleles were cut separately from the gel, re-amplified, and the nucleotide sequence determined by direct sequencing.

RESULTS

Burrow mates defend an exclusive territory throughout the year. In addition to the main burrow, which often has a central position in the territory, a pair's territory contains from one to seven smaller burrows (Sommer, 2000). A map with burrow and territory distribution is given in Figure 7.1.

To investigate whether limited food resources, and therefore high energetic and physiological costs of reproduction, may explain the male's assistance in rearing young (**'resource defence hypothesis'**), range sizes of animals with offspring were compared with those without offspring (Figure 7.2). Depending on whether all-purpose territories are defended throughout the year or a larger area is required to ensure vital resources during reproduction, home range sizes should be stable or increase after birth of offspring. In *Hypogeomys*, the reproductive season starts with the onset of the rainy season (around December). In December, animals with young (two males, three females) used significantly smaller areas than animals without young (two males, two

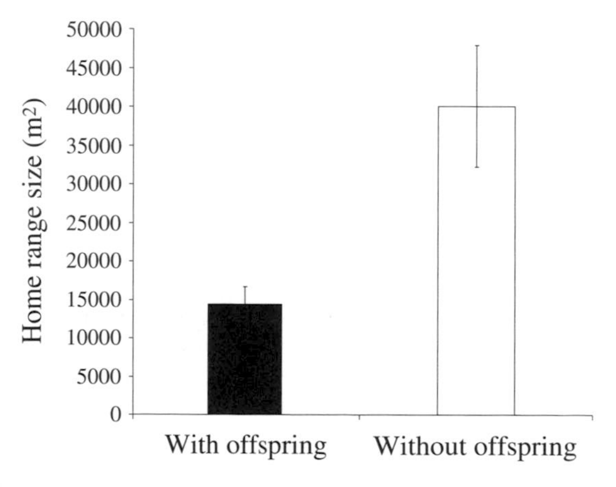

Figure 7.2. Home ranges (m^2) (mean ± SE) of animals with and without offspring at the onset of the reproductive period in December (modified from Sommer, 1997).

females) (Figure 7.2, *t*-test: $N = 9, t = 3.13, P = 0.04$), which suggests that enough food was available and males may not have been necessary to ensure vital food resources during reproduction (Sommer, 1997).

Confidence of paternity seems to be very high for *Hypogeomys* males. Only two extra-pair young were found through genetic analyses of the 48 parent–offspring trios (4%) (Sommer & Tichy, 1999; Sommer, unpublished data).

While a male and a female stay together until one mate dies (Sommer, 1998), after a mate's death, new pair bonds are formed within days or weeks. According to current theory, this renders current offspring potential targets for infanticide (Hausfater & Hrdy, 1984; Labov *et al.*, 1985; van Schaik & Kappeler, 1993). In three cases the new male's paternity (one example is given in Figure 7.3) and in one case the new female's maternity was genetically excluded (Sommer & Tichy, 1999, unpublished data). During ten years of fieldwork, neither female nor male replacement was observed to have a negative effect on the welfare of already existing offspring.

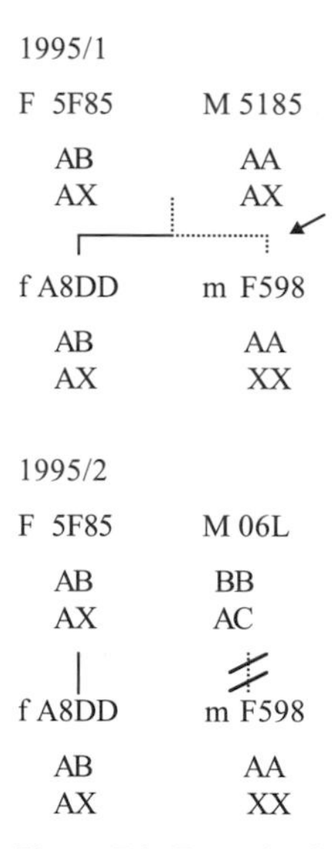

Figure 7.3. Example of male tolerance of unrelated offspring revealed by MHC-DQA and DRB gene inheritance (modified from Sommer & Tichy, 1999). After predation, but during the mating season, male 5185 was replaced by 06L. Offspring F598 was born in 1995/1 but was first trapped half a year later in 1995/2. Although it was probably sired by male 5185 and female 5F84, it was tolerated by the unrelated new male 06L. M, adult male; F, adult female; m, male offspring; f, female offspring;....., questionable paternity; //, paternity refuted by SSCP; ←, paternity confirmed by SSCP. Genotypes at the DQA and DRB locus are indicated below the identification number (modified from Sommer & Tichy, 1999).

Predation

In both 1995 and 1996, a predation peak occurred at the beginning of the dry season (June 1995 and May 1996, respectively). During these periods, more than 20% of the animals under study were predated (Sommer, 2000). The predation impact on *H. antimena* during periods of high predation pressure was sex- and age-specific (Figure 7.4). Significantly more naive offspring than adult animals were taken (Fisher's exact test: $P = 0.006$). In 1995, 57.1% and in 1996, 50.0% of the radio-collared offspring were predated between April and June. Though the sex ratio of radio-collared males and females was balanced, no female was predated during this period, whereas, on average, 24.0% of the tagged males were killed (Fisher's exact test: $P = 0.058$) (Sommer, 2000).

To investigate why offspring are more vulnerable to predation during this period, movement behaviour before and during the predation peak was analysed. As usual in mammals, *Hypogeomys* offspring move more as they grow (ANOVA: month term: $F_{1,27} = 0.86$, n.s.; group term: $F_{5,27} = 0.96$, n.s.; Figure 7.5*a*), and they use larger home ranges within the parental territory (Wilcoxon two-sample test: $T = 0, N = 5, P < 0.05$; Figure 7.5*b*), both of which increase the distance to safe burrows (Sommer, 2000). But even during the predation peak, offspring still moved significantly less than adults (Kruskal–Wallis test: mean distance: offspring = 318 m ± 147 ($N = 19$), males = 434 m ± 113 ($N = 18$), females = 403 m ± 108 ($N = 16$), $P < 0.03$).

Sex-specific behaviour

Home ranges were significantly reduced with increasing predation risk (Wilcoxon test for pairwise comparison: $N = 12$, April–May 1995: n.s., May–June 1995: $Z = -3.05$, $P < 0.001$) (Figure 7.6*a*). The reduction in home range sizes was similar in both sexes (Mann–Whitney *U*-test: n.s.) (Figure 7.6*b*).

The use of the forest by animals sharing burrows located in dense forest but still near an open forest path (5 m wide) was investigated at the beginning of

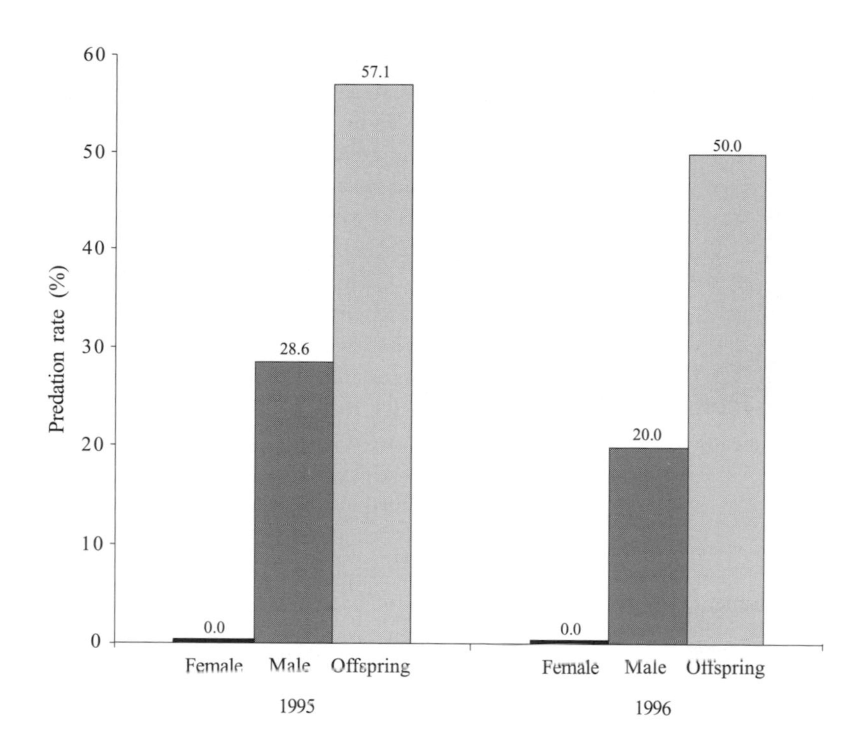

Figure 7.4. Sex- and age-specific predation at the beginning of the dry season. The number of predated individuals and radio-collared individuals for females 0/8 (1995); 0/8 (1996); for males 2/7 (1995); 2/10 (1996); and for offspring 4/7 (1995); 3/6 (1996).

the rainy season (November/December 1995, no predation of tagged animals) and at the end of the rainy season (May 1996, high predation of tagged animals) (Figure 7.7). In both pairwise comparisons of the same individuals, and in comparisons of animals of the same sex (one male and one female were replaced after predation of the former mate), all animals of both sexes used significantly more open paths in November/ December than in May (χ^2 test: F66A: $N_{95/2} = 46$, $N_{96/1} = 41$, $\chi^2 = 16.4$, $P < 0.001$; 10R: $N_{95/2} = 49$, $N_{96/1} = 57$, $\chi^2 = 15.6$, $P < 0.001$; 17R – F598 + B521: $N_{95/2} = 54$, $N_{96/1} = 67$, $\chi^2 = 16.0$, $P < 0.001$; 14L – 5913: $N_{95/2} = 49$, $N_{96/1} = 34$, $\chi^2 = 10.4$, $P = 0.001$). Latrines were also found more often on open paths in November/December (mean 38.7 ± 3.8 latrines per 800 m path length, $N = 6$ sample days) than in May (2 latrines per 800 m path length, $N = 15$ sample days).

Does *Hypogeomys* show any sex-specific behavioural responses during times of high predation impact? Males and females did exhibit some behavioural differences when their offspring were at highest risk of predation. The distance between the pair increased during their nocturnal activities during the predation peak (ANOVA: period: $F_{1,184} = 19.55$, $P < 0.0001$; group: $F_{4,184} = 5.27$, $P < 0.0001$) (Table 7.1). While males travelled further (ANOVA: period: $F_{1,27} = 9.46$, $P = 0.005$; group: $F_{4,27} = 0.81$, n.s.), female travel showed no significant changes (ANOVA: period: $F_{1,26} = 3.01$, n.s.; group: $F_{4,26} = 1.81$, n.s.) (Table 7.1).

Table 7.1. *Behavioural traits of* Hypogeomys *males and females before (April) and during (May) the predation peak in 1996. Medians and quartiles are given*

		April 1996 (before predation peak)	May 1996 (during predation peak)
Distance between	M–F	37 (19, 70)	78 (40, 121)
family	M–O	60 (36, 102)	78 (51, 125)
members (m)	F–O	61 (29, 88)	108 (52, 132)
Travel distance	M	321 (196, 392)	424 (354, 486)
per night (m)	F	355 (178, 459)	418 (302, 504)

M, male; F, female; O, offspring.
From Sommer, 2000.

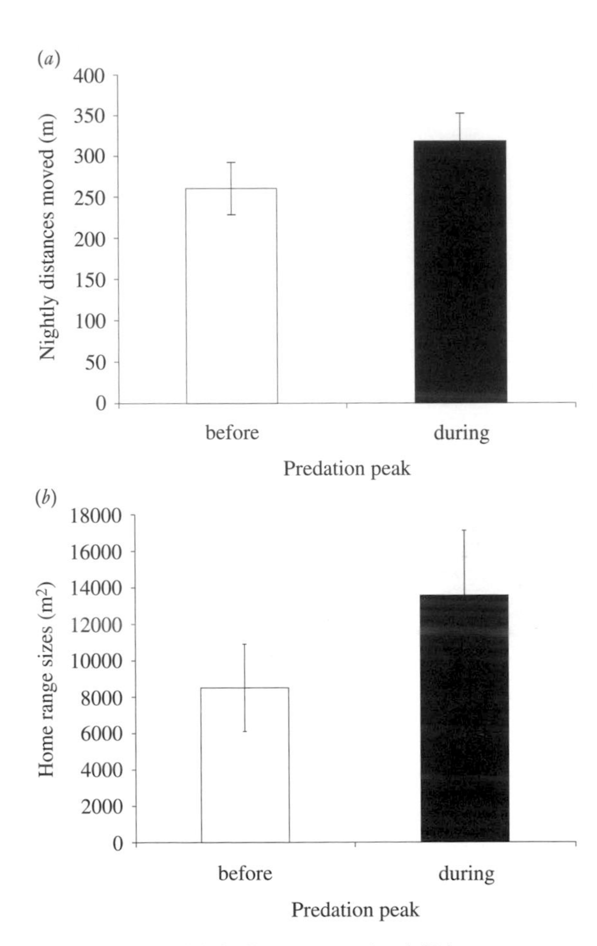

Figure 7.5. (a) Nightly distances moved and (b) home range sizes (m^2) of offspring before and during the predation peak (1996). Means ± SE are given.

At the same time, while males and females increased their distance from their offspring, males stayed closer to them than did females (ANOVA: females: period: $F_{1,74} = 9.54$, $P = 0.003$; group: $F_{2,74} = 1.72$, n.s.; males: period: $F_{1,74} = 5.16$, $P = 0.03$; group: $F_{2,74} = 1.81$, n.s.) (Table 7.1). The interaction term between month (before/during the period of high predation) and pair or individual (group) was not significant in any of the analyses (Sommer, 2000).

DISCUSSION

Hypogeomys antimena lives in obligate monogamy. A male and a female stay together until one mate dies. In *Hypogeomys*, confidence of paternity seems to be very high for males; only two extra-pair young were found in genetic analyses of 48 parent–offspring trios (4%) (Sommer & Tichy, 1999; Sommer, unpublished data). However, the exclusion probability of only two loci is low (0,43) and the number of identified extra-pair young might increase with the inclusion of additional loci. The circumstances of the extra-pair young remain unclear. For males, the main potential benefit of extra-pair copulations (EPC) is an increase in reproductive success, while the risk of cuckoldry seems to be the only cost sufficiently important to modify male reproductive behaviour. In females, the most likely cost of EPCs is reduced paternal care. Genetic benefits, on the other hand, are the most widely accepted explanation for females seeking EPCs (Birkhead & Møller, 1992; Petrie & Kempenaers, 1998). Another potential benefit would be a decreased risk of infanticide (Hrdy, 1977).

High degrees of paternity certainty in monogamous birds are often associated with high levels of male care (Birkhead & Møller, 1992). For example, in the colonial breeding, socially monogamous Great Northern diver (common loon, *Gavia immer*), where pairs produce large young in which they invest heavily to secure offspring survival, no extra-pair young were found (Piper *et al.*, 1997). Similarly, in lesser kestrels (*Falco naumanni*), extreme dependence on male provisioning can result in genetic monogamy (Villarroel *et al.*, 1998). By comparison, in the pair-living, fat-tailed dwarf lemur (*Cheirogaleus medius*), where biparental care is essential for successful reproduction, a high rate of extra-pair paternity (44%) was found. This study indicated that females do not seem to run the risk of reduced paternal care, and that superior genetic quality of the males might be crucial to female choice (Fietz *et al.*, 2000; Fietz, chapter 14).

Resource defence

Males do not seem to be constrained into monogamy by female dispersion (Emlen & Oring, 1977; Rutberg, 1983). The length of male night journeys within their territories indicates that males could easily cover the range of more than one female. However, polygyny was never observed, although in one case an adult female and two adult males shared one burrow and territory (Sommer, 1998). Both sexes defend exclusive territories throughout the year irrespective of food abundance and presence of young. Only burrow mates have overlapping

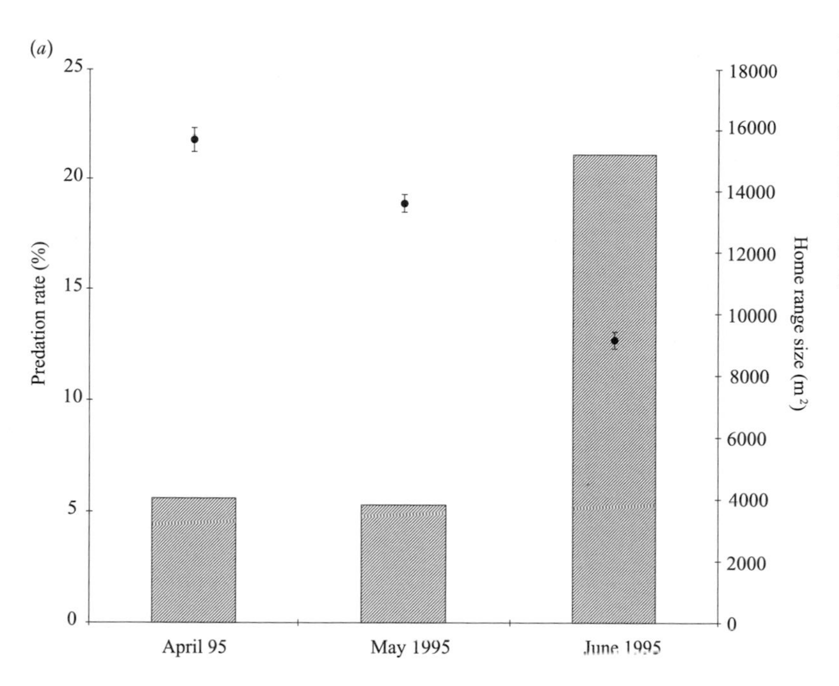

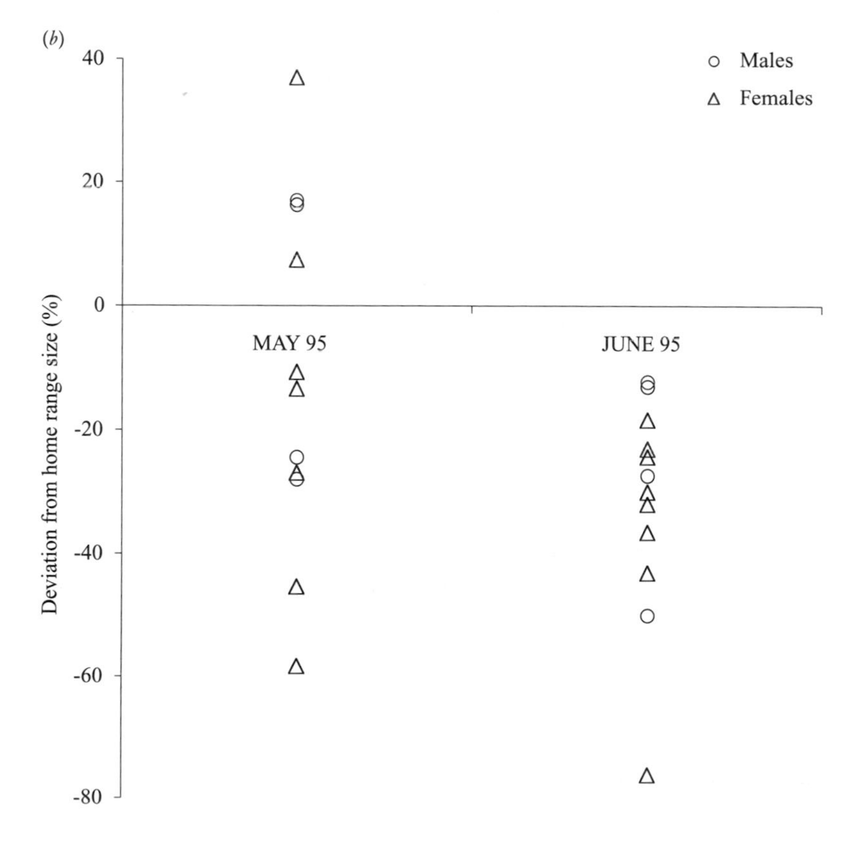

Figure 7.6. (a) Predation rates (shaded bars, %) and home range sizes (m^2, $N = 12$ individuals, means ± SE) in 1995. (b) Sex-specific deviation (%) from home range size (m^2) before (May compared with April) and during (June compared with May) the predation peak.

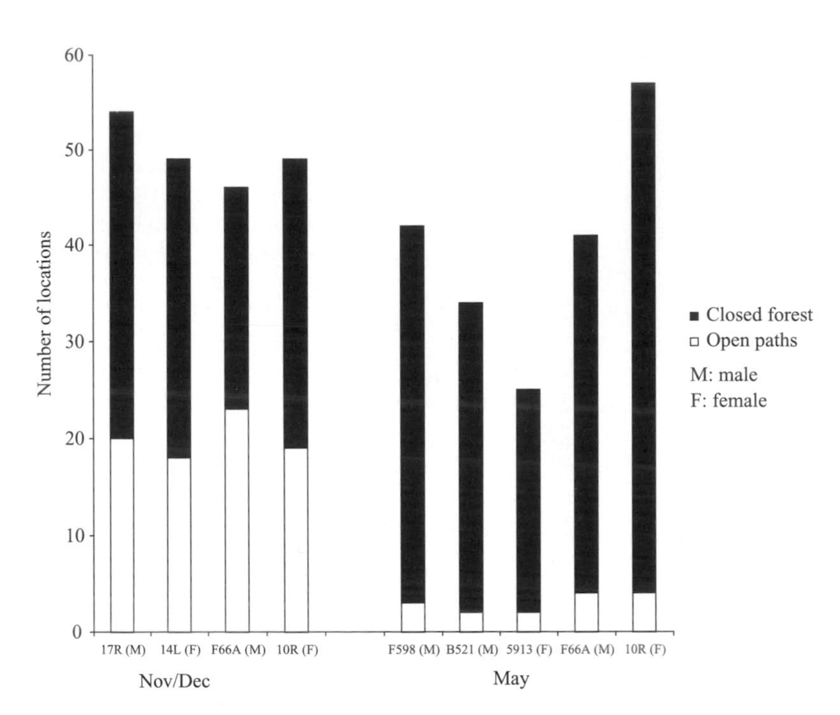

Figure 7.7. Number of localizations of radio-collared animals on open paths and in closed forest in November/December 1995 (no predation of tagged animals) and in May 1996 (high predation of tagged animals).

ranges. Within the couple's territory, females seem to range independently of their partners, scent-mark the pair's territory, and chase away intruders (but do so less than males). Consequently, explaining monogamy in terms of male control of female movements, i.e., a 'mate-guarding strategy' (as suggested for dik-diks: Brotherton & Manser, 1997) does not seem appropriate in the case of *Hypogeomys*.

Home ranges of both sexes are generally larger during the extended dry season (September/October), before the birth of young, than at the beginning of the rainy and reproductive season (December) (Cook *et al.*, 1991; Sommer, 1997). In December, animals with young used significantly smaller areas within their territories than animals without young. No correlation between territory size and the body mass of inhabitants was found (Sommer, 1998). During the study's ten years, reproduction took place in all territories, independent of the area's total size, which suggests that not only individuals in 'rich' ranges reproduce. It might be that the parts of the forest that cannot fulfil year-round demands are not inhabited, which would explain the patchy distribution of *H. antimena* within the dry, deciduous forest on Madagascar's western coast. *Hypogeomys* reproduction (at least lactation and weaning) takes place, as in most other mammal species in the dry deciduous forest of Madagascar (Garbutt, 1999), during the rainy season when food is probably not limited. Deciduous species produce their new leaf flush and flowers rich in nectar and/or fleshy parts just before or at the beginning of the rainy season. Furthermore, during this time, the protein content of fruits of several plant species is high, compared with the mean value of most tropical fruits (Hladik, 1980). Since 1992, no female or male was ever observed carrying and providing food to offspring. Offspring start eating herbivorous food at the age of four weeks when they first leave the burrow during the night. So far, there are no indications that the mating system is constrained by food resources (**'resource defence hypothesis'**). This is also supported by the behaviour of female offspring. Female young show a delayed dispersal, despite reaching adult body mass at the age of one year, and they use the resources of the parental territory for one more reproductive season. Parental body mass was not influenced by the presence of female offspring (Sommer, 1998). The function of the delayed dispersal of female offspring still remains unclear. Possible explanations are that this delay serves some helper function, or that there are limited dispersal options (e.g., burrows) because of habitat saturation. The fact that captive *H. antimena* mature within their first year (R. Reyes, personal communication) suggests that yearlings in the field are sexually suppressed while remaining in their family group.

Burrows are essential for raising offspring, for protection against predation, and for shelter against heat during the day. A pair's territory usually contains one to seven smaller burrows. The main burrow system, which is usually used by the whole family, is about 5 m in length, the surface is covered with bar soil, and it sometimes has up to seven entrance holes. Smaller burrows often have only a single entrance hole and rarely show signs of recent digging. Smaller burrows were never known to become 'main burrows'. If smaller burrows are located closer to the territory border where neighbouring animals might try to enter, they can be temporarily used or as shelter during thunderstorms, to escape predators, and serve as interim homes for dispersing offspring. Digging new burrows seldom occurs. After the death of resident animals, burrows are sometimes occupied temporarily by neighbouring animals but are inevitably lost once new immigrants arrive (Sommer, 2000). Thus, burrows might be a critical resource for *Hypogeomys* (**'resource defence hypothesis'**).

Infanticide avoidance

Though *Hypogeomys* couples usually stay together until one mate dies, predated mates are replaced within a few days to weeks. If male replacements occur during the reproductive period, current theory predicts that the new male residents would benefit by committing infanticide. As a consequence, these males would eliminate any kind of parental investment in offspring they had not sired and would reduce the time that elapsed before females entered their next fertile period, hence increasing their own reproductive rate (Hausfater & Hrdy, 1984; Labov *et al.*, 1985; van Schaik & Kappeler, 1993). Two of the predicted conditions in *Hypogeomys*, e.g., probability is zero or close to zero that the male sired the infant, and the infanticidal male has an increased probability of siring the next offspring, might fit the infanticide hypothesis. However, the third condition, that the mother can be fertilized earlier than if the infant had lived, is only partly met. *Hypogeomys* is a seasonal breeder and only reproduces during a short time span. Females give birth during the rainy season to two consecutive litters (usually a singleton). The third condition would be met only if infanticide occurred early enough during the reproductive period so that the female returned to ovarian cycling sooner than she would have if the infant had lived. This would then make infanticide profitable for the male. In three cases in this study, the exclusion of the new male as the parent (and in one case, the exclusion of the new female as the parent) could be confirmed by genetic data. In the former case the oestrous condition of the already present female is unknown. Neither the female nor the male replacement affected the welfare of already existing offspring (Sommer & Tichy, 1999, unpublished data). So far, there is no support for the **'infanticide avoidance hypothesis'**, and therefore the need of males to protect their progeny against infanticidal conspecifics. However, the first four weeks of life, before offspring leave the burrow, remain somewhat mysterious. This is, however, probably the period during which offspring are most likely to be killed by infanticidal males.

Predation avoidance

One possible male care contribution is reducing the predation risk to offspring (**'predation avoidance hypothesis'**). This hypothesis was investigated by analysing the behavioural traits of males and females in response to predation pressure from the two natural predator species (*Cryptoprocta ferox*, Carnivora: Viverridae; *Acrantophis dumerili*, Boidae). In both 1995 and 1996, a *Hypogeomys* predation peak occurred at the beginning of the dry season. In 1995, 57.1% and in 1996, 50.0% of the radio-collared offspring were predated between April and June. Though the sex ratio of radio-collared males and females was balanced, no female was predated during this period, whereas on average, 24.0% of the tagged males were killed (Sommer, 2000).

The transition from rainy to dry season is the time when insectivorous mammals (*Tenrec ecaudatus*, *Setifer setosus*, *Echinops telfarii*) and snakes (including *A. dumerili*) begin to hibernate. The insectivores are common prey items during the rainy season for both *C. ferox* and *A. dumerili*, whereas lemurs and *Hypogeomys* are the most widely consumed prey of *C. ferox* in the Kirindy Forest during the dry season. Like felids, *C. ferox* opportunistically takes small as well as large prey (Rasoloarison *et al.*, 1995; Goodman *et al.*, 1997). Though the main predation impact took place at almost the same time during consecutive years, it is still unclear whether *Hypogeomys* can predict when a fossa (*C. ferox*) is present.

Predators should select prey items that are the most profitable in terms of energy intake per unit of handling time (for a review see Stephens & Krebs, 1986). Therefore, they should focus on the young, old, sick (and infirm), or otherwise disabled individuals of the prey population. The impact of predation on the prey population also depends on frequency of contact with the predator. Considerable evidence exists for male-biased predation in sexually dimorphic, polygynous species and is considered to be the cost associated with male ornaments and with intensive mate-acquisition behaviour, all of which leads to an increased probability of encountering predators (e.g., Fitzgibbon, 1990; Clutton-Brock, 1991; Magnhagen, 1991; Owen-Smith, 1993). However, in monogamous mammals that are sexually monomorphic in size or appearance this should not be the case (and so far has not been documented). In *H. antimena*, male-biased predation cannot be related to sexual dimorphism because this species shows no sexual differences in any body measurements (Sommer, 1996, 1997, 1998), unlike findings in other studies (Magnhagen, 1991; Owen-Smith, 1993). An alternative explanation for sex-biased predation might be that other sex-specific cues may appeal to a predator. Neither predator species of *H. antimena* engages in energy-consuming hunting, and both will probably attack any prey they encounter as long as the energy gain is beneficial. The risk of encounter with these predators thus depends on the prey's knowledge of and experience with the predator and its activity level (for a review see Stephens & Krebs, 1986; Norrdahl & Korpimäki, 1998). Therefore, the age-specific predation rate is not surprising. Offspring in most species are more vulnerable to predation because they lack anti-predator skills, and the risk often increases with the intensity of their explorations (e.g., Fitzgibbon & Fanshawe, 1989; Dickman *et al.*, 1991; Curio, 1993). And as with most mammals, *Hypogeomys* offspring become more mobile as they mature, using increasingly larger home ranges within the parental territory. This in turn extends the distance they must travel to reach safe burrows, making them more vulnerable to predation (Sommer, 2000). It might be assumed that with increasing experience offspring might learn to react more effectively if a predator is in the territory.

Ecological factors

Although there were significant reductions in home ranges during periods of high predation risk, no sex-specific reductions were recorded. Smaller home ranges might facilitate earlier recognition of patrolling predators, use of alarm calls, and increase the likelihood of reaching a safe burrow. However, other coinciding ecological effects cannot be ruled out as causes in the reduction of home range size.

During periods of high predation risk, *Hypogeomys* more often moved along open paths and used latrines to mark territory borders, than in times of low risk. These behavioural changes could be coincidental or the result of other, unidentified ecological constraints that might also be correlated with predation risk. The use of open paths during the rainy season might be an advantage because rodents are more likely to see snakes that are only active during the rainy season under the higher illumination of moonlight (Bouskila, 1995). This might decrease the risk of boa predation during this period. On the other hand, *Cryptoprocta ferox* also prefers to travel along paths and roads in the forest (personal observation; C. Hawkins, personal communication). However, the high impact of *C. ferox* predation on *Hypogeomys* started at the beginning of the dry season when insectivorous mammals and snakes were becoming inactive. At this time, *Hypogeomys* shifted its travel and marking activities to closed forest areas. Kangaroo rats (*Dipodomys merriami*) also showed seasonal shifts in microhabitat use. They preferred open areas in summer, but shifted towards closed areas in winter. The results were compatible with the distribution of the risk of snakes (Brown *et al.*, 1988). The same behaviour was indicated by another study of *D. merriami* and *D. deserti*, where the microhabitat use (open desert vs. bushes) was a trade-off between the risk of owl and snake predation (Bouskila, 1995). Since *C. ferox* marks by scent, the mere odour may be sufficient to trigger an anti-predatory response in *H. antimena*. Responses to predator odours have been reported from experimental data for other rodent species (Jedrzejewski *et al.*, 1993).

Sex-specific behaviour and predator encounters

Whereas the nightly travel distance of males and females did not differ before the period of high predation of offspring, males increased their movement intensity during high predation periods. This may suggest that males move more in order to detect predators and to protect offspring. Males' behavioural responses could lead to a higher frequency of encounters with predators, which decreases survival prospects. Alternative explanations for increased male movements could be pressure from other groups or the search for females. But it seems unlikely that all males would increase their movements in response to these other pressures: at this time of the year (April–June) the female oestrous period is probably over, and male offspring dispersal, which is associated with increased competition, usually takes place at the beginning of the reproductive period, around December (Sommer, 1998, 2001).

In contrast to males, females maintained their usual activity levels during the period of high offspring predation; their distance from their offspring was also greater. As a consequence of the sex-differentiated behaviours, the distance between pair-mates also increased during the predation peak, the females' behaviour probably leading to fewer encounters with predators, which may increase the females' survival (Lima & Dill, 1990; Magnhagen, 1991; Norrdahl & Korpimäki, 1998; Sommer, 2000). Therefore, one explanation for the higher predation of males when offspring were at high risk could be that through vigilant behaviour males expose themselves to higher predation risk while trying to reduce offspring mortality.

Hypogeomys males invest in current offspring at the possible cost of higher predation risk, while females protect their residual reproductive value. But the higher male predation risk might be balanced by sexually different life history traits: males can start breeding when they are a year old; females mature at the age of two years (Sommer, 2001). In the field, the survival of adults declined dramatically between the ages of four and five years, with only 4% reaching the age of seven. During each reproductive cycle, the reproductive value of offspring increases as they mature because of increased survival probability. Thus, for animals with parental care, the expectation is that parental risk-taking will increase with time spent with one particular brood (summarized by Magnhagen, 1991). This strategy can be evolutionarily stable only if the fitness benefits gained by this intensive male parental care outweigh the risks of losing future reproductive opportunities. Otherwise *Hypogeomys* males would be expected to adopt another strategy (Trivers, 1972; Kleiman & Malcolm, 1981). Male-biased predation can also be interpreted as a result of a failed male defence against predators. One explanation for this failure might be a recently increased predator density to which *Hypogeomys* has not yet adapted. And perhaps this explains the alarming population decline that has recently been observed throughout its remaining geographic range, probably a consequence of increasing habitat degradation (Sommer & Hommen, 2000; Ganzhorn *et al.*, 2001; Sommer *et al.*, 2002*b*).

For *Hypogeomys* males, the universe of possible anti-predator behaviours is limited. Offspring are too big to be carried around (in contrast to, for example, marmosets), and elevated spots suitable for surveying a larger area, unlike those available to meerkats (examples are summarized by Kleiman & Malcolm, 1981; Woodroffe & Vincent, 1994), rarely exist. Acting as protector of those at-risk offspring means offering oneself as potential prey, distracting the predator's attention from the more vulnerable young. However, unlike in birds, this does not seem to be a successful anti-predator strategy for *Hypogeomys* because the chances that any of the rats will escape a predator attack may be lower if the predator is close by. The safest strategy for adult and young rats to escape a *C. ferox* attack might be to run back to the burrow. Another possibility for reducing predation risk might be to patrol the territory in order to identify predators in enough time to send alarm calls warning other family members. Besides contact calls, *Hypogeomys* uses alarm calls (a shrill whistle) both when predators are present and when rats from neighbouring territories are in the vicinity. These calls can be easily detected by the human ear from distances of 50 m or more, depending on the density of the vegetation. With their long (>6 cm), rabbit-like ears, the rats can presumably hear the calls over much greater distances, and it can be assumed that family members at least are always within hearing distance. Animals react to alarm calls with increased vigilance. Instead of walking on all fours, they sit upright on their hind feet and their ears move slightly. Based on observations of trapped families released in front of their burrows, young rats react more sensitively to any noise (e.g., cracking of small dry branches, alarm calls in the neighbourhood) than do the adults, and upon hearing these sounds immediately run into the parental burrow (Sommer, 2000).

Males more than females perform patrols along the territorial borders, and although they vary their routes, criss-crossing the home range was not observed. Though mates use the same latrines, 60% of all observed visits were made by males and 34.3% by females. In 5.7% of the visits, mates accompanied each other ($N = 36$) (Sommer, 1998). Therefore, it remains unclear whether patrolling is primarily conducted as a part of predator avoidance or is predominantly performed in order to defend the territory against intruding rats and only secondarily as an element of predator monitoring. As the season progresses, infant survival might be higher because young have themselves begun to learn how to detect and avoid predators, independent of male parental behaviour. Despite sex differences in the territorial defence behaviour, other factors, for example different foraging strategies of males and females, might also contribute to the sex-specific differences in the predation rate.

CONCLUSION

Hypogeomys antimena is a socially monogamous rodent species. Pair mates remain together until one partner dies. It seems that solitary parents cannot successfully rear offspring. Of 112 family units studied between 1992 and 1999, only four consisted of a solitary parent with one offspring. Only one of these offspring, which was about a year old and weighed about 1050 g when its father was killed, survived and dispersed (Sommer, 1998, unpublished data). This may be indirect evidence of the importance of male care. Because of the difficulty of quantifying many of the confounding ecological variables, several questions remain unanswered and further data collection is necessary. And whether the defence of the critical 'resource burrow' ('resource defence' monogamy) or the 'predation risk' (need for biparental care) was the catalyst for the evolution of monogamy in *H. antimena* remains difficult to determine.

Acknowledgements

This study was supported by the 'Commission Tripartite' of the Malagasy Government, the Laboratoire de Primatologie et des Vertébrés de l'Université d'Antananarivo, the Ministère pour la Production Animale et des Eaux et Forêts, the Centre de Formation Professionnelle Forestière de Morondava, B. Rakotosamimanana, R. Rasoloarison, and L. Razafimanantsoa. Financial support was provided by Landesgraduiertenförderung (LGFG), German Academic Exchange Service (DAAD), Stifterverband für die Deutsche Wissenschaft, WWF International and Madagascar, and the German Science Foundation (DFG, SO 428/1-1, SO 428/1-3). The encouraging support by J. U. Ganzhorn is gratefully acknowledged. J. U. Ganzhorn, B. Kempenaers, other editors, and two anonymous referees provided very helpful comments on the manuscript.

References

Albignac, R. (1973). *Mammifères carnivores. Faune de Madagascar*, Volume 21. Paris and Antananarivo: ORSTOM/CNRS.

Anzenberger, G. (1992). Monogamous social systems and paternity in primates. In *Paternity in Primates: Genetic Tests and Theories*, ed. R. D. Martin, A. F. Dixson & E. J. Wickings, pp. 203–24. Basel: Karger.

Barlow, G. W. (1984). Patterns of monogamy among teleost fishes. *Archiv für Fischereiwissenschaft*, **35** (supplement 1), 75–123.

Birkhead, T. R. & Møller, A. (1992). *Sperm Competition in Birds: Evolutionary Causes and Consequences*. London: Academic Press.

Bouskila, A. (1995). Interactions between predation risk and competition: a field study of kangaroo rats and snakes. *Ecology*, **76**, 165–78.

Brotherton, P. N. M. & Manser, M. B. (1997). Female dispersion and the evolution of monogamy in dik-dik. *Animal Behaviour*, **54**, 1413–24.

Brotherton, P. N. M. & Rhodes, A. (1996). Monogamy without biparental care in a dwarf antelope. *Proceedings of the Royal Society of London, Series B*, **263**, 23–9.

Brown, J. S., Kotler, B. P., Smith, R. J. & Wirtz W. O. II (1988). The effects of owl predation on the foraging behavior of heteromyid rodents. *Oecologia*, **76**, 408–15.

Caro, T. M. (1986). The functions of stotting in Thomson's gazelles: some tests of the hypotheses. *Animal Behaviour*, **34**, 663–84.

Clutton-Brock, T. H. (1989). Mammalian mating systems. *Proceedings of the Royal Society of London, Series B*, **236**, 339–72.

(1991). *The Evolution of Parental Care*. Princeton, New Jersey: Princeton University Press.

Cockburn, A. (1988). *Social Behaviour in Fluctuating Populations*. London: Croom Helm.

Cook, J. M., Trevelyan, R., Walls, S. S., Hatcher, M. & Rakotondraparany, F. (1991). The ecology of *Hypogeomys antimena*, an endemic Madagascan rodent. *Journal of Zoology, London*, **224**, 191–200.

Curio, E. (1993). Proximate and developmental aspects of antipredator behaviour. *Advanced Study of Behaviour*, **22**, 135–238.

Desy, E. A., Batzli, G. O. & Liu, J. (1990). Effects of food and predation on behaviour of prairie voles: a field experiment. *Oikos*, **58**, 159–68.

Dickman, C. R. (1992). Predation and habitat shift in the house mouse, *Mus domesticus*. *Ecology*, **73**, 313–22.

Dickman, C. R., Predavec, M. & Lynam, A. J. (1991). Diffferential predation of size and sex classes of mice by the barn owl, *Tyto alba*. *Oikos*, **62**, 67–76.

Dunbar, R. I. M. & Dunbar, E. P. (1980). The pairbond in klipspringer. *Animal Behaviour*, **28**, 219–29.

Emlen, S. T. & Oring, L. W. (1977). Ecology, sexual selection, and the evolution of mating systems. *Science*, **197**, 215–23.

Fietz, J., Zischler, H., Schwiegk, C., Tomiuk, J., Dausmann, K. H. & Ganzhorn, J.U. (2000). High rates of extra-pair young in the pair-living fat-tailed dwarf lemur *Cheirogaleus medius*. *Behavioral Ecology and Sociobiology*, **49**, 8–17.

Fitzgibbon, C. D. (1990). Antipredator strategies of immature Thomson's gazelles: hiding and the prone response. *Animal Behaviour*, **40**, 846–55.

Fitzgibbon, C. D. & Fanshawe, J. (1989). The condition and age of Thomson's gazelles killed by cheetahs and wild dogs. *Journal of Zoology, London*, **218**, 99–107.

Ganzhorn, J. U. & Sorg, J.-P. (1996). Ecology and economy of a tropical dry forest in Madagascar. *Primate Report*, **46–1**, 1–382.

Ganzhorn, J. U., Porter, P., Lowry, I. I., Schatz, G. E. & Sommer, S. (2001). Madagascar: one of the world's hottest biodiversity hotspot on its way out. *Oryx*, **35**, 346–8.

Garbutt, N. (1999). *Mammals of Madagascar*. East Sussex: Pica Press.

Girman, D. J. (1996). The use of PCR-based single-stranded conformation polymorphism analysis (PCR-SSCP) in conservation genetics. In *Molecular Genetic Approaches in Conservation*, ed. T. B. Smith & R. K. Wayne, pp. 167–82. New York: Oxford University Press.

Glaw, F. & Vences, M. (1994). *A Fieldguide to the Amphibians and Reptiles of Madagascar*, 2nd Edition. Cologne: M. Vences & F. Glaw Verlags GbR.

Goodman, S. M., Langrand, O. & Rasolonandrasana, B. P. N. (1997). The food habits of *Cryptoprocta ferox* in the high mountain zone of the Andringitra Massif, Madagascar (*Carnivora, Viverridae*). *Mammalia*, **61**, 185–92.

Gubernick, D. J. & Teferi, T. (2000). Adaptive significance of male parental care in a monogamous mammal. *Proceedings of the Royal Society of London, Series B*, **267**, 147–50.

Hausfater, G. & Hrdy, S. B. (ed.) (1984). *Infanticide: Comparative and Evolutionary Perspectives*. New York: Aldine de Gruyter.

Hladik, C. M. (1980). The dry forest of the west coast of Madagascar: Climate, phenology, and food available for prosimians. In *Nocturnal Malagasy Primates: Ecology, Physiology and Behaviour*, ed. P. Charles-Dominique, H. M. Cooper, C. M. Hladik, E. Pages, G. F. Pariente, A. Petter-Rousseaux, J. J. Petter & A. Schilling, pp. 3–40. New York: Academic Press.

Hrdy, S. B. (1977). *The Langurs of Abu*. Cambridge, Massachusetts: Harvard University Press.

Hughes, C. (1998). Integrating molecular techniques with field methods in studies of social behavior: a revolution results. *Ecology*, **79**, 383–99.

Jedrzejewski, W., Rychlik, L. & Jedrzejewski, B. (1993). Responses of bank voles to odours of seven species of predator – experimental data and their relevance to natural predator–vole relationships. *Oikos*, **68**, 251–7.

Kleiman, D. G. (1977). Monogamy in mammals. *Quarterly Review of Biology*, **52**, 39–69.

Kleiman, D. G. & Malcolm, J. R. (1981). The evolution of male parental investment in mammals. In *Parental Care in Mammals*, ed. D. Gubernick & P. Klopfer, pp. 347–87. New York: Plenum Press.

Komers, P. E. (1996). Obligate monogamy without parental care in Kirk's dikdik. *Animal Behaviour*, **51**, 131–40.

Komers, P. E. & Brotherton, P. N. M. (1997). Female space use is the best predictor of monogamy in mammals. *Proceedings of the Royal Society of London, Series B*, **264**, 1261–70.

Labov, J. B., Huck, U. W., Elwood, R. W. & Brooks, R. J. (1985). Current problems in the study of infanticidal behavior of rodents. *Quarterly Review of Biology*, **60**, 1–20.

Lack, D. (1968). *Ecological Adaptions for Breeding in Birds*. London: Methuen.

Law, J. C., Facher, E. A. & Deka, A. (1996). Nonradioactive single-strand conformation polymorphism analysis with application for mutation detection in a mixed population of cells. *Analytical Biochemistry*, **236**, 373–5.

Lessa, E. P. & Applebaum, G. (1993). Screening techniques for detecting allelic variation in DNA sequences. *Molecular Ecology*, **2**, 119–29.

Leighton, D. R. (1987). Gibbons: territoriality and monogamy. In *Primate Societies*, ed. B. B. Smuts, D. L. Cheney, R. M. Seyfarth, R. W. Wrangham & T. T. Struhsaker, pp. 135–45. Chicago: University of Chicago Press.

Lima, S. L. & Dill, L. M. (1990). Behavioural decisions made under the risk of predation: a review and prospectus. *Canandian Journal of Zoology*, **68**, 619–40.

Magnhagen, C. (1991). Predation risk as a cost of reproduction. *Trends in Ecology and Evolution*, **6**, 183–6.

Mohr, C. O. (1947). Table of equivalent populations of North American small mammals. *American Midland Naturalist*, **37**, 223–49.

Norrdahl, K., Korpimäki, E. (1998). Does mobility affect risk of predation by mammalian predators? *Ecology*, **79**, 226–32.

Orita, M., Suzuki, Y., Sekiya, T. & Hayashi, K. (1989). Rapid and sensitive detection of point mutations and DNA polymorphisms using the polymerase chain reaction. *Genomics*, **9**, 408–12.

Owen-Smith, N. (1993). Comparative mortality rates of male and female kudus: the costs of sexual dimorphism. *Journal of Animal Ecology*, **62**, 428–40.

Petrie, M. & Kempenaers, B. (1998). Extra-pair paternity in birds: explaining variation between species and populations. *Trends in Ecology and Evolution*, **13**, 52–8.

Piper, W. H., Evers, D. C., Meyer, M. W., Tischler, K. B., Kaplan, J. D. & Fleischer, R. C. (1997). Genetic monogamy in the common loon (*Gavia immer*). *Behavioral Ecology and Sociobiology*, **41**, 25–31.

Rasoloarison, R., Rasolonandrasana, B. P. N., Ganzhorn, J. U. & Goodman, S. M. (1995). Predation on vertebrates in the Kirindy forest, western Madagascar. *Ecotropica*, **1**, 59–65.

Rutberg, A. T. (1983). The evolution of monogamy in primates. *Journal of Theoretical Biology*, **104**, 93–112.

Sommer, S. (1996). Ecology and social structure of *Hypogeomys antimena*, an endemic rodent of the dry deciduous forest in western Madagascar. In *Biogéographie de Madagascar*, ed. W.R. Lourenco, pp. 295–302. Paris: Editions de l'Orstom.

(1997). Monogamy in *Hypogeomys antimena*, an endemic rodent of the deciduous dry forest in western Madagascar. *Journal of Zoology, London*, **241**, 301–14.

(1998). Populationsökologie und -genetik von *Hypogeomys antimena*, einer Endemischen Nagerart im Trockenwald Westmadagaskars. Ph.D. thesis, University of Tübingen. Göttingen: Cuvillier Verlag.

(2000). Sex specific predation rates on a monogamous rat (*Hypogeomys antimena*, Nesomyinae) by top predators in the tropical dry forest of Madagascar. *Animal Behaviour*, **59**, 1087–94.

(2001). Reproductive ecology of the endangered monogamous Malagasy giant jumping rat, *Hypogeomys antimena*. *Mammalian Biology*, **66**, 111–15.

(2003). Natural history of the Malagasy Giant Jumping Rat, *Hypogeomys antimena*. In *The Natural History of Madagascar*, ed. S. Goodman & J. Benstead. Chicago: University of Chicago Press (in press).

Sommer, S. & Hommen, U. (2000). Modelling the effects of life history traits and changing ecological conditions on the population dynamics and persistence of the endangered Malagasy giant jumping rat (*Hypogeomys antimena*). *Animal Conservation*, **4**, 333–43.

Sommer, S. & Tichy, H. (1999). MHC-Class II polymorphism and paternity in the monogamous *Hypogeomys antimena*, the endangered, largest endemic Malagasy rodent. *Molecular Ecology*, **8**, 1259–72.

Sommer, S., Schwab, D. & Ganzhorn, J. U. (2002). MHC diversity of endemic Malagasy rodents in relation to geographic range and social system. *Behavioral Ecology and Sociobiology*, **51**, 214–21.

Sommer, S., Toto Volahy, A. & Seal, U. S. (2002). A population and habitat viability assessment for *Hypogeomys antimena*, the largest endemic rodent of Madagascar. *Animal Conservation*, **5**, 263–73.

Stephens, D. W. & Krebs, J. R. (1986). *Foraging Theory*. Princeton, New Jersey: Princeton University Press.

Trivers, R. L. (1972). Parental investment and sexual selection. In *Sexual Selection and the Descent of Man 1871–1971*, ed. B. Campell, pp. 136–79. Chicago: Aldine.

van Schaik, C. P. & Dunbar, R. I. M. (1990). The evolution of monogamy in large primates: a new hypothesis and some crucial tests. *Behaviour*, **115**, 30–62.

van Schaik, C. P. & Kappeler, P. M. (1993). Life history, activity period and lemur systems. In *Lemur Social Systems and their Ecological Basis*, ed. P. M. Kappeler & J. U. Ganzhorn, pp. 241–60. New York: Plenum Press.

van Schaik, C. P. & van Noordwijk, M. A. (1989). The special role of male *Cebus* monkeys in predation avoidance and its effects on group composition. *Behavioral Ecology and Sociobiology*, **24**, 265–76.

Villarroel, M., Bird, D. M. & Kuhnlein, U. (1998). Copulatory behaviour and paternity in the American kestrel: the adaptive significance of frequent copulations. *Animal Behaviour*, **56**, 289–99.

Westneat, D. F. & Sargent, R. C. (1996). Sex and parenting: the effects of sexual conflict and parentage on parental strategies. *Trends in Ecology and Evolution*, **11**, 87–91.

Wittenberger, J. F. & Tilson, R. L. (1980). The evolution of monogamy: hypotheses and evidence. *Annual Review of Ecology and Systematics*, **11**, 197–232.

Woodroffe, R. & Vincent, A. (1994). Mother's little helpers: patterns of male care in mammals. *Trends in Ecology and Evolution*, **9**, 294–7.

Wynne-Edwards, K. E. (1987). Evidence of obligate monogamy in the Djungarian hamster, *Phodopus campbelli*: pup survival under different parenting conditions. *Behavioral Ecology and Sociobiology*, **20**, 427–37.

CHAPTER 8

Social polyandry and promiscuous mating in a primate-like carnivore: the kinkajou (*Potos flavus*)

Roland Kays

INTRODUCTION

The class Mammalia includes a diverse array of societies, from solitary species that rarely meet except for mating, to species so gregarious that they are rarely out of contact with their many group mates. Within this continuum, social monogamy is more likely to evolve in the less gregarious lineages, but remains a rare strategy across the class (Kleiman, 1977; van Schaik & Kappeler, chapter 4).

However, the mammalian class is dominated by solitary social systems that are rarely studied in detail (Eisenberg, 1981). Some have suggested that the secretive nature of many mammals has made them difficult to study, and that 'solitary mammal' has become a default term used to classify many of these elusive species. Leyhausen (1965) recognized this: 'the main reason why so many mammals are said to be solitary seems to be that they can only be shot one at a time'. Indeed, modern techniques such as radio telemetry, night vision, and molecular analyses have revealed surprising complexity and sociality among many species previously classified as solitary.

Among the Carnivora, telemetry work has found male associations in slender mongooses (*Galerella sanguinea*: Rood, 1989) and northern raccoons (*Procyon lotor*: Gehrt & Fritzell, 1998). An ambitious cadre of night-loving primatologists has revealed an amazing diversity of sociality among nocturnal primates (Kappeler, 1997; Bearder, 1999; Sterling *et al.*, 2000). This includes nightly interactions between Indonesian spectral tarsiers (*Tarsius spectrum*: Gursky, 2000), permanent, bisexual sleeping groups and home range overlap in the fat-tailed dwarf lemur (*Cheirogaleus medius*: Fietz, 1999), and female sleeping groups and home range overlap in the grey mouse lemur (*Microcebus murinus*: Radespiel, 2000). These recent discoveries have led to a re-evaluation of the evolution of primate sociality, and reclassification of the dispersed social organization classification of species that spend much of their active time alone, but maintain complex social networks (Müller & Thalmann, 2000). Pair-based social systems are more likely to have evolved in these secretive, small social groups, and many 'new' socially monogamous species are likely to be discovered as we turn a more probing scientific eye to these reclusive mammals.

In addition to living in small groups, socially monogamous species tend to share life history traits such as sexual monomorphism, a long life span, and substantial parental care requirements (Kleiman, 1977; van Schaik & Kappeler, chapter 4). They are arboreal, fruit-eating members of the raccoon family (Carnivora, Procyonidae) with an ecology and behaviour convergent with many primate species (Kays, 2001). This includes specializing in a diet of large ripe fruits, relatively low predation risk, and a dispersed social system with parallels seen in primate groups including *Ateles*, *Daubentonia*, *Pan*, and *Pongo*. Kinkajous (*Potos flavus*) are long-lived, have little sexual dimorphism, and show relatively long gestation and lactation periods (Table 8.1).

Thus, kinkajous have many traits typical of socially monogamous species, and may be expected to live solitarily, or in some form of monogamous social system. However, field studies have found that their behaviour varies from these systems in surprising ways. In this chapter I will first describe what is known about the social organization of this nocturnal, arboreal species and then discuss an evolutionary hypothesis for this system, highlighting the roles of resources, mate access, and parental investment.

Table 8.1. *Vital statistics of kinkajous from the literature and from recent field work in Panama*

	Species-wide[a]	This study[b]
Weight	2–4.6 kg	Av. 3.09 kg
Sexual dimorphism	Only in some skull measures	Only in teeth (*t*-tests on 8 males and 6 females). Mass: Male 3.17 kg, Female 3.00 kg, d.f. = 12, $P = 0.25$ Body length: Male 941.9 mm, Female 916.7 mm, d.f. = 12, $P = 0.14$ Lower canines: Male 12.6 mm, Female 11.2, d.f. = 12, $P = 0.02$ Upper canines: Male 10.4 mm, Female 9.3 mm, d.f. = 12, $P = 0.007$
Litter size	1–2	1
Diet	Fruit, nectar, leaves; occasionally insects	90–99% fruit; remaining nectar and leaves; no insects or other prey
Density	12–74/km^2	*c.* 12/km^2
Activity	Nocturnal and arboreal	Nocturnal and arboreal
Habitat	Variety of tropical forest types	Tropical moist forest, mix of primary forest and *c.* 70-year-old secondary growth
Lifespan	20–40 years in captivity	Unknown
Parental care	Female only	Female only
Oestrus length	17 days	Unknown
Female receptive period	2 days	*c.* 2–3 days
Breeding seasonality	Absent and/or varies across sites	May and June breeding peak; some females may mate in other times
Gestation length	*c.* 4 mo	*c.* 4 mo
Lactation length	*c.* 4 mo	*c.* 4 mo
Dispersal bias	Unknown	Female
Territorial	Yes and no	Yes

[a] From Poglayen-Neuwall, 1976; Ford & Hoffmann, 1988.
[b] From Kays & Gittleman, 1995, 2001; Kays, 1999*a*; Kays *et al.*, 2000.

SITE, TIME, AND SPECIES

Kinkajous are 2–4 kg members of the Procyonidae (Carnivora), closely related to olingos (*Bassaricyon* spp.), and distantly related to other members of the family (e.g., raccoons, coatis: Decker & Wozencraft, 1991). Table 8.1 summarizes their basic natural history and general reproductive behaviour. While they have been reported to eat substantial amounts of insects (Redford *et al.*, 1989), my study population was almost completely frugivorous, which is probably more typical of populations across their range (Naveda, 1992; Julien-Laferrière, 1999; Kays, 1999*a*). Kinkajous live in a variety of forested habitats from Mexico to Brazil, and are completely nocturnal and arboreal. Although males are on average slightly heavier, the only significant sexual dimorphisms found are in the skull and teeth.

My field research was conducted over 19 months, from 1993 to 1996, in the lowland, tropical, moist forest of Parque National Soberanía in the Republic of Panama (22 100 ha; 9° 9′ N, 79° 44′ W). Work was centred on the trail network of the 104-ha Limbo research plot (Robinson *et al.*, 2000). The Limbo plot consisted of a mix of forest types and was dominated by older secondary growth (*c.* 60–120 years old), but also included remnant patches of tall forest (*c.* 400 years old: Karr, 1971). This population included four non-overlapping social groups; some observations are also reported from a fifth social group observed in preliminary work 6 km from the main study site (Kays & Gittleman, 1995).

Twenty-five kinkajous were caught in 192 total captures using 50 Tomahawk live traps (32 × 32 × 102 cm). Traps were baited with banana for 1292 trap nights and hung in trees 4–25 m above the ground using a hoistable trap design (Kays, 1999*b*). Ten kinkajous were fitted with radio collars that were marked with a unique pattern of coloured reflective tape. Fifteen kinkajous were fitted with a similar reflective identification collar without a radio transmitter.

Marked kinkajous habituated quickly and could be followed and observed without obvious disturbance. Radio-collared animals were followed on 74 half-nights for 380 hours (including four females for 156 h, four adult males for 169 h, and two subadult males for 54 h), either from the point when an individual animal left its sleeping den at dusk until midnight, or from midnight until an individual entered its sleeping den at dawn. Animals could not always be seen clearly because of understorey vegetation and darkness. Therefore, sporadic visual observation, falling fruit, and sounds were used in combination to determine a focal animal's behaviour. Generally, kinkajous could be seen directly while they were feeding or resting in a tree. As individuals moved through the trees, however, they could only be seen intermittently, and therefore telemetry was used to follow travelling animals. In addition to direct observation, 74 (2.4% of total) nocturnal animal locations were fixed with the triangulation of telemetric bearings.

The behaviour of a focal animal was continuously recorded as feeding, resting, travelling, social, or unknown. Feeding was defined as eating or actively searching a fruiting tree for food. Travelling was defined as moving between trees. An animal was classified as resting if it stopped travelling for >1 minute while not eating or being social. Social behaviour included activities that brought two or more animals <1 m from each other for ≥ 1 minute. Certain silent social behaviours, such as grooming, resting in contact, or scent marking, were often difficult to detect when kinkajous were hidden behind vegetation. Some of these behaviours were probably missed and recorded as rest or unknown, and may be underrepresented in the results.

An allogrooming bout was defined as one animal grooming another by licking and biting at the fur, and was considered to have ended when the animals separated for >1 min. Fights between kinkajous were designated by loud screams, aggressive chases, and combative physical contact. Any instance of two or more kinkajous simultaneously feeding from the same fruit patch was considered to be group feeding. A fruiting patch usually consisted of a single fruiting tree or vine, but occasionally included two fruiting plants if their canopies were adjacent. Each time a kinkajou fed on fruit, flowers, or leaves the following were recorded: plant species, location, diameter at breast height (DBH) of the plant, and a categorical estimation of the number of fruits or flowers on the ground under the plant (<10, <50, <100, <500, <1000, >1000).

Home range size was calculated with the minimum convex polygon (MCP) method using the WILDTRAK computer program, Version 2 (Todd, 1995). To exclude outlying points from an MCP and prevent a few extreme and atypical points from contributing a large additional area, a 95% MCP was used by selecting 95% of the data points lying closest to the arithmetic mean centre of the range. When calculating the 95% MCP home ranges, an independence interval of 70 min was used to avoid temporal autocorrelation of consecutive locations. This is approximately the amount of time it would take a kinkajou to cross a home range (Doncaster & Macdonald, 1997). Observed patterns of grooming, group feeding, and group denning were compared using non-parametric chi-squared tests.

I analysed the variation in 11 newly developed microsatellite loci to assess the degree of kinship within and between four social groups totalling 25 kinkajous (Kays *et al.*, 2000). I used the Queller and Goodnight (1989) index of relatedness (R) calculated by the computer program RELATEDNESS 5.0 (Goodnight, 1998) to estimate kinship. To evaluate paternity, I used both the probability of paternity exclusion value following Chakraborty *et al.* (1988) and Chakravarti and Li (1983), as well as the likelihood method (Meagher, 1986), as implemented in the program CERVUS 1.0 (Marshall *et al.*, 1998). The mean relatedness of male and female kinkajous was evaluated with a two-sample randomization test (de Ruiter & Geffen, 1998), using the program RT 2.0 (Manly, 1991). In this test, the observed mean difference was compared with the means of 5000 random samplings of the same set of relatedness values.

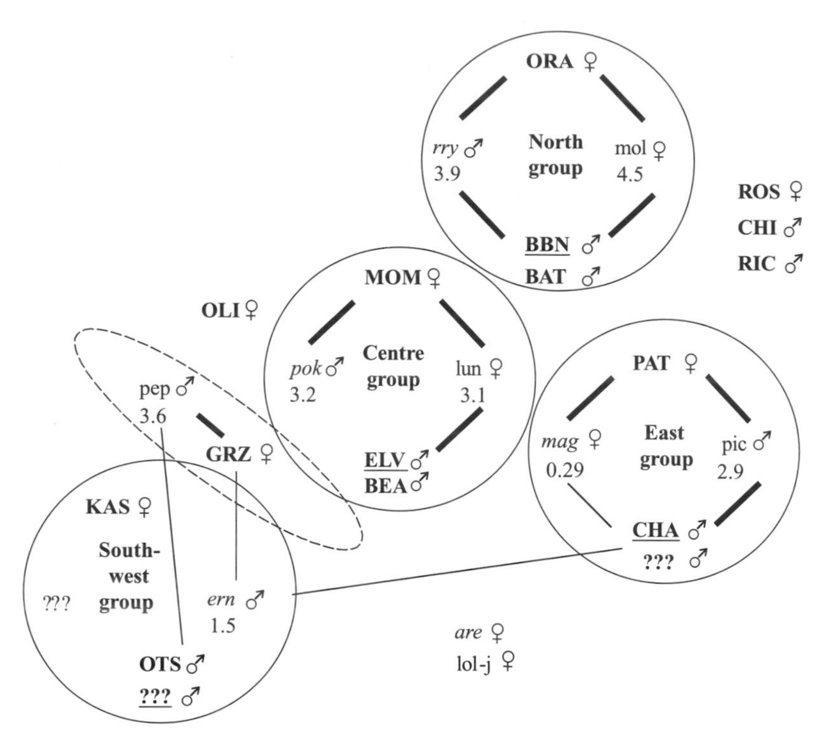

Figure 8.1. Schematic of social group structure and genetic parentage for kinkajous in the Limbo plot. Three-letter codes indicate marked individuals. Capital letters refer to adults, lower case to subadults (*italics*) or juveniles. Question marks indicate individuals that were observed but never captured. Circles represent approximate group home range boundaries and are not to scale. The broken ellipse represents the home range of a female and her pup that bordered two groups, but did not associate with a social group. Individuals outside the circles were not radio-collared and could not be assigned to a group or home range. Underlined males were dominant in their group. Thick lines show parentage at 95% confidence level and thin lines show parentage at 80% confidence levels. Numbers below offspring labels are LOD scores (the natural logarithm of the product of parentage likelihood ratios across all loci). See Kays *et al.* (2000) for details.

KINKAJOU SOCIAL ORGANIZATION

While early anecdotal records of kinkajou social behaviour often reported groups of animals feeding together (e.g., Goldman, 1920), a systematic survey found them most often as solitary animals (Walker & Cant, 1977). Following animals through the forest without radio collars has proved almost impossible, except in full moonlight (Forman, 1985), and collars wrapped in reflective tape have significantly improved behavioural observations (Kays & Gittleman, 1995). Using spotlights and radio collars, preliminary observational studies of a few individuals documented a mostly solitary lifestyle punctuated with regular social interactions (Forman, 1985), and led the way for more detailed studies (Julien-Laferrière, 1993; Kays & Gittleman, 1995).

My field and genetic study of a population of 25 kinkajous in central Panama has revealed more of the details of how these social interactions form the base of an unusual social organization, and is the source for the description below (Kays, 1999*c*; Kays *et al.*, 2000; Kays & Gittleman, 2001). At least one female (GRZ) reproduced outside the typical social group structure (see below). This was also suspected for other females from the periphery of the study site, although no detailed behavioural data were collected to test this assertion.

The social organization of kinkajous in Panama centred on a social group of two males, one female, one subadult, and one juvenile (Figure 8.1). These individuals all overlapped in home range, regularly shared day dens, and socialized around dens and fruiting trees. Neighbouring social groups had almost no overlap in

Table 8.2. *Relationships between kinkajou group feeding and fruit patch size from 162 observed feeding bouts. Fruits under the tree were estimated as <10, <50, <100, <500, <1000, >1000*

Fruit patch size	[a] Patches fed in by kinkajous of three group sizes mean ± SD (N)			
	1	2	3+[b]	P-value
DBH cm	53 ± 40 (120)	67 ± 37 (26)	110 ± 33 (6)	0.0007
Number of fruits under tree	496 ± 545 (145)	668 ± 594 (28)	1187 ± 458 (8)	0.0015

[a] One-way ANOVA.

[b] There was one feeding group of four animals (Kays, 1999*a*, *c*; Kays *et al.*, 2000; Kays & Gittleman, 2001).

range, and rarely interacted; border fights between males (Julien-Laferrière, 1993; Kays & Gittleman, 1995) or between females (Kays & Gittleman, 2001) were rare. These strict group territories are probably marked with specialized scent glands (Poglayen-Neuwall, 1966). Some females were apparently not members of any social group, and raised offspring in home ranges that fitted between the territories of group females. The one non-group female studied in detail overlapped somewhat (15–44%) with neighbouring group males, but did not socialize with them on a regular basis during the course of this study, and never shared a day den with them (Kays & Gittleman, 2001).

Kinkajous typically moved through their home range and foraged alone; 97% of travel between fruiting trees was alone ($N = 145$ trips) and 80% of foraging bouts ($N = 153$ bouts) were solitary. However, they regularly met up with members of their social group, especially at larger fruiting trees where feeding competition is less important (Table 8.2). Males fed in larger groups than females, both when data were pooled (females, $N = 83$, mean $= 1.08 \pm 0.32$; males, $N = 101$, mean $= 1.26 \pm 0.59$; T-value $= -2.38$, d.f. $= 182$, $P = 0.018$) and when data were considered for individuals (MANOVA; females, $N = 4$, mean $= 1.05$; males, $N = 5$, mean $= 1.44$; $F = 21.0$; $P < 0.0001$). For group members, at least 70% of their days were spent sleeping in the same tree den with other animals, in groups of two to five.

Allogrooming was observed on 38 occasions, for a total of 254 min. This is representative of typical behaviour, and excludes four nights when copulations were observed. The mean length of all bouts was 6.44 min (SD 5.44), the longest a 28-min bout between an adult male and a subadult male. Allogrooming was most frequent at large (>50 cm DBH) fruiting trees ($N = 19$ grooming bouts), but also occurred in smaller fruiting trees ($N = 4$) and near day dens just after dusk ($N = 9$). Affiliative behaviours were generally observed between all group members (Figure 8.2). Patterns of social grooming, group denning, and group feeding show strong juvenile female but weak juvenile male bonds, reflecting the absence of male parental care. While the two adult males occasionally groomed each other, this was less than expected if grooming was conducted at random, and short fights were observed between group males on six occasions in 224 hours of observation (not including nights with copulations). Adult and subadult males groomed each other more than expected.

With this adult male/subadult male grooming bias, the presence of two adult males in the social group, and the paucity of interactions with the female, the social interactions in kinkajou groups are distinctively male-centred; this probably reflects the patrilineal nature of the groups. Although long-term data are not available to show a patrilineal inheritance of group territories, this idea is supported by both behavioural and genetic evidence of female-biased dispersal. Both marked subadult females dispersed during the study, while none of the three subadult males did. Furthermore, males were more related to their male neighbours than females were to their female neighbours (average male relatedness $= 0.118 \pm 0.25$, $N = 7$; average female relatedness $= -0.02 \pm 0.31$, $N = 8$; $P < 0.05$, two-sample randomization test) – suggesting that males were living among family while females had dispersed away from their natal groups.

Kinkajous were observed mating on four nights. On two nights, the two group males competed over access to a non-group female; all these fights were won by the dominant male, who also received all copulation observed. In contrast, the subordinate male did not contest copulations between the group female and the dominant male, but followed quietly behind. At the end of the second night of oestrus the dominant male stopped

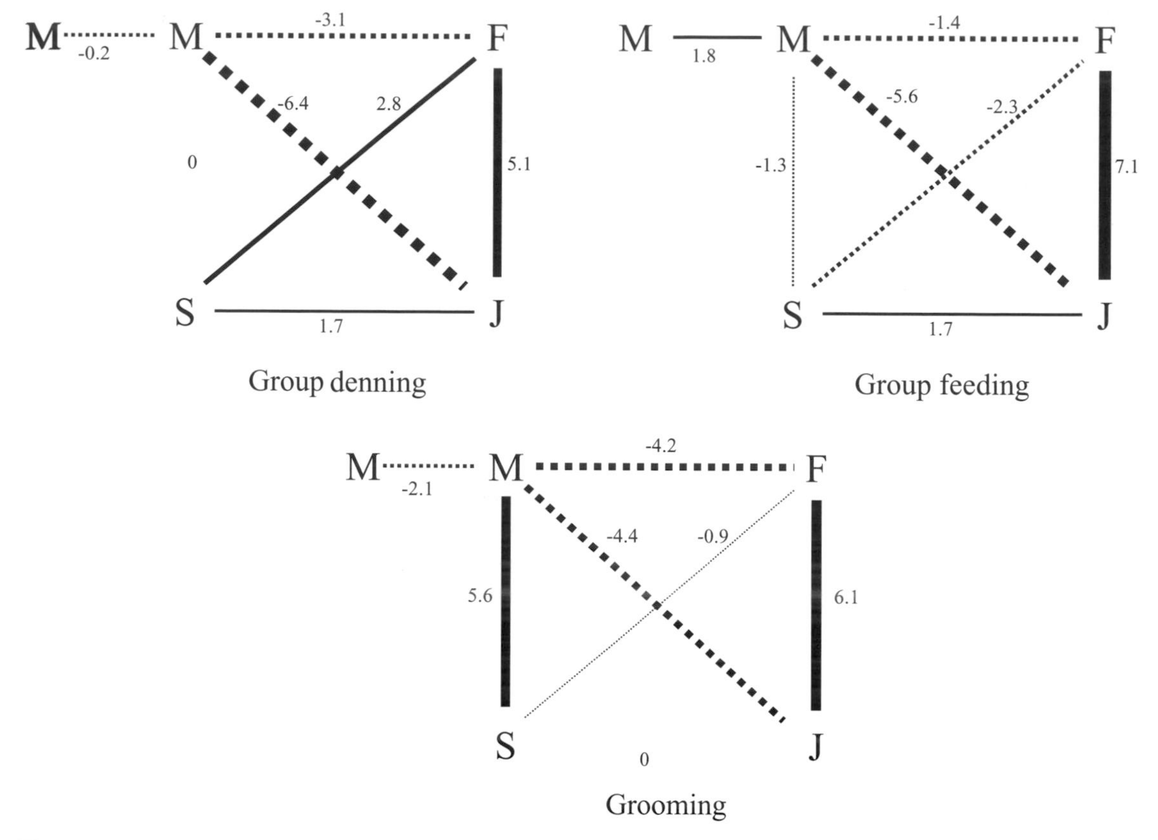

Figure 8.2. Social relationships between group members as measured by residuals from non-parametric chi-squared tests using expected values calculated from kinkajou group structure and focal animal sample size. Grooming and feeding relationships were significant at $P < 0.005$, group denning was not significantly different than would be expected if they were at random. M, male; F, female; J, juvenile; S, subadult. Values for interactions between group members and males are averaged between the two group males, except for the relationship between the two group males themselves, which are represented by connection between the two Ms. Numbers are residuals. Solid lines show interactions that occurred more than expected; broken lines show interactions that occurred less than expected. Data are from 59 kinkajou group dennings, 39 group feeding bouts, and 38 grooming bouts on nights without copulations. For details see Kays & Gittleman (2001).

guarding the group female (presumably to feed) and the subordinate immediately copulated with her. A genetic analysis of paternity supported the idea that the dominant male monopolized fertilizations (Figure 8.1).

It is difficult to make a rigid classification of the social organization observed in Panamanian kinkajous, and it is constructive to consider the various levels independently. Patterns of spatial overlap, grouping, and sociality indicate a polyandrous system with one group female and two group males; however, not all females live in groups. Furthermore, the handful of mating observations suggests male promiscuity; a more detailed description of this mating system (e.g., as extra-pair copulations, or random matings) will require more field observations. Genetic data from three social groups show a family structure typical of monogamous systems, with both the juvenile and subadult typically sired by the dominant male of the group. However, this genetic story is complicated by the presence of non-group females with pups sired by neighbouring group males.

While some details of this unusual social organization remain very unclear, the evolution of the basic group structure and patterns of social interaction deserve discussion. I will do this by considering how these patterns relate to the investment in offspring, the procurement of food by females, and access to mates for males.

INVESTMENT IN OFFSPRING

With the charge of livebirth and lactation, females are the dominant caregivers in most mammal species. Only rarely are the needs of offspring so great that help from males is also needed, and most mammal species have female-biased parental care (Clutton-Brock, 1991). Kinkajous fit this mammalian pattern, and only females have been observed fulfilling parental duties such as carrying young between day dens and fruit trees (in captivity: Poglayen-Neuwall, 1976; in the wild: Kays & Gittleman, 2001). Young pups were 'parked' (Kays & Gittleman, 2001), rather than carried by adults during typical feeding in fruit trees, as in many nocturnal, arboreal primates (Charles-Dominique, 1977). Older pups travelled with their mother between feeding trees, and sometimes fed alongside group males in larger fruit trees. Although males were never seen carrying young or obviously 'baby-sitting', they showed no aggression towards pups, regularly shared fruiting trees and day dens, and occasionally played with them. This system provides few opportunities for the males to contribute substantial parental care, and most of the costs of raising offspring seem to be associated with gestation and lactation.

Reproduction is no small investment for a female kinkajou, as she is either pregnant or lactating for about two-thirds of the year (Table 8.1), translating into a higher energetic cost than most carnivores face. Female kinkajous fall well above regression lines in plots comparing female investment in offspring for the Carnivora (brain size vs. gestation length; body mass vs. gestation length; body weight vs. mass at birth: Gittleman, 1994*a*, *b*, personal communication). Feeding with juveniles in fruiting trees may also reduce her food intake per feeding tree because of increased feeding competition. This suspected increased feeding competition may be reflected in the fact that females travel more than males (in metres and minutes: Kays, 1999*c*), presumably to visit more fruiting trees. Because of their extreme frugivory (90% of feeding bouts, 99% scats: Kays, 1999*a*), females may have difficulty not only in maintaining total caloric intake, but also in satisfying the need for specific fruit species that provide protein and other nutrients that are often rare in fruits (e.g., selection of nutritious figs: O'Brien *et al.*, 1998; Kays, 1999*a*; Wendeln *et al.*, 2000). The sexual difference in the importance of food is further reflected in the social relationships between kinkajous.

THE IMPORTANCE OF FRUIT TO FEMALES

Fruit is seasonal in most habitats, and shows an annual cycle in central Panama with a low in production towards the end of the rainy season (November) and a peak at the beginning of the rainy season (May) (Croat, 1978; Kays, 1999*a*). Kinkajou body weight fluctuated two months behind these food resources, suggesting that fruit is a limiting resource, especially for females (Figure 8.3). The lag time between fruit abundance and weight change may actually be somewhat shorter, as

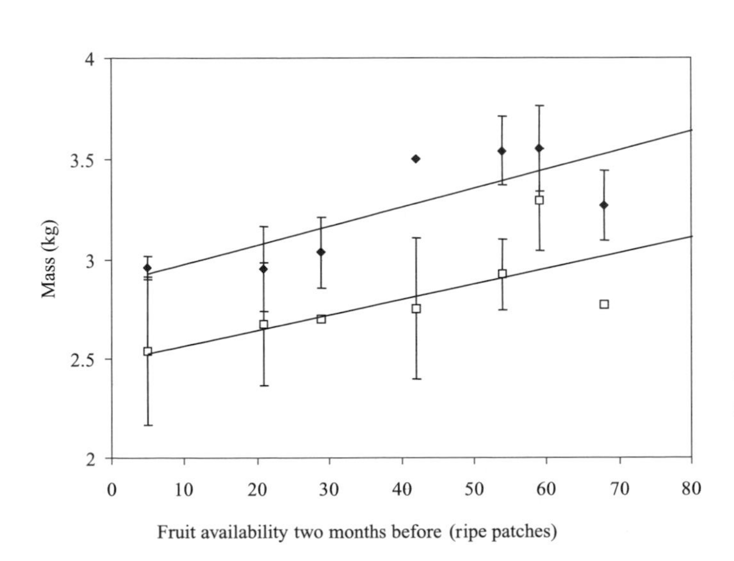

Figure 8.3. Correlation between body mass and number of edible fruit patches available two months earlier. Males, Spearman $r = 0.75$, $P = 0.0521$; females, Spearman $r = 0.893$, $P = 0.0068$. Fallen fruit was censused twice per month along 11.4 km of trail. Weights were averaged for same-sex animals captured in the same month (N = 1–4 individuals, ±SD). See Kays (1999*a*, *b*) for details.

these fruits were censused after they fell to the ground, and not when they were ripe and available to arboreal animals.

Further evidence of the importance of fruit to kinkajous comes from correlations between fruit abundance and kinkajou activity budgets (Kays, 1999*c*). Variation in activity patterns reflected seasonal changes in fruit abundance. Females responded to changes in food availability by travelling more and eating less as fruit increased. Overall, adjustments by males to seasonal fruit changes were not significant. However, outside the main breeding season, when males are less 'distracted' by females, males responded to increased fruit abundance by eating less and resting more. Females may be travelling more in search of high quality fruit species with which to meet high nutritional (e.g., calcium: O'Brien *et al.*, 1998) as well as energetic needs. Because they have lower reproductive costs, males are not under such nutritional stress, and appear to use surplus time offered by high fruit abundance to increase resting and social time (Kays, 1999*c*).

Other behavioural differences also support the hypothesis that females have higher nutritional needs, and that feeding competition for ripe fruit is more important to females than males. Females reduced feeding competition by eating in adult feeding groups that were smaller than those of males (Kays & Gittleman, 2001). This is also reflected in an overall lower sociality by females compared with males. Females spent marginally less time in social behaviours such as grooming and resting in contact with other kinkajous (females average 1.4 % social, males average 4.4% social, d.f. = 7, $P = 0.070$). Furthermore, a female's sociality was directed towards her pup, and not to other adult members of the social group (Figure 8.2). Unburdened by reproductive costs, male kinkajous have lower energetic needs, and thus can accommodate the added feeding competition associated with sociality. This may also be related to the female bias in dispersal (Kays *et al.*, 2000) if females do not tolerate sharing fruiting trees with their adult offspring.

Neither the relative importance of food to female kinkajous nor their minimal sociality is unusual. Typically, the distribution of food determines the distribution of female mammals in an area, which, in turn, determines the distribution of males (Bradbury & Vehrencamp, 1977; Emlen & Oring, 1977; Wrangham, 1980; Clutton-Brock, 1989). Thus a priority for the female is to secure an exclusive territory small enough to defend, but large enough to supply her with food throughout the year. Given the unpredictability of finding ripe fruit (Croat, 1978; O'Brien *et al.*, 1998; Wright *et al.*, 1999), her foraging area must be large enough to encompass both the spatial and temporal patchiness of tropical fruit (Carr & Macdonald, 1986).

With their arboreal and nocturnal habits, kinkajous have few predators and a low predation risk (Kays, 1999*c*). Without the strong predation-related selection pressures to live in groups, as experienced by diurnal, arboreal mammals (i.e., Primates: Isbell, 1994; Hill & Dunbar, 1998), female kinkajous can afford to exclude other females from their territories, thus reducing feeding competition. Excluding males is probably impractical because of the males' equal to slightly larger body size, larger canines, and a shared carnivorous ancestry (Table 8.1). Lower female gregariousness reduces scramble competition for food, and should allow them to survive on fewer feeding trees per night, thus allowing smaller territories to be defended. Like kinkajous, the large arboreal aye-ayes have a low predation risk and females are solitary, using exclusive ranges (Sterling, 1993; Sterling & Richard, 1995). Although more social than female kinkajous or aye-ayes, the highly frugivorous spider monkeys and chimpanzees have similar sexual differences in sociality, also presumably caused by scramble competition for ripe fruit patches (Wrangham, 2000). The distribution of female kinkajous in medium-sized (average 22 ha), non-overlapping home ranges, combined with the species' low predation risk, sets the stage for a unique distribution of males (Kays & Gittleman, 2001).

MALES ACCESS TO MATES

Theory predicts that if the home ranges of a species of female mammal overlap, or are very small, single males may be able to overlap and monopolize more than one breeding female (Clutton-Brock, 1989), and larger female ranges would prevent males from monopolizing more than one female. In areas where females have medium-sized home ranges males may use a mixed strategy, monopolizing one female within their own social group and keeping tabs on nearby non-group females. This is one possible explanation for the behaviour of Panamanian kinkajous, where the dominant male associated with both a group female and one or more territorial non-group females (Kays & Gittleman, 2001). Female ranges are hypothesized to be too large for a male to defend two complete ranges, yet small enough that one male may overlap at least partially with more than one female range.

The maximum size of a male's territory is probably set by a combination of the distance a kinkajou can travel in one night, and the amount of effort needed to keep a territory well marked. Given the tight abutment and non-overlap of neighbouring social group territories in Panama, the boundaries are probably well marked with scent (Kays & Gittleman, 2001). Because it is so difficult to observe in the field, there is little detail on natural kinkajou scent marking. However, we do know that kinkajous have very specialized scent glands (Poglayen-Neuwall, 1966), and that the time and energy needed to mark an area are probably not trivial (Gosling, 1990). The combination of two male kinkajous in a social group is surprising, and the reasons for it are not obvious (Kays & Gittleman, 2001). One possibility is that they cooperate in territorial scent marking, allowing them to mark and defend an area that totally encompasses one group female, and also provides access to neighbouring non-group females.

The dominance of certain males appears to prevent the monopolization of non-group females by subordinate males, thus preventing them from breaking off to form a breeding pair with neighbouring non-group females. This was most obvious when the dominant male from the centre group repeatedly trounced the subordinate male in fights over an oestrous non-group female. Although subordinate males were not found to be the parents of any young kinkajous, one was observed to copulate once with the female of his group after the dominant male halted his pursuit of her halfway through the night. Thus, reproductive opportunities appear to be rare, but not non-existent, for subordinate males.

The choice of which female to socialize with on a regular basis as part of a social group may be influenced by female quality, since the non-group female was older and smaller than the group females. On the other hand, this older, more experienced female may have been better at excluding males from her territory, thus reducing her feeding competition. Unfortunately, little detail is available on other non-group females to make more quantitative comparisons.

CONSIDERING ALTERNATIVES

These adaptive explanations for the evolution of kinkajou social organization fit well with the somewhat limited amount of behavioural data from the field. However, there are few compelling analyses to show any benefit of grouping. This is often the case in elusive species, especially if sociality is cryptic or newly evolved (Rood, 1989; Woodroffe & Macdonald, 1993). Nonetheless, the hypotheses described above offer the best explanation for the available data, and the alternative evolutionary scenarios have little or no support.

The costs and benefits selecting for solitary female kinkajous seem clear cut, and converge with the low female sociality seen in other arboreal mammals with low predation risk (e.g., *Daubentonia madagascariensis*: Sterling, 1993; Sterling & Richard, 1995; *Pan paniscus*: Wrangham, 2000). A more gregarious existence would increase the costs associated with feeding competition and jeopardize a kinkajou mother's ability to sustain long gestation and lactation periods. There also seems to be little advantage to joining groups for protection from predators. Kinkajou body size, together with their arboreal and nocturnal habits, essentially makes them too large for owls, too nocturnal for diurnal raptors, and too arboreal for large cats. Thus, predation risk is low, which is supported by their slow reproductive rate (species with low reproductive rates could not sustain high predation rates: Hill & Dunbar, 1998) and the fact that females pay no heed to moonlight (nocturnal predators are more efficient on bright nights: Bowers & Dooley, 1993; Kays, 1999*c*). Finally, there seems to be little risk of infanticide to pressure females to join groups (van Schaik, 1996; van Schaik & Kappeler, 1997). Indeed, some females raised young independently of any group, and very young juveniles were typically 'parked' unprotected while their mothers foraged further away (Kays & Gittleman, 2001). There is probably little advantage to females to overlap in daily home range with two males since this would increase feeding competition. However, no female carnivores are known to exclude males from their range, and this may not be an alternative for female kinkajous: general avoidance and low sociality may be the best alternatives. Thus, a more gregarious female existence makes no sense given the high nutritional costs it would entail, and the apparent insignificance of the additional protection grouping could provide from predation or infanticide.

Considering the alternatives for male behaviour is more complicated. Given a distribution of solitary females, males might rove through many female territories in search of fertile females (e.g., Ostfeld, 1985; Sandell, 1989), defend an area large enough to encompass two or more female territories (e.g., Emmons, 1988), or match up with single females in a classic monogamous system (e.g., Brotherton *et al.*, 1997).

Males may be expected to rove in search of females if they can survive in unfamiliar areas and recognize the presence of an oestrous female. With their low predation risk, the increased riskiness of unfamiliar areas would probably not be significant for roving male kinkajous (Kays, 1999*c*; Brotherton & Komers, chapter 3). However, finding food in unfamiliar areas may be a serious hindrance to roving animals, given the seasonal and spatial patchiness of ripe fruit. Roving males are more likely to find a reproductive female if there is a predictable, short, synchronized breeding season, or if females widely advertise their condition (Brotherton & Komers, chapter 3). While there was a two-month seasonal breeding peak in Panamanian kinkajous, this was not tightly synchronized, and mating was also observed five months after the typical season (Kays & Gittleman, 2001). Given the obvious vaginal swelling of oestrous females (Poglayen-Neuwall, 1976), their reproductive condition may not be easily concealed. However, it is unclear how widely it is advertised through scent, and roving males may be unlikely to pick up on their weakly seasonal, short (*c.* 2 day) receptive period.

Kinkajou males have not been found to rove widely. In fact, radio-tracking data show them to be territorial, never crossing into other groups' territories (Kays & Gittleman, 2001). However, males occasionally travel beyond the home range of their group female, apparently to meet up with neighbouring non-group females. The ranges of non-group females are overlapped to some degree by males from different groups and are likely be visited by numerous neighbouring group males. Thus, kinkajou male pairs appear to be following a somewhat mixed strategy, marking and occupying a core area surrounding their group-female, but also roving in surrounding areas to keep track of non-group females, and mate with them if the opportunity arises. The importance of being familiar with the location and seasonality of fruiting trees may limit the extent of male roving, keeping them tied to a core home range.

Many territorial male carnivores defend large areas that allow them to overlap with more than one female in a polygynous system (Sandell, 1989). This requires that females' ranges are small enough for a male to be able to defend an area encompassing more than one of them. In addition, some males must be strong enough to either exclude other male competitors from these large areas or maintain dominance over sympatric males. As sympatric kinkajou males have a clear-cut dominance relationship, the size of female home ranges probably limits their ability to defend more than one (Kays & Gittleman, 2001). The size of male territories is probably limited by the difficulty of arboreal travel and the expense of scent-marking boundaries in a rainy environment (Gosling, 1990). However, if female home ranges within parts of the kinkajous' geographic range are smaller, contain more food, or are less seasonal, males may be able to defend two females.

Given the territorial nature of males, why do they not focus on a single female in a monogamous sense? Social monogamy is seen in species where males are needed to prevent infanticide, help with parental care, or where males constantly guard their mates (Brotherton & Komers, chapter 3; van Schaik & Kappeler, chapter 4). Male kinkajous do not help with parental care, and infanticide seems unimportant in their social evolution (see above). Also, given the high cost of feeding competition in small fruiting trees, constant mate guarding may not be a viable option. Even within the female's two- to three-day receptive period, I found that males only guarded for the first half of the night, before hunger (apparently) forced individuals to go their separate ways in search of food (Kays & Gittleman, 2001). Instead, coalitions of males may mark, exclusively use, and presumably defend an area that contains the resources needed by a female. This resource defence, rather than mate defence hypothesis is also supported by the presence of female-biased dispersal in kinkajous (Kays *et al.*, 2000). Resource defence is rare among mammals, and thought to be related to a resource defence mating system (Greenwood, 1980).

The size of female home ranges in Panama appears to offer male kinkajous an option somewhere between classical polygyny and monogamy. Female ranges seem too large for a male to overlap completely with more than one, yet small enough that males can overlap completely with one, and partially with other non-group females. The cooperation of two males may be needed to scent-mark such a large area, or a subordinate male may simply be accepted because the costs of excluding him are greater than the benefits. Male coalitions may also be encouraged by the relatedness of at least some male pairs (Kays *et al.*, 2000). Access to non-group females may not be possible in parts of their geographic range where females have larger home ranges because of lower food abundance, or higher seasonality. In these cases males may overlap with only a single female, in a monogamous social system.

CONCLUSION

Surprising sociality is now being discovered in many species of elusive mammals that were previously classified as 'solitary'. Flexible social monogamy is most likely in species that are not highly gregarious, and should be expected in some of these species with dispersed social systems. Kinkajous spend most of their active time alone, yet regularly interact in small social groups. The high parental investment of females makes finding sufficient fruit food a priority, and females live a largely solitary existence in order to minimize feeding competition. Without the energetic burden of parental investment, feeding competition is less important for males, and they have more complex social interactions. Male pairs appear to defend an area large enough to allow them to monopolize the breeding of one group female, and also take advantage of mating opportunities with neighbouring non-group females through limited roving. This results in a system that is socially polyandrous with a monogamous or promiscuous mating pattern.

With their carnivoran heritage, it is most parsimonious to consider that kinkajous evolved from a solitary ancestor, and that their newly evolved sociality and group life are adaptations to their derived niche as arboreal fruit-eaters. In the course of this evolution, they have converged with other arboreal frugivores, most notably with dispersed foragers in the Primates. This suggests a universality to the adaptive evolutionary links between a number of ecological and behavioural traits. For example, kinkajous appear to converge with a number of primate species (e.g., *Ateles*, *Daubentonia*, *Pan*, *Pongo*), where low predation risk and a diet of patchy arboreal foods (typically, ripe fruit) have selected for low levels of gregariousness, and dispersed sociality is biased toward males (Chapman *et al.*, 1995; Sterling, 1993; Sterling & Richard, 1995; van Schaik, 1999; Wrangham, 2000). Since many of these unrelated, large, arboreal frugivores are also among the minority of mammals having female-biased dispersal, it is likely that these ecological traits, or the resulting dispersed social system, also select for this dispersal strategy (Greenwood, 1980; Pusey & Packer, 1987), although the mechanisms behind this remain unclear.

Acknowledgements

Many thanks to U. Reichard and C. Boesch for inviting me to a very well-organized and productive workshop. Additional thanks to the workshop participants and attending students for insightful comments and suggestions. Gracias to INRENA for allowing me to work in their beautiful National Park, and to the Smithsonian Tropical Research Institute for funding and other support in Panama. Generous support was also provided by a predoctoral fellowship from the National Science Foundation, the Science Alliance and Department of Ecology and Evolutionary Biology at the University of Tennessee, the National Geographic Society, the American Museum of Natural History, Eagle Creek and The Sharper Image. Thanks to R. Azipure, C. Carassco, C. Foster, J. Kays, N. Kays, C. Krieger, L. Slatton, N. Smythe, D. Staden, and J. Young for help in the field. D. Robinson and T. Robinson expertly created the trail network at the Limbo research area and allowed me to work there. J. L. Gittleman served as an excellent advisor during this research, and provided the comparative data on kinkajou maternal investment.

References

Bearder, S. K. (1999). Physical and social diversity among nocturnal primates: a new view based on long term research. *Primates*, **40**, 267–82.

Bowers, M. A. & Dooley, J. L. (1993). Predation hazard and seed removal by small mammals – microhabitat versus patch scale effects. *Oecologia*, **94**, 247–54.

Bradbury, J. W. & Vehrencamp, S. L. (1977). Social organisation and foraging in emballonurid bats. *Behavioral Ecology and Sociobiology*, **2**, 1–17.

Brotherton, P. N. M., Pemberton, J. M., Komers, P. E. & Malarky, G. (1997). Genetic and behavioural evidence of mongamy in a mammal, Kirk's dik-dik (*Madoqua kirkii*). *Proceedings of the Royal Society of London, Series B*, **264**, 675–81.

Carr, G. M. & Macdonald, D. W. (1986). The sociality of solitary forager: a model based on resource dispersion. *Animal Behaviour*, **34**, 1540–79.

Chakraborty, R., Meagher, T. R. & Smouse, P. E. (1988). Parentage analysis with genetic markers in natural populations. I. The expected proportion of offspring with unambiguous paternity. *Genetics*, **118**, 527–36.

Chakravarti, A. & Li, C. C. (1983). The effect of linkage on paternity calculations. In *Inclusion Probabilities in Parentage Testing*, ed. R. H. Walker, pp. 411–20. Arlington, Virginia: American Association of Blood Banks.

Chapman, C. A., Wrangham, R. W. & Chapman, L. J. (1995). Ecological constraints on group size: an analysis of spider monkey and chimpanzee subgroups. *Behavioural Ecology and Sociobiology*, **36**, 59–70.

Charles-Dominique, P. (1977). *Ecology and Behaviour of Nocturnal Primates*. London: Duckworth.

Clutton-Brock, T. H. (1989). Mammalian mating systems. *Proceedings of the Royal Society of London, Series B*, **236**, 339–72.

(1991). *The Evolution of Parental Care.* Princeton, New Jersey: Princeton University Press.

Croat, T. B. (1978). *Flora of Barro Colorado Island*. Stanford, Califonia: Stanford University Press.

Croat, T. B., de Ruiter, J. R. & Geffen, E. (1998). Relatedness of matrilines, dispersing males and social groups in long-tailed macaques (*Macaca fascicularis*). *Proceedings of the Royal Society of London, Series B*, **265**, 79–8.

Decker, D. M. & Wozencraft, W. C. (1991). Phylogenetic analysis of recent procyonid genera. *Journal of Mammalogy*, **72**, 42–55.

Doncaster, C. P. & Macdonald, D. W. (1997). Activity patterns and interactions of red foxes (*Vulpes vulpes*) in Oxford city. *Journal of Zoology, London*, **241**, 73–87.

Eisenberg, J. F. (1981). *The Mammalian Radiations.* Chicago: University of Chicago Press.

Emlen, S. T. & Oring, L. W. (1977). Ecology, sexual selection, and the evolution of mating systems. *Science*, **197**, 215–23.

Emmons, L. H. (1988). A field study of ocelots (*Felis pardalis*) in Peru. *Revue d'Ecologie (Terre et Vie)*, **43**, 133–58.

Fietz, J. (1999). Monogamy as a rule rather than exception in nocturnal lemurs: the case of the fat-tailed dwarf lemur, *Cheirogaleus medius*. *Ethology*, **105**, 259–72.

Ford, L. S. & Hoffmann, R. S. (1988). *Potos flavus*. *Mammalian Species*, **321**, 1–9.

Forman, L. (1985). Genetic Variation in Two Procyonids: Phylogenetic, Ecological and Social Correlates. Ph.D. thesis, New York University, New York.

Gehrt, S. D. & Fritzell, E. K. (1998). Resource distribution, female home range dispersion and male spatial interactions: group structure in a solitary carnivore. *Animal Behaviour*, **55**, 1211–27.

Gittleman, J. L. (1994*a*). Are the pandas successful specialists or evolutionary failures? *Bioscience*, **44**, 456–64.

(1994*b*). Female brain size and parental care in carnivores. *Proceedings of the National Academy of Sciences of the USA*, **91**, 5495–7.

Goldman, E. A. (1920). Mammals of Panama. *Smithsonian Miscellaneous Collections*, **69**, 1–579.

Goodnight, K.F. (1998). *RELATEDNESS 5.0*. available from http://gsoft.smu.edu/GSoft.html

Gosling, L. M. (1990). Scent-marking by resource holders: alternative mechanisms for advertising the costs of competition. In *Chemical Signals in Vertebrates*, Volume 5, ed. D. W. Macdonald, D. Müller-Schwarze & S. E. Natynczuk, pp. 315–28. Oxford: Oxford University Press.

Greenwood, P. J. (1980). Mating systems, philopatry, and dispersal in birds and mammals. *Animal Behaviour*, **28**, 1140–62.

Gursky, S. (2000). Sociality in the Spectral Tarsier, *Tarsius spectrum*. *American Journal of Primatology*, **51**, 89–101.

Hill, R. A. & Dunbar, R. I. M. (1998). An evaluation of the roles of predation rate and predation risk on selective pressures on primate grouping behaviour. *Behaviour*, **135**, 411–30.

Isbell, L. A. (1994). Predation on primates: Ecological patterns and evolutionary consequences. *Evolutionary Anthropology*, **3**, 61–71.

Julien-Laferrière, D. (1993). Radio-tracking observations on ranging and foraging patterns by kinkajous (*Potos flavus*) in French Guiana. *Journal of Tropical Ecology*, **9**, 19–32.

(1999). Foraging strategies and food partitioning in the neotropical frugivorous mammals *Caluromys philander* and *Potos flavus*. *Journal of Zoology, London*, **247**, 71–80.

Kappeler, P. M. (1997). Determinants of primate social organization: comparative evidence and new insights from Malagasy lemurs. *Biological Reviews*, **72**, 111–51.

Karr, J. R. (1971). Structure of avian communities in selected Panama and Illinois habitats. *Ecological Monographs*, **41**, 207–33.

Kays, R. W. (1999*a*). Food preferences of kinkajous (*Potos flavus*): a frugivorous carnivore. *Journal of Mammalogy*, **80**, 589–99.

(1999*b*). A hoistable arboreal mammal trap. *Wildlife Society Bulletin*, **27**, 298–300.

(1999*c*). The Solitary Group Life of a Frugivorous Carnivore: Ecology, Behavior, and Genetics of Kinkajous (*Potos flavus*). Ph.D. thesis, University of Tennessee, Knoxville.

(2001). Kinkajou. In *The Encyclopedia of Mammals*, ed. D. W. Macdonald, pp. 92–3. Oxford: Oxford University Press.

Kays, R. W. & Gittleman, J. L. (1995). Home range size and social behavior of kinkajous (*Potos flavus*) in the Republic of Panama. *Biotropica*, **27**, 530–4.

(2001). The social organization of the kinkajou *Potos flavus* (Procyonidae). *Journal of Zoology*, **253**, 491–504.

Kays, R. W., Gittleman, J. G. & Wayne, R. K. (2000). Microsatellite analysis of kinkajou social organization. *Molecular Ecology*, **9**, 743–51.

Kleiman, D. G. (1977). Monogamy in mammals. *Quarterly Review of Biology*, **52**, 39–69.

Leyhausen, P. (1965). The communal social organisation of solitary mammals. *Symposium of the Zoological Society of London*, **14**, 249–63.

Manly, B. F. J. (1991). RT: a Program for Randomisation Testing. West Inc. 2003 Central Ave. Cheyenne, Wyoming, 82001, USA.

Marshall, T. C., Slate, J., Kruuk, L. E. B. & Pemberton, J. M. (1998). Statistical confidence for likelihood-based paternity inference in natural populations. *Molecular Ecology*, **7**, 639–55.

Meagher, T. R. (1986). Analysis of paternity within a natural population of *Chamaelirium letueum*. I. Identification of most-likely male parents. *American Naturalist*, **128**, 199–215.

Müller, A. E. & Thalmann, U. (2000). Origin and evolution of primate social organisation: a reconstruction. *Biological Reviews of the Cambridge Philosophical Society*, **75**, 405–35.

Naveda, J. A. S. (1992). Historia Natural y Ecologia del Cuchicuchi (*Potos flavus*: Carnivora), en Barlovento, Estado Miranda, Venezuela. Thesis para Licenciado en Biologia, Universidad Central de Venezuela.

O'Brien, T. G., Kinnaird, M. F., Dierenfeld, E. S., Conklin-Brittain, N. L., Wrangham, R. W. & Silver, S. C. (1998). What's so special about figs? *Nature*, **392**, 668.

Ostfeld, R. S. (1985). Limiting resources and territoriality in microtine rodents. *American Naturalist*, **126** 1–15.

Poglayen-Neuwall, I. (1966). On the marking behavior of the kinkajou (*Potos flavus* Schreber). *Zoologica*, **51**, 137–41.

(1976). Zur Fortpflanzungsbiologie und Jugendentwicklung von *Potos flavus* (Schreber 1774). *Der Zoologische Garten*, **46**, 237–83.

Pusey, A. E. & Packer, C. (1987). Dispersal and Philopatry. In *Primate Societies*, ed. B. B. Smuts, D. L. Cheney, R. M. Seyfarth, R. W. Wrangham & T. T. Struhsaker, pp. 250–66. Chicago: University of Chicago Press.

Queller, D. & Goodnight, K. (1989). Estimating relatedness using genetic markers. *Evolution*, **43**, 258–75.

Radespiel, U. (2000). Sociality in the Gray Mouse Lemur (*Microcebus murinus*) in Northwestern Madagascar. *American Journal of Primatology*, **51**, 21–40.

Redford, K. H., Stearman, A. M. & Trager, J. C. (1989). The kinkajou (*Potos flavus*) as a myrmecophage. *Mammalia*, **53**, 132–4.

Robinson, W. D., Brawn, J. D., & Robinson, S. K. (2000). Forest bird community structure in central Panama: influence of spatial scale and biogeography. *Ecological Monographs*, **70**, 209–35.

Rood, J. P. (1989). Male associations in a solitary mongoose. *Animal Behaviour*, **38**, 725–8.

Sandell, M. (1989). The mating tactics and spacing patterns of solitary carnivores. In *Carnivore Behavior, Ecology, and Evolution*, Volume 1, ed. J. L. Gittleman, pp. 164–82. Ithaca, New York: Cornell University Press.

Sterling, E. J. (1993). Patterns of range use and social organization in aye-ayes (*Daubentonia madagascariensis*) on Nosy Mangabe. In *Lemur Social Systems and their Ecological Basis*, ed. P. M. Kappeler & J. U. Ganzhorn, pp. 1–10. New York: Plenum Press.

Sterling, J. E. & Richard, A. F. (1995). Social organization in the aye-aye (*Daubentonia madagascariensis*) and perceived distinctiveness of nocturnal primates. In *Creatures of the Dark: The Nocturnal Prosimians*, ed. L. Altermann, G. A. Doyle & M. K. Izard, pp. 439–51. New York: Plenum Press.

Sterling, E. J., Nguyen, N. & Fashing, P. J. (2000). Spatial patterning in nocturnal prosimians: a review of methods and relevance to studies of sociality. *American Journal of Primatology*, **51**, 3–19.

Todd, I. A. (1995). *Wildtrak II: The Integrated Approach to Home Range Analysis*. Oxford University.

van Schaik, C. P. (1996). Social evolution in primates: the role of ecological factors and male behaviour. *Proceedings of the British Academy*, **88**, 9–31.

(1999). The socioecology of fission-fusion sociality in orangutans. *Primates*, **40**, 69–86.

van Schaik, C. P. & Kappeler, P. M. (1997). Infanticide risk and the evolution of male-female association in primates. *Proceedings of the Royal Society of London, Series B*, **264**, 1687–94.

Walker, P. L. & Cant, J. G. (1977). A population survey of kinkajous (*Potos flavus*) in a seasonally dry tropical forest. *Journal of Mammalogy*, **58**, 100–2.

Wendeln, M. C., Runkle, J. R. & Kalko, E. K. V. (2000). Nutritional values of 14 fig species and bat feeding preferences in Panama. *Biotropica*, **32**, 489–501.

Woodroffe, R. & Macdonald, D. W. (1993). Badger sociality – models of spatial grouping. *Symposium of the Zoological Society of London*, **65**, 145–69.

Wrangham, R. W. (1980). An ecological model of female-bonded primate groups. *Behaviour*, **75**, 262–97.

(2000). Why are male chimpanzees more gregarious than mothers? A scramble competition hypothesis. In *Primate Males: Causes and Consequences of Variation in Group Composition*, ed. P. Kappeler, pp. 248–58. Cambridge: Cambridge University Press.

Wright, S. J., Carrasco, C., Calderon, O. & Paton, S. (1999). The El Niño Southern Oscillation, variable fruit production, and famine in a tropical forest. *Ecology*, **80**, 1632–47.

CHAPTER 9

Monogamy correlates, socioecological factors, and mating systems in beavers

Lixing Sun

INTRODUCTION

Mating systems are among the central topics in the study of animal social organization. Monogamy is loosely defined and used to refer to the situation in which a mating pair stays together exclusively for at least one breeding season (e.g., Fuentes, 1999). Obviously, this definition applies only to social, not genetic, monogamy. In the following discussion, unless otherwise indicated, I use the term monogamy to refer to social monogamy.

Kleiman (1977) was the first to summarize mating systems in mammals and found that fewer than 3% of mammalian species were socially monogamous (also see Rutberg, 1983; Kinzey, 1987). This was in sharp contrast to birds, where about 90% of the species were traditionally believed to be socially monogamous (Lack, 1968). This general impression was prevalent for decades until a recent flurry of data revealed that many of these socially monogamous birds and mammals routinely engage in extra-pair copulations (EPCs) (see Birkhead & Møller, 1995; Fuentes, 2002). Several chapters in this volume provide some fresh data about EPCs in mammals, which are less studied than birds in this respect.

In mammals, parental investment is extremely skewed towards females owing to the time- and energy-consuming aspects of mammalian female gestation and lactation. As a result, male mammals are selected for reproducing more polygynously than males in other groups of animals. Males provide parental care in fewer than 10% of mammalian species (Woodroffe & Vincent, 1994). However, even male parental care does not necessarily lead to social monogamy, and based on a few socially monogamous species that have been genetically screened (see Ribble, chapter 5; Fietz, chapter 14), strict genetic monogamy appears to be much rarer in mammals. Social monogamy in mammals, though infrequent, appears to exist mostly in rodents, primates, and canids (e.g., Emlen & Oring, 1977; Kleiman, 1977; Getz & Carter, 1996; Reichard, chapter 13).

Primatologists have contributed a great deal to our understanding of mammalian mating systems. Data based primarily on situations in primates have revealed several shared elements in socially monogamous species, for example, prolonged association and strong pair bonding between two adults, sexual monomorphism, strong territoriality, joint territorial defence, and presumable adult–offspring genetic relatedness of 0.5 (Kleiman, 1977; Wittenberger & Tilson, 1980; Rutberg, 1983; van Schaik & Dunbar, 1990; Anzenberger, 1992). Examining these common elements uncovered in monogamous primates can be highly useful because they may be potentially valuable predictors of social monogamy. Nevertheless, how reliable these elements are in predicting monogamy, whether the predictability can be extrapolated into non-primate mammals, and why these elements correlate with social monogamy, have yet to be investigated.

Beavers (*Castor canadensis*, the North American beaver and *C. fiber*, the Eurasian beaver) are among the few non-primate mammals that show a typical pattern of social monogamy (Svendsen, 1989) and also demonstrate, without exception, the 0.5 genetic relatedness (which is rarely studied in mammals) of these monogamy correlates found in primates. Hence, they provide us a superb opportunity to examine closely whether there is a causational relationship between monogamy and monogamy-related elements in groups other than primates.

In this chapter, I will systematically examine the features in beavers that are related to social monogamy. I intend to answer the following four questions.

1 What are the characteristics associated with social monogamy in beavers?

2 Does EPC or polygynous mating exist in beavers?
3 What are the causational relationships between social monogamy and the characteristics believed to be associated with it?
4 How do immediate socioecological factors cause a switch between mating systems?

SITE, TIME, AND SPECIES

Most of the data, results, and facts discussed here are based on published or unpublished data or results from the beaver population at Allegany State Park in western New York State, USA. This population of the North American beaver has been intensively studied since 1984. The beaver population had not been trapped or managed for a period that extended for about two decades prior to and through my study in the 1990s. In some areas, the population density reached 1.14 families per kilometre of stream length, the highest ever reported (Müller-Schwarze & Schulte, 1999). Since 1984, beavers in all accessible areas of the park were live-trapped and ear-tagged for individual identification in spring, summer, and/or autumn every year. By 1995, more than 300 beavers had been trapped and tagged, and about 70–80% of the individual animals could be identified. In the core study area, where 22 beaver families were living, all individuals were identifiable (Sun & Müller-Schwarze, 1997). Behavioural observations began in 1985 and have continued every summer since then. With numerous field assistants participating in intensive trapping and observation at 30–50 beaver settlements each year, a large data set of meticulous records has been assembled. These records include information about beaver morphological measurements, family composition, family member replacements, lengths of pair bonds, dispersal events, territorial behaviour, and numerous other behavioural and ecological data. Some details about the habitat, population density, intercolony distance, family size, family composition and body size are summarized and compared with other populations. This information can be found in a paper by Müller-Schwarze and Schulte (1999).

The two beaver species (*C. canadensis* and *C. fiber*) are similar in size and live in family units consisting of an adult pair and their offspring (see Müller-Schwarze & Sun, 2003). Adult males and females form long-term pairs, and typically live and breed monogamously in exclusive territories (Svendsen, 1989), where they build dams and lodges and feed on woody and herbaceous plants. They mate in winter and females give birth in early summer. A typical family has a pair of adults plus a varying number of young from both the current year and the previous year. The average family size ranges from four at Allegany State Park (Müller-Schwarze & Schulte, 1999) and Alaska (Boyce, 1974) to more than eight in Massachusetts (Brooks *et al.*, 1980) and Nevada (Busher *et al.*, 1983). Young beavers, both male and female, disperse from their natal family at the age of two (Sun *et al.*, 2000). Beavers typically stay in the same settlement until food plants in the accessible areas are depleted. The details of the life of the Eurasian beaver are less clear but are presumably similar to those of its North American counterpart. Most of the unpublished data and observations about the Eurasian beaver used in my discussion were kindly provided by F. Rosell, whose research is based on a population in Norway.

CHARACTERISTICS ASSOCIATED WITH MONOGAMY

Sexual monomorphism

Beavers are sexually monomorphic in body morphology. There is no statistically significant difference in body size between adult males and females in *C. canadensis* (see Jenkins & Busher, 1979). In my own study (unpublished data), adults (> 3.5 years old) show no sexual dimorphism in the five major body measurements: body weight, body length, chest girth, tail length, and hind foot length. The only exception is that females have a slightly wider tail than males. In adult (>3 years old) *C. fiber*, females are actually slightly heavier than males, but there is no difference in body length, tail length, or tail width (F. Rosell, personal communication). Smith and Jenkins (1997) measured seasonal variation in body size and found no sexual dimorphism in any age class.

Lack of sexual dimorphism in body size in beavers is in consensus with the observation that almost all of the currently recognized socially monogamous species are sexually monomorphic. The underlying reason for this is the presumably lower opportunity for sexual selection (Arnold & Wade, 1984) to diversify the body size between males and females, i.e., lack of a disruptive selection force on the body size of the two sexes.

Food, nutrition, and energy expenditure

Beavers are generalist herbivores (Jenkins & Busher, 1979). It is well known that they eat tree bark, but they also consume substantial amounts of other woody and herbaceous plants, including aquatic plants when available. Although beavers prefer deciduous trees, particularly poplars (*Populus*) and willows (*Salix*), over conifers (Fryxell & Doucet, 1993), almost all tree species growing near beaver-settled areas are included in their diets (e.g., Jenkins, 1975; Heidecke, 1989; Yu *et al.*, 1994). Here, I focus on trees because tree bark is the only food that is available in high latitude areas in winter and thus is the limiting factor for nutritional requirements.

Though abundant and readily available, tree bark is low in nutritional value. In *C. canadensis*, an adult needs 760–850 kcal daily for maintenance (Stephenson, 1956; Pearson, 1960), the equivalent of 0.7–0.9 kg of fresh aspen (*Populus tremuloides*). The estimate for *C. fiber* also comes close to this number (Yu *et al.*, 1994; Nolet *et al.*, 1995). Beavers have to consume a much larger amount of less nutritious dietary plants in order to sustain growth. Even if conditions are unrealistically ideal, with an abundance of favourite trees and a high tree regrowth rate, a beaver still needs a large area to sustain itself for a few years in the same habitat. Relatively even distribution of food resources and low nutritional values of food items, together with low mobility, limit the size of beaver social units to a single family. Young offspring consume a substantial amount of food critical for wintering, resulting in a larger loss of body mass for both adults and yearlings compared with beaver families without young from the current year (Smith & Jenkins, 1997). Beaver family size appeared to be positively correlated with habitat quality (measured as availability of deciduous woody plants within the home range of resident beavers), and at marginal habitats, beaver families tended to be smaller with a higher relocation rate (Sun, unpublished data). Therefore, food resources appear to be limiting: either there is not enough food, or beavers have to shorten their period of residence in any one settlement, especially if numerous adults and their offspring live in the same area. For this reason, resource defence polygyny, frequently seen in mammals due to male monopoly of resources, is not possible in beavers.

In order to avoid predators, water is vital for beavers, and they build dams and lodges to minimize their exposure on land. They may also construct canals to facilitate escape and transportation of food items, such as tree branches. Diets of low nutritional value force beavers to allocate a substantial proportion of their time to foraging, yet arduous construction work and ensuing maintenance tasks also require a large expenditure of energy and time. Such energy and time constraints, especially in families with young of the year, make it difficult for adult beavers to leave their settlements to actively search for EPC opportunities.

Obviously, offspring are constrained by food requirements. The number of offspring born appears to be positively correlated with the quality of habitat, especially food (Smith & Jenkins, 1997; author's personal observation). Adult males play a pivotal role in food provisioning, guarding, and comforting newborn young (Wilsson, 1971; Patenaude, 1983), although other male parental contributions are not clear. It seems that males may gain a good deal in terms of reproductive success by assisting females in raising offspring. This complies with the conditions of the mate assistance hypothesis for monogamy (Kleiman, 1977; Wittenberger & Tilson 1980).

Seasonality of reproduction and physical limitation

The mating season of *C. canadensis* is from late January to early March, peaking in mid-February (Hodgdon & Hunt, 1966; Bergerud & Miller, 1977), when beaver ponds are normally deeply frozen in high latitude areas. In western New York, beaver ponds remained deeply frozen in four of the five winters during the period of this study, between 1991 and 1996. Thus, beaver movements in winter are physically limited to the lodge and to the free water underneath the ice. Beavers need a well-consolidated settlement with sufficient food stored (which they hoard near the lodge during the previous autumn) and free water beneath the ice that gives them access to the food cache. In the winter, lodges are vital for warmth and protection against predators (Müller-Schwarze & Sun, 2003). Seeking mating opportunities without a relatively permanent residential base during the mating season would be extremely risky. Therefore, alternative reproductive strategies such as roaming, frequently seen in other mammalian species, are not an option for beavers in high latitude areas.

Heavy male parental investment theoretically makes it important for males to guard their mates to ensure their paternity. Strong seasonality and short duration of female oestrus (only 12–24 hours for each female in *C. fiber*: Doboszynska & Zurowski, 1983) make it feasible for males to concentrate their mate-guarding efforts within a small window of time. This situation is similar to synchronized female oestrus, which favours monogamy because few EPC opportunities would be left for males and males would lose little by being monogamous (Orians, 1969). Thus, even in warmer areas without ice in winter, EPC should be infrequent because of the likelihood of mate guarding by males and the strong reproductive seasonality of females. Unfortunately, little information about these behavioural aspects is available from beaver populations living in these areas.

Dispersal patterns

Both male and female beavers disperse, typically at the age of two (Sun *et al.*, 2000). Females tend to disperse further (10.15 ± 2.42 [SE] km) than males (3.49 ± 0.86 km), a rare instance in mammals. Unlike the female philopatry commonly seen in rodents such as marmots (e.g., Armitage, 1991), where several females of breeding age stay together, both female and male beavers disperse and settle individually. Thus, breeding females do not form groups nor are they close enough so that two or more females can be contained in one male's home range. This female distribution pattern makes male resource defence polygyny unlikely. Under this situation, monogamy may prevail as shown in some primates (Wrangham, 1980; Rutberg, 1983; van Schaik & van Hooff, 1983). Ribble (1992) observed a similar dispersal pattern in a small rodent species, the California mouse (*Peromyscus californicus*), where females tend to disperse farther than males. Interestingly, most individuals of this species also breed monogamously.

Territoriality and pair bonding

Beavers live in discrete family units, each consisting of an adult male and female and their offspring from current and previous years (see Wilsson, 1971; Jenkins & Busher, 1979). Adults are highly territorial (see Jenkins & Busher, 1979; Rosell & Nolet, 1997), with both sexes participating in marking and territorial defence. Since water and food resources are vital yet limiting to beavers (Müller-Schwarze & Sun, 2003), the exclusive function of a beaver territory appears to be to provide access to and use of these crucial resources. Settled beavers would allow their dispersing younger siblings to stay temporarily in their lodges for as long as an entire year when no vacant settlement was available for dispersers to colonize (Sun, unpublished data).

Despite an extremely high density, beaver territories (not different from home ranges in size: see Müller-Schwarze & Sun, 2003) at Allegany State Park do not overlap. The mean distance between settlements in this population is 1.11 km, whereas in other populations studied, the mean nearest neighbour distances range from 0.51 km to 1.59 km (Müller-Schwarze & Schulte, 1999). Compared with a typical territory span of 0.1 km along the stream, the length of unsettled stream section between neighbours is roughly one magnitude longer. To travel through the unsettled area between neighbouring families, beavers have to expose themselves to a much higher risk of predation. This risk can also be estimated at least one magnitude higher than when beavers are in their own territory – the unsettled areas are long, contain less water, and provide no structures that can be used as immediate shelter against predators such as wolves (*Canis lupus*), coyotes (*C. latrans*), black bears (*Ursus americanus*), and cougars (*Felis concolor*) (Mech, 1966; Allen, 1979; Engelhart & Müller-Schwarze, 1995). Hence, seeking EPC opportunities in neighbouring beaver families would incur additional costs, the amount depending on the distance and availability of water. These costs, in turn, would offset some of the benefits of EPC and discourage beavers from pursuing this alternative reproductive strategy.

Beavers demonstrate long-term pair bonding between the adult male and female, defined as a male and female exclusively living together in the same family. Svendsen (1989) estimated that the bonding period in *C. canadensis* could be as long as eight years, with the average bonding period 2.5 ± 1.5 [SD] years. My own study (unpublished data) in New York showed that pair bonds lasted 4.3 ± 1.7 years on average, with the longest at least nine years. Serial monogamy, with one individual paired with one mate at a time but with several mates over several years, is the dominant pattern in *C. canadensis*, although the reason for mate replacement is not clear. If 12 years is the longevity of pair bonds in the wild and the pair bond starts at two years of age (Müller-Schwarze & Sun, 2003), a beaver in New York in its lifetime should, on average, have 2.3 mates compared with 4.0 mates in Ohio (based on data from Svendsen, 1989).

Both the adult male and female in a family exclude invading conspecifics of either sex. Despite the fact that beavers are nocturnal and direct encounters between residents and invaders are rare (but for which no quantitative data are available), I personally observed both adult male and female resident *C. canadensis* aggressively pursuing and expelling intruding beavers. In *C. fiber*, F. Rosell (personal communication) observed that both resident males and females bluffed intruding conspecifics by using sticks to splash the water. Thus, beavers demonstrate numerous important social elements in monogamy: explicit territoriality, joint display (visual, olfactory, and acoustic) in defending their territory, and repulsion of conspecific individuals of both sexes.

Kin recognition

Beavers are capable of recognizing various degrees of relatives and family members. They use their keen olfactory sense to recognize kin based on anal gland and/or castor sac secretions, and behave favourably towards relatives and aggressively toward strangers (Sun & Müller-Schwarze, 1997, 1998*a*, *b*, 1999; Sun *et al.*, 2000). In beavers, the anal gland secretion contains a genetically controlled kinship pheromone that allows beavers to determine relatedness by phenotype-matching. This enables them to determine the relatedness of an individual even if they have never met before (Sun & Müller-Schwarze, 1998*a*). There are at least two major benefits to having such a highly developed recognition capability. This skill helps relatives to disperse successfully and maintains an optimal breeding system (Sun & Müller-Schwarze, 1997). Therefore, even if EPCs do occur, the resulting offspring should be detectable by the cohabiting adult male and it would presumably be treated unfavourably. Infanticide is a phenomenon among numerous mammalian species (see Hausfater & Hrdy, 1984; van Schaik & Janson, 2001), and might also be expected in beavers, although there is no empirical evidence yet. The ability to identify kin accurately should be a strong enough force to deter females from engaging in EPC, and consequently should favour monogamy, although the behavioural and evolutionary relationship between the kin recognition ability and EPC or polygyny is still not understood.

CASES OF POLYGYNOUS MATING

Social polygyny, consisting of one male and two females, has been observed in a few cases in beavers. Hammond (1943) first reported that two pregnant *C. canadensis* females in North Dakota were found in the same settlement. Another case involving two lactating females in one settlement was observed in Newfoundland (Bergerud & Miller, 1977). Wheatly (1993) found two pregnant females in the same settlement in the taiga of southern Manitoba. The most surprising finding was in Little Valley, Nevada, where three of four settlements in 1974 and all four settlements in 1975 contained two lactating females (Busher *et al.*, 1983). In *C. fiber*, F. Rosell (personal communication) observed three cases where two breeding females were living in the same settlement. However, except for Busher *et al.*'s (1983) report, details on the length of cohabitation and tentative reasons for two females to live with a single male are not available for the reports mentioned above. Therefore, it is difficult to determine whether the social polygyny in these cases is also genetic polygyny, despite a strong indication in that direction.

My study on *C. canadensis* at Allegany State Park in western New York State revealed that in the 35 beaver families that I trapped and observed intensively over a four-year period (1992–95), three of them (8.57%) had two breeding females cohabiting with one adult male. In two of these three cases, two females lived with one male in the same family for a year, and both females were observed regularly participating in lodge and dam maintenance and bringing food to newborn young inside their lodges. The remaining case involved a large beaver family (11 members) in a lake. The two females and one male lived in the same family for at least two years.

One on-going project in my laboratory is the investigation of the paternity of beaver families using eight microsatellite primers to clarify whether EPC or genetic polygyny happens in beavers, and if so, at what frequency.

ARE RELATIONSHIPS BETWEEN MONOGAMY AND MONOGAMY ELEMENTS CAUSATIONAL?

Beavers demonstrate numerous characteristics considered consistent with monogamy on the social level. Yet, anecdotal reports, more detailed information from Busher *et al.*'s (1983) study and my own, have shown that social and perhaps genetic polygyny can occur, thus contradicting the traditional view. Therefore, a lack of sexual dimorphism in body size, widely dispersed females, long-term pair bonds, strong territoriality with both males and females participating in defence, keen kin recognition ability, and resource and other physical limitations unfavourable for polygamous mating, do not guarantee a socially or genetically monogamous mating system in the two species. Hence, beavers show a certain degree of plasticity and, thus, variation, in their mating systems. It is, at the very least, unclear whether there is any causational relationship between social monogamy and any of the characteristics examined in this chapter. These characteristics appear more likely to be side effects rather than prerequisites for both social and sexual monogamy, as implied by Wickler and Seibt (1983). They should more appropriately be called monogamy correlates.

WHAT ARE THE EFFECTS OF SOCIOECOLOGICAL FACTORS ON RODENT MONOGAMY?

In order to achieve the highest individual reproductive success, animals, particularly vertebrates, can be highly plastic in their mating behaviour in response to immediate socioecological conditions (e.g., Lott, 1991). Whether a mating strategy is adaptive or not is determined by the socioecological environment that an animal is experiencing. Here, I have grouped the demographic parameters with ecological factors in my discussion.

Variations in mating systems and the socioecological factors that determine them in wild vertebrates have been extensively reviewed by Lott (1991). For those rodents in which monogamous mating has been documented, most show variations in their mating systems depending on specific socioecological conditions. For example, bushy-tailed woodrats (*Neotoma cinerea*) can form monogamous or polygynous mating systems depending on the size of the breeding habitat – rock outcrops (Escherick, 1981). In prairie voles (*Microtus ochrogaster*), temperature is the key to mating systems: monogamous mating is less frequent when temperatures are low in winter than when temperatures are high in other seasons (Getz *et al.*, 1987). Females' tolerance determines mating patterns in yellow-bellied marmots (*Marmota faviventris*): monogamy is present only when dominant females are intolerant of the presence of other females, those that are typically not closely related. Otherwise, the mating system is more likely to be polygynous, with one male and several closely related females (Armitage, 1991). In Gunnison's prairie dog (*Cynomys gunnisoni*), the distribution pattern of food determines the mating pattern. The number of monogamous mating pairs increases when food distribution is uniform but decreases when food distribution is patchy (Travis *et al.*, 1995). Monogamy in the Malagasy giant jumping rat (*Hypogeomys antimena*) is determined by high predation pressure (Sommer, 2000). The only rodent that has yet to show any variation in social and genetic monogamy is *Peromyscus californicus* (Ribble, chapter 5). The failure to detect a variation in the mating system of this species may be because most of the available information is from one intensively studied population in a fairly uniform habitat.

In *C. canadensis*, the likely polygynous mating in Little Valley, Nevada, was attributed to the shortage of suitable habitats, resulting in a locally high population density (Busher *et al.*, 1983). This may also be the reason for the occasional polygynous mating that I observed at Allegany State Park, New York, where population density could be as high as 1.14 families per km stream length in certain areas, compared with the density range of 0.40–1.09 per km stream length reported in other populations (Müller-Schwarze & Schulte, 1999). A high population density would be likely to result in a shorter distance between different beaver settlements. This in turn would make between-family interactions more frequent than when population density is low. In such a situation, beavers would increase their defences against invasion from neighbours by intensifying territorial marking (Houlihan, 1989).

These are some case studies illustrating how, in rodents, socioecological conditions can trigger switching between monogamy and polygamy, at least from the social, if not genetic, perspective. Other mating systems in other groups of vertebrates under different socioecological conditions are extensively reviewed by Lott (1991).

Orians (1969) proposed the influential model of the polygyny threshold to show how immediate ecological conditions can influence individual females' decisions to mate either monogamously or polygynously. This idea was later incorporated in Emlen and Oring's (1977) compelling argument about the role of socioecological factors in determining mating systems in animals. Since then, a substantial body of literature has evolved, showing that variations in mating systems exist among populations of the same species and among individuals of the same population in a wide variety of vertebrate species (see, e.g., Lott, 1991). These studies convincingly demonstrate that numerous animals belonging to vastly different evolutionary lineages can and do frequently switch between different mating systems in response to immediate socioecological conditions. In birds and mammals, available studies have shown little evidence for any phylogenetic constraints on mating systems (e.g., Sillén-Tullberg & Møller, 1993; Di Fiore & Rendall, 1994; Temrin & Tullberg, 1995).

FURTHER DISCUSSION

Because of unequal investment in male and female gametes (anisogamy), males are selected for mating with as many females as possible while females are selected for successfully raising as many offspring as possible (Bateman, 1948). The presence of social and genetic monogamy in mammals is truly remarkable because both require a specific and stringent set of socioecological conditions that make a male and female stay together and mate exclusively for at least an entire breeding season. Since the biological reasons for males and females to engage in EPC are entirely different (see Alcock, 1997), social and genetic monogamy are delicate equilibria. They hinge on the male–female concurrence of exclusive breeding after weighing two *very* different sets of considerations for the male and female, respectively. Any perturbation resulting in the reversal of the cost:benefit ratio favouring monogamy for individuals of either sex, due to changes in socioecological conditions, would work against the continuation of this social and genetic monogamy. Switching between mating systems should be viewed as an adaptive reproductive strategy, which tracks the dynamics of socioecological conditions to achieve the highest reproductive success.

Important socioecological factors influencing monogamy include parental care, infanticide, mate guarding, sexual dimorphism, female distribution patterns, food distribution patterns and availability, predation pressure, habitat type, spatial behaviour, dispersal patterns, and many others. These factors strongly affect the strengths of both natural and sexual selection, which ultimately determine which mating system is optimal under a specific condition at a specific time. The importance of some of these factors in shaping mating systems, especially monogamy, is substantially treated in this volume.

Selection operates predominantly at the individual level (though it can work at other levels under some specific situations: see Keller, 1999) and therefore, how animals mate is an individually-based reproductive strategy rather than a species-specific character, and should be flexible enough to adapt to different socioecological conditions. Hence, for a species as a whole, variations in mating systems are the norm rather than the exception. This is by no means to say that there is no pattern. On the contrary, different conspecific individuals under similar socioecological conditions tend to adopt the same mating strategy. This forms a statistical pattern, with the modal mating pattern being the one used by most individuals. A species-specific, long-term, monogamous mating system can exist only when all individuals of the species adopt the same mating strategy, and when it is more beneficial than any other alternative. This could happen when fluctuations in the socioecological conditions over time and space do not reverse the cost:benefit ratio favouring the monogamous mating system. Obviously this condition is difficult to meet for all individuals of a species, especially in birds and mammals where most populations are highly subdivided spatially, and experience vastly different sets of socioecological conditions over time. Changes, mild or significant, in these conditions do happen in space and time, resulting in frequent reversals of the cost:benefit ratio of monogamy for either or both sexes.

Therefore, long-term, species-specific monogamy is rare. Indeed, studies not just of beavers but also of many other animals have revealed that variations in mating systems are routine, and can be readily found in different populations of the same species, in different individuals of the same population, or in the same individuals under different socioecological conditions (e.g., Lott, 1991; Owens & Bennett, 1997; Fuentes, 2000; Reichard, chapter 13). All these point to one conclusion: mating is an individualized strategy determined by the immediate socioecological conditions and, thus, it is flexible. The case of beavers illustrates that even in species demonstrating so many of the characters associated with monogamy, polygynous mating can be frequent under certain socioecological conditions.

Acknowledgements
I would like to thank U. Reichard and C. Boesch for the opportunity to participate in stimulating and congenial discussions with many leading researchers during the Monogamy

Workshop at the Max-Planck-Institute. This article benefited a great deal from discussions with A. Fuentes and four anonymous reviewers. C. Straub also provided valuable input into an earlier draft.

References

Alcock, J. (1997). *Animal Behavior: An Evolutionary Approach*, 6th Edition. Sunderland, Massachusetts: Sinauer Associates.

Allen, D. L. (1979). *Wolves of Minong: Their Vital Role in a Wild Community*. Ann Arbor, Michigan: University of Michigan Press.

Anzenberger, G. (1992). Monogamous social systems and paternity in primates. In *Paternity in Primates: Genetic Tests and Theories*, ed. R. D. Martin, A. F. Dixson & E. J. Wichlings, pp. 203–24. Basel: Karger.

Armitage, K. M. (1991). Social and population dynamics of yellow-bellied marmots: results from long-term research. *Annual Review of Ecology and Systematics*, **22**, 379–407.

Arnold, S. J. & Wade, M. J. (1984). On the measurement of natural and sexual selection: theory. *Evolution*, **38**, 709–19.

Bateman, A. J. (1948). Intrasexual selection in *Drosophila*. *Heredity*, **2**, 349–68.

Bergerud, A. T. & Miller, D. R. (1977). Population dynamics of Newfoundland beaver. *Canadian Journal of Zoology*, **55**, 1480–92.

Birkhead, T. R. & Møller, A. P. (1995). Extra-pair copulation and extra-pair paternity in birds. *Animal Behaviour*, **49**, 843–8.

Boyce, M. S. (1974). Habitat ecology of an unexploited population of beavers in interior Alaska. *Proceedings of Worldwide Furbearers Conference*, **1**, 155–86.

Brooks, R. P., Fleming, M. W. & Kennelly, J. J. (1980). Beaver colony responses to fertility control: evaluating a concept. *Journal of Wildlife Management*, **44**, 568–75.

Busher, P. E., Warner, R. J. & Jenkins, S. H. (1983). Population density, colony composition, and local movements in two Sierra Nevadan beaver populations. *Journal of Mammalogy*, **64**, 314–18.

Di Fiore, A. & Rendall, D. (1994). Evolution of social organization: a reappraisal for primates by using phylogenetic methods. *Proceedings of the National Academy of Sciences of the USA*, **91**, 9941–5.

Doboszynska, T. & Zurowski, W. (1983). Reproduction of the European beaver. *Acta Zoologica Fennica*, **174**, 123–6.

Emlen, S. T. & Oring, L. W. (1977). Ecology, sexual selection and the evolution of mating systems. *Science*, **197**, 215–23.

Engelhart, A. & Müller-Schwarze, D. (1995). Responses of beaver (*Castor canadensis*, Kuhl) to predator chemicals. *Journal of Chemical Ecology*, **21**, 1349–64.

Escherick, P. C. (1981). Social biology of the bushy-tailed woodrat, *Neotoma cinerea*. *University of California Publication in Zoology*, **110**, 1–132.

Fryxell, J. M. & Doucet, C. M. (1993). Diet choice and the functional response of beavers. *Ecology*, **74**, 1297–306.

Fuentes, A. (1999). Re-evaluating primate monogamy. *American Anthropologist*, **100**, 890–907.

(2000). Hylobatid communities: changing views on pair bonding and social organization in hominoids. *Yearbook of Physical Anthropology*, **43**, 33–60.

(2002). Patterns and trends in primate pair bonds. *International Journal of Primatology*, **24** (in press).

Getz, L. L. & Carter, S. C. (1996). Prairie-vole partnerships. *American Scientist*, **84**, 56–62.

Getz, L. L., Hofmann, J. E. & Carter, C. S. (1987). Mating system and population fluctuations of the prairie vole, *Microtus ochrogaster*. *American Zoologist*, **27**, 909–20.

Hammond, M. C. (1943). Beaver on the Lower Souris Refuge. *Journal of Wildlife Management*, **7**, 316–21.

Hausfater, G. & Hrdy, S. B. (ed.) (1984). *Infanticide: Comparative and Evolutionary Perspectives*. New York. Aldine de Gruyter.

Heidecke, D. (1989). Ökologische Bewertung von Biberhabitaten. *Säugertierkundliche Mitteilungen*, **3**, 13–28.

Hodgdon, K. W. & Hunt, J. H. (1966). *Beaver Management in Maine*. Game Division, Bulletin 3. Augusta: Maine Department of Inland Fisheries and Game.

Houlihan, P. W. (1989). Scent Mounding by Beaver (*Castor canadensis*): Functional and Semiochemical Aspects. M.Sc. thesis, State University of New York: Syracuse.

Jenkins, S. H. (1975). Food selection by beavers: a multidimensional contingency table analysis. *Oecologia*, **21**, 157–73.

Jenkins, S. H. & Busher, P. E. (1979). *Castor canadensis*. *Mammalian Species*, **120**, 1–8.

Keller, L. (ed.) (1999). *Levels of Selection in Evolution*. Princeton, New Jersey: Princeton University Press.

Kinzey, W. G. (1987). Monogamous primates: a primate model for human mating systems. In *The Evolution of Human Behavior: Primate Models*, ed. W. G. Kinzey, pp. 87–104. Albany: State University of New York.

Kleiman, D. G. (1977). Monogamy in mammals. *Quarterly Review of Biology*, **52**, 39–69.

Lack, D. (1968). *Ecological Adaptations for Breeding in Birds*. London: Methuen.

Lott, D. F. (1991). *Intraspecific Variation in the Social Systems of Wild Vertebrates*. Cambridge: Cambridge University Press.

Mech, L. D. (1966). *The Wolves of Isle Royale*. Fauna of the National Parks of the United States, Fauna Series, No. 7. Washington, DC: US Government Printing Office.

Müller-Schwarze, D. & Schulte, B. A. (1999). Behavioral and ecological characteristics of a "climax" population of beaver (*Castor canadensis*). In *Beaver Protection, Management and Utilization in Europe and North America*, ed. P. Busher & R. Dzieciolowski, pp. 161–77. New York: Kluwer Academic/Plenum.

Müller-Schwarze, D. & Sun, L. (2003). *The Beaver: Natural History of a Wetlands Engineer*. Ithaca, New York: Cornell University Press.

Nolet, B. A., van der Veer, P. J., Evers, E. G. J. & Ottenheim, M. M. (1995) A linear programming model of diet choice of free-living beavers. *Netherlands Journal of Zoology*, **45**, 315–37.

Orians, G. H. (1969). On the evolution of mating systems in birds and mammals. *American Naturalist*, **103**, 589–603.

Owens, I. P. F. & Bennett, P. M. (1997). Variation in mating system among birds: ecological basis revealed by hierarchical comparative analysis of mate desertion. *Proceedings of the Royal Society of London, Series B*, **264**, 1103–10.

Patenaude, F. (1983). Care of the young in a family of wild beavers (*Castor canadensis*). *Acta Zoologica Fennica*, **174**, 121–2.

Pearson, A. M. (1960). A Study of the Growth and Reproduction of the Beaver (*Castor canadensis*; Kuhl) Correlated with the Quality and Quantity of some Habitat Factors. M.Sc. thesis, University of British Columbia, Vancouver.

Ribble, D. O. (1992). Dispersal in a monogamous rodent, *Peromyscus californicus*. *Ecology*, **73**, 859–66.

Rosell, F. & Nolet, B. A. (1997). Factors affecting scent marking behavior in the European beaver. *Journal of Chemical Ecology*, **23**, 673–89.

Rutberg, A. T. (1983). The evolution of monogamy in primates. *Journal of Theoretical Biology*, **104**, 93–112.

Sillén-Tullberg, B. & Møller, A. P. (1993). The relationship between concealed ovulation and mating systems in anthropoid primates: a phylogenetic analysis. *American Naturalist*, **141**, 1–25.

Smith, D. W. & Jenkins, S. H. (1997). Seasonal change in body mass and size of tail in northern beavers. *Journal of Mammalogy*, **78**, 869–76.

Sommer, S. (2000). Sex-specific predation on a monogamous rat, *Hypogeomys antimena* (Muridae: Nesomyinae). *Animal Behaviour*, **59**, 1087–94.

Stephenson, A. B. (1956). Preliminary Studies on Growth, Nutrition and Blood Chemistry of Beavers. M.Sc. thesis, University of British Columbia, Vancouver.

Sun, L. & Müller-Schwarze, D. (1997). Sibling recognition in the beaver: a field test for phenotype matching. *Animal Behaviour*, **55**, 493–502.

(1998*a*). Anal gland secretion codes for relatedness in the beaver, *Castor canadensis*. *Ethology*, **104**, 917–27.

(1998*b*). Anal gland secretion codes for family membership in the beaver. *Behavioral Ecology and Sociobiology*, **44**, 199–208.

(1999). Chemical signals in the beaver: one species, two secretions, many functions? In *Advances in Chemical Signals in Vertebrates*, ed. R. E. Johnston, D. Müller-Schwarze & P. W. Sorensen, pp. 281–8. New York: Kluwer Academic/Plenum.

Sun, L., Müller-Schwarze, D. & Schulte, B. A. (2000). Dispersal pattern and effective population size of the beaver. *Canadian Journal of Zoology*, **78**, 393–8.

Svendsen, G. E. (1989). Pair formation, duration of pair-bonds, and mate replacement in a population of beavers (*Castor canadensis*). *Canadian Journal of Zoology*, **67**, 336–40.

Temrin, H. & Tullberg, B. S. (1995). A phylogenetic analysis of the evolution of avian mating systems in relation to altricial and precocial young. *Behavioral Ecology*, **6**, 296–307.

Travis, S. E., Slobodchikoff, C. N. & Keim, P. (1995). Ecological and demographic effects on intraspecific variation in the social system of prairie dogs. *Ecology*, **76**, 1794–803.

van Schaik, C. P. & Dunbar, R. I. M. (1990). The evolution of monogamy in large primates: a new hypothesis and some crucial tests. *Behaviour*, **115**, 30–62.

van Schaik, C. P. & Janson, C. H. (ed.) (2001). *Infanticide by Males and its Implications*. Cambridge: Cambridge University Press.

van Schaik, C. P. & van Hooff, J. A. R. A. M. (1983). The ultimate causes of primate social systems. *Behaviour*, **85**, 91–117.

Wheatley, M. (1993). Report of two pregnant beavers, *Castor canadensis*, at one beaver lodge. *Canadian Field-Naturalist*, **107**, 103.

Wickler, W. & Seibt, U. (1983). Monogamy: an ambiguous concept. In *Mate Choice*, ed. P. Bateson, pp. 33–52. Cambridge: Cambridge University Press.

Wilsson, L. (1971). Observation and experiments on the ethology of the European beaver (*Castor fiber* L.). *Viltrevy*, **8**, 115–266.

Wittenberger, J. F. & Tilson, R. L. (1980). The evolution of monogamy: hypotheses and evidence. *Annual Review of Ecology and Systematics*, **11**, 197–232.

Woodroffe, R. & Vincent, A. (1994). Mother's little helpers: patterns of male care in mammals. *Trends in Ecology and Evolution*, **9**, 294–7.

Wrangham, R. W. (1980). An ecological model of female-bonded primate groups. *Behaviour*, **75**, 262–300.

Yu, C., Lu, H., Shao, W., Jia, C. & Zhen, R. (1994). The feeding habit of beavers in Xinjiang. *Scientia Silvae Sinicae*, **30**, 519–24.

CHAPTER 10

Social monogamy and social polygyny in a solitary ungulate, the Japanese serow (*Capricornis crispus*)

Ryosuke Kishimoto

INTRODUCTION

Among ungulates, monogamy is generally said to have evolved in a small number of species inhabiting forests or thickets (Estes, 1974; Geist, 1974; Jarman, 1974; Leuthold, 1977). In such habitats, these species have adapted to selective browsing, and their food habits have resulted in spacing, territoriality, solitary living, or living as single male–female pairs (Jarman, 1974). In those species where pair members spend comparatively long periods of time together, such as Kirk's dik-dik (*Madoqua kirkii*: Hendrichs & Hendrichs, 1971; Hendrichs, 1975; Tilson & Tilson, 1986; Brotherton & Rhodes, 1996; Komers, 1996; Brotherton & Manser, 1997), and the klipspringer (*Oreotragus oreotragus*: Dunbar & Dunbar, 1980), social monogamy has been well documented. Genetic monogamy has also been proven in Kirk's dik-dik, (Brotherton *et al.*, 1997). However, because interactions between males and females are very rare, it has not yet been demonstrated that solitary ungulates are socially monogamous. If in fact they are socially monogamous, why do they exhibit this type of monogamy, and how do they maintain the pair bond?

A solitary ungulate, the Japanese serow (*Capricornis crispus*) is a good species to use as a case study to answer these questions. The recent increase in the population of Japanese serows, which has been strictly protected in Japan since 1955, now facilitates behavioural observation in the field. In fact, preliminary studies of the social organization of Japanese serows began in the 1970s (e.g., Akasaka & Maruyama, 1977; Kiuchi *et al.*, 1978, 1979, 1986; Haneda *et al.*, 1979; Sakurai, 1981), and were followed by detailed and long-term studies in the 1980s and 1990s (e.g., Ochiai, 1983*a*, *b*, 1993; Kishimoto, 1987, 1989*a*; Kishimoto & Kawamichi, 1996). Most of these studies presumed social monogamy, including social polygyny, based on observations of the home range overlaps of male–female units of the species, although so far there have been no genetic studies of the mating system(s) of this species. The aim of this chapter is to review these studies of social monogamy and social polygyny in Japanese serows, and to argue that this solitary ungulate usually exhibits social monogamy.

SITE, TIME, AND SPECIES

Over a period of seven years (1979–85), I studied the social organization of Japanese serows by direct observation of 159 individuals occupying a study site that covers 320 ha in Akita, Japan. I collected data for a total of 1057 days, usually between March and December of each of those years (Kishimoto, 1987, 1989*a*; Kishimoto & Kawamichi, 1996). Ochiai (1983*a*, *b*, 1993) also conducted a detailed study of Japanese serows, this work carried out at a site covering 120 ha in Kusoudomari, Japan, from 1976 to 1992. These studies provide data that support the assumptions of the characteristics of Japanese serow social organization because most, if not all, residents in the study areas were identified each year, and the sites were almost completely encompassed by territories of both sexes. My descriptions of the social monogamy and social polygyny of Japanese serows are based on these studies, supplemented by observations from other studies conducted at additional sites in Japan.

The two study sites that are the focus of this chapter, Akita and Kusoudomari, were located within montane areas of northern Japan, and are about 200 km apart. Both sites were primarily covered by forests, although Akita had some patchily distributed, clear-cut logging areas. Both sites were part of larger areas that included widespread populations of Japanese serows, although half of the Kusoudomari site fronted on the sea. The snow cover at both sites is no more than 1 m in depth, remaining approximately three months of the year, between December and March.

Japanese serows weigh about 35 kg, although some adults can weigh as much as 50 kg or more (Miura, 1986). This species exhibits so little sexual dimorphism that accurate sexing in the field is usually possible only by observing the external genitalia (Kishimoto, 1988). The external appearance of large preorbital glands is indistinguishable between the sexes, although the average weight of the glands is a little heavier in males than in females (Kodera *et al.*, 1982). A browser, the main diet of the Japanese serow consists of browsed leaves and dwarf bamboos (Takatsuki & Suzuki, 1984; Suzuki & Takatsuki, 1986; Takatsuki *et al.*, 1988; Ochiai, 1999), and the botanical composition of this diet is similar for males and females (Suzuki & Takatsuki, 1986). There have been no obvious predators of Japanese serows since the wolf (*Canis lupus*) became extinct in Japan in 1905.

Both male and female serows become sexually mature at about 2.5–3 years of age (Sugimura *et al.*, 1981; Tiba *et al.*, 1988), and in the field four age classes are categorized by horn shape (Kishimoto, 1988): (i) infants, (ii) yearlings, (iii) subadults (two years old), and (iv) adults (three years old or more). Morphological studies of serow ovaries and fetuses indicate that peak fertilization occurs from late October to early November (Sugimura *et al.*, 1983; Kita *et al.*, 1987). Based on behavioural studies, the period between September and November was defined as the rutting season because association between adult males and adult females is conspicuously more frequent in these months than during the rest of the year (Ochiai, 1983*a*; Kishimoto 1989*a*). After a 210–220-day gestation period (Ito, 1971; Komori, 1975), females give birth to single infants, almost always between May and June (Kishimoto, 1989*b*; Ochiai, 1993).

The Japanese serow has been defined as a solitary ungulate because the combined sightings of single animals (71.7%, $N = 523$: Sakurai, 1981; 76.1%, $N = 305$: Hanawa *et al.*, 1980; 79.3%, $N = 3259$: Kishimoto, 1989*a*; 71.8%, $N = 2397$: Ochiai, 1993), and single females accompanied by a single infant (12.0%: Kishimoto, 1989*a*; 16.2%: Ochiai, 1993), were quite high. Infants left their mothers to become solitary within one year after birth, although they were weaned some months before their departure (Kishimoto, 1989*b*). Only 5.5% and 5.1% of sightings were of adult male–female pairs, either accompanied by or without offspring (Kishimoto, 1989*a* and Ochiai, 1993, respectively). Groups with four animals, usually an adult male–female pair accompanied by two offspring between the ages of 0 and two years, were rarely observed (0.1%: Kishimoto, 1989*a*; 0.2%: Ochiai, 1993), and thus far there have been no verifiable sightings of groups with more than four animals.

SPATIAL ORGANIZATION OF HOME RANGES

Sociological studies of Japanese serows have usually been conducted by direct observation using binoculars or a spotting scope. Because behavioural interactions were rarely observed, social organization was presumed principally by distribution of annual home ranges, which are obtained by connecting the outermost periphery of all the locations of individuals observed each year. In this section, I will describe intrasexual territoriality and social mating units presumed by this method. First, I will focus on the social mating units of territory holders. Then, because some mature daughters are tolerated within their mothers' home ranges, I will focus on social mating units that include the mature daughters, and on territory establishment of offspring. Lastly, I will consider the results of recent radio-telemetry studies, some of which revealed larger home ranges than those detected through direct observation.

Serow territoriality

Both the Akita (Kishimoto, 1989*a*) and Kusoudomari (Ochiai, 1993) study sites were almost completely encompassed by adult home ranges of the same sex each year, and the same-sex ranges were regularly spaced with little overlap (Figure 10.1). These adults were regarded as territory holders because they defended their ranges through territorial chases against other individuals of the same sex. The average home range size for territorial males was 15.2 ha (range 1.7–35.2 ha, $N = 53$) in Akita and 13.6 ha (range 7.5–22.4 ha, $N = 4$) in Kusoudomari (Ochiai, 1983*b*), and for females 10.4 ha (range 2.3–23.4 ha, $N = 62$) in Akita and 10.3 ha (range 4.3–14.7 ha, $N = 4$) in Kusoudomari (Ochiai, 1983*b*). These average sizes were similar to those of other study areas (e.g., 24.4 ha, $N = 6$ males, 15.7 ha, $N = 4$ females: calculated from Kiuchi *et al.*, 1986; 18.4 ha, $N = 6$ males, 11.9 ha, $N = 7$ females: Haneda *et al.*, 1979). The average rates of the same-sex range overlaps were usually low (see Table 10.2), although considerable range overlaps

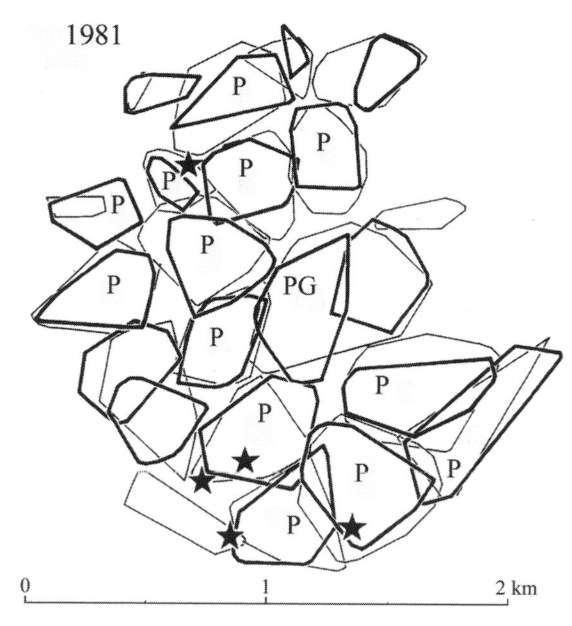

Figure 10.1. Distribution of annual home ranges of resident adults of Japanese serow in Akita (from Kishimoto & Kawamichi, 1996). Thick lines delineate female ranges, thin lines male ranges. Ranges of sexually mature offspring within parental ranges and a few non-territorial males are not shown. Stars indicate locations of territorial chases. P, social pairs; PG, socially polygynous groups. To analyse home range sizes and social units, home ranges surrounded by other ranges belonging to the same sex and/or bounded by valleys on at least three sides were selected and identified as P or PG. Areas not consistently observed over a long enough period to determine whether they were home ranges were omitted from this analysis.

were occasionally observed in connection with territorial intrusion or replacement. In Akita, while a total of 32 males during the study period were regarded as territory holders, only five adult males, who usually occupied territorial boundary areas for at least two but no more than six years, were regarded as non-territorial males.

Both sexes usually occupy their territories over a period of consecutive years (Table 10.1). In Akita, although the entire period of territorial occupancy (from establishment to loss) was usually not known, during the study period the longest duration was over five years for three of 32 male territories. For four of 24 female territories, occupancy lasted throughout the seven-year study period and, generally, females held territories for longer than males. In Kusoudomari, the longest period of territory occupancy was 13 years for a male–female pair (Ochiai, 1993).

Intrasexual encounters between territory holders and neighbours are very rare. In Akita, only 12 encounters were observed between territorial males, seven between territorial males and non-territorial males, two between territorial females, and five between territory holders and neighbouring offspring of 1–3 years of age. All the encounters resulted in aggressive chases, with the exception of one encounter between territorial males where the males avoided each other. These aggressive chases occurred on the periphery of the home ranges (stars in Figure 10.1), and the length of chases to expel intruders usually exceeded 50 m. In Kusoudomari, only 12 encounters between adult males were observed, of which ten resulted in aggressive chases and two resulted in one-sided retreats; only six encounters between adult females were observed, of which four resulted in aggressive chases and two resulted in one-sided retreats (Ochiai, 1993). Furthermore, when chases went deep into neighbouring ranges, Ochiai (1993) observed them switch to the original chaser being chased, a reversal of dominance. Territorial chases were also observed in other study areas (Akasaka & Maruyama, 1977; Haneda *et al.*, 1979; Sakurai, 1981). Based on my occasional witness of territory holders with some injury, actual fights with horn contact between territory holders of the same sex may occur. This kind of fight, however, has never been observed, probably because, if it were to happen at all, it would occur very quickly.

The small degree of range overlaps and the small number of intrasexual encounters suggest that neighbours of the same sex avoid each other. This may be the result of both scent marking with preorbital

Table 10.1. *Territorial occupancy periods over a seven-year span in Akita*

	Period (years)							
	>6	5	4	3	2	1	<1	Total
Territorial males	0	3	4 (1)	7	2	10 (2)	6 (3)	32 (6)
Territorial females	4	5	4	6	1	4	0	24

Numbers indicate minimum length of territorial occupancy; numbers in parentheses indicate territorial occupancy from establishment to loss.
From Kishimoto & Kawamichi, 1996.

Table 10.2. *Average size of annual home ranges and average degree of home range overlap in territory holders in Akita*

				Average rate of home range overlaps (% ± SD range)			
				Intrasexual		Intersexual	
		N	Average size of annual home ranges ha ± SD Range	Largest overlap	2nd largest overlap	Largest overlap	2nd largest overlap
Social monogamy	Male	43	13.1 ± 6.6	13.5 ± 15.8	2.6 ± 3.8	66.7 ± 19.6	8.6 ± 15.3
			1.7–28.9	0–69.9	0–17.8	30.1–100.0	0–88.5
	Female	43	10.7 ± 4.9	7.8 ± 12.6	1.4 ± 2.5	80.5 ± 22.0	9.3 ± 14.5
			2.5–23.4	0–68.2	0–9.9	23.6–100	0–68.3
Social polygyny	Male	9	25.4 ± 6.9	13.9 ± 11.5	2.2 ± 2.0	52.2 ± 12.0	23.0 ± 9.0
			16.6–35.2	0–40.0	0–4.8	37.7–79.9	11.4–36.9
	Female A	9	14.2 ± 3.4	13.0 ± 13.2	5.0 ± 6.4	91.2 ± 7.6	17.2 ± 22.5
			8.4–20.6	0–46.4	0–20.2	75.2–100.0	0–73.5
	Female B	9	6.7 ± 3.4	23.8 ± 2.71	1.4 ± 1.9	90.6 ± 9.0	1.8 ± 3.3
			3.8–12.7	0–76.9	0–6.1	71.2–100.0	0–8.7

In socially polygynous groups, the degree of male home range overlap is larger for the ranges of female A than for the ranges of female B. Average degree of home range overlap for individual territory holders is analysed for the largest overlaps and the second largest overlaps; other smaller overlaps are omitted.

glands as well as territorial chases: territory holders frequently scent-mark with these glands, while non-territorial males and family offspring that have not yet established their territories scent-mark very little (Kishimoto, 1986). There has been no evidence that serows scent-mark along territory boundaries either with preorbital glands, horn rubbing, drops, or urination (Haneda *et al.*, 1979; Ochiai, 1983*b*; Baba *et al.*, 1996), unlike dik-diks, who mark the boundaries of their territories with dung piles (Hendrichs & Hendrichs, 1971).

Social mating units of territory holders

Because territory holders are tolerant of the opposite sex, the home ranges of opposite-sex adults overlap considerably, and mating units could be recognized by the patterns of intersexual range overlaps. In Akita, one territorial male range could almost completely overlap one or two female territorial ranges (Kishimoto, 1989*a*). As social mating units, male–female pairs and polygynous groups with two females were defined by these two overlapping patterns (Figure 10.1). The average range size of polygynous males was nearly twice that of paired males (Table 10.2, Student's t-test: $t = 4.70$, two-tailed $P < 0.001$). This was because ranges of polygynous males covered the ranges of two dispersed females, although one polygynous female's range tended to be much smaller than the other female's range (the average range size was significantly different between Female A and Female B in Table 10.2, Student's t-test: $t = 4.35$, $P < 0.001$). This may be one reason why some males occupied additional female ranges to become socially polygynous. In each year of the study, 9–13 pairs and 1–4 polygynous groups were observed, and the average annual proportion of polygynous territorial males was 18.6% (range 7.1–28.6%, $N = 7$ years). A total of 48 social mating units, including 39 pairs and nine polygynous groups, were observed during the study period (Table 10.3). Although the entire duration was known for only 23 of the 48 mating units, eight of the 39 pairs lasted more than three years, whereas none of the polygynous groups lasted this long, suggesting that territorial males had difficulty maintaining polygynous groups for long periods. The longest duration of a social pair was more than five years.

In Kusoudomari, of 36 social mating units, 23 were pairs, eight were polygynous groups with two females, and five were polygynous groups with three females (Ochiai, 1993). In the polygynous groups with three

Table 10.3. *Duration of pairs and polygynous groups over a seven-year span in Akita*

	Duration (years)							
	>6	5	4	3	2	1	<1	Total
Pairs	0	1	2 (1)	5 (1)	4 (1)	10 (5)	17 (11)	39 (18)
Polygynous groups	0	0	0	0	1	5 (3)	3 (2)	9 (5)

Numbers indicate minimum length of social mating units; numbers in parentheses indicate durations of social mating units from establishment to abandonment.
From Kishimoto & Kawamichi, 1996.

females, one female range either overlapped considerably one other female range or the other two female ranges, probably in some cases because of a process of territory establishment by female offspring (K. Ochiai, personal communication). The longest duration of a social pair was over 13 years for one pair, and duration for other social mating units was usually several years (Ochiai, 1993).

Male–female pairs were also found in other study sites such as Kasabori, where Akasaka and Maruyama (1977) observed family home ranges with an adult male–female pair and their offspring, Asahi Mountains (Kiuchi *et al.*, 1979, 1986), Nagano (Haneda *et al.*, 1979), and Sobo-Katamuki (Ono *et al.*, 1984). Haneda and Furihata (1978), however, reported that a subordinate adult male was tolerated within one female home range of a polygynous group with two females, and Sakurai (1981) also reported that a subordinate resident was tolerated within the home range of a male–female pair. These subordinate residents may have been mature offspring since some offspring remained within their mothers' ranges for over three years (see below).

Changes in social mating units

In Akita, changes in social mating units during the study period resulted either from the disappearance of territory holders and/or the transformation of home range boundaries (Table 10.4). A total of 21 territorial males and seven territorial females disappeared. Of the 18 paired males that disappeared, 11 and four were replaced by a male newcomer and a neighbouring paired male, respectively. When three polygynous males disappeared, male newcomers each took over one of the females in each of the polygynous groups to become a new pair, because the other female of the group remained solitary or disappeared. Of the seven territorial females that disappeared, three paired females were replaced by their mature daughters. At the time of their disappearance two females had no mature daughters in their ranges, and neighbouring territorial females expanded their ranges: one expansion occurred within a polygynous group, so that the mating unit changed from a polygynous group to a pair; the other coincidentally occurred in two neighbouring pairs where a paired male and a neighbouring paired female disappeared within two months, resulting in the fusion of two pairs into one pair. For the other two females without mature daughters, the type of mate changes are unknown.

Changes in social mating units by transformation of home ranges were observed in five pairs and four polygynous groups (Table 10.4; see Figure 10.2 for examples). Two paired males absorbed a neighbouring mature daughter during the process of her territory establishment. Three polygynous groups were intruded upon by male newcomers, resulting in the groups reforming into two pairs. One polygynous group was intruded upon by a neighbouring paired male, changing the social mating units from a polygynous group to a pair and from a pair to a polygynous group, respectively. Two female newcomers intruded on paired females' ranges to establish their territories, and consequently the pairs changed to polygynous groups.

In Kusoudomari, 12 changes of social mating units were observed (Ochiai, 1993). Three pairs changed to new pairs, replacing paired males or a paired female; three pairs changed to polygynous groups with two females, either by young females joining or by the absorption of a neighbouring female; and one pair changed to a polygynous group with three females by absorbing

Table 10.4. *Changes in social mating units following the disappearance of territory holders and/or transformation of home ranges in Akita*

Event	Change	New status	*N*
Disappearance of paired male	Pair to new pair	Replacement by male newcomer	11
	Pair to polygyny	Absorption of widowed female by neighbouring paired male	3
	Two pairs to pair	Fusion of two pairs into one pair	1[a]
	Unknown	New mate not found by widowed female	2
		Disappearance of widowed female	1
Disappearance of polygynous male	Polygyny to pair	Replacement by male newcomer and disappearance of one widowed female	2
		Replacement by male newcomer and new mate not found by one widowed female	1
Disappearance of paired female	Pair to new pair	Replacement by her mature daughter	3
	Two pairs to pair	Fusion of two pairs into one pair	1[a]
	Unknown	Disappearance of male mate	1
Disappearance of a polygynous female	Polygyny to pair	Expansion of the other polygynous female's range	1
	Unknown	Disappearance of male mate	1
Transformation of male home range	Pair to polygyny	Absorption of neighbouring mature daughter by paired male upon her establishment of territory	2
	Polygyny to two pairs	Intrusion into a polygynous female's range by male newcomer	3
	Inversion of pair and polygyny	Absorption of a polygynous female by neighbouring paired male	1
Transformation of female home range	Pair to polygyny	Intrusion by female newcomer into range of a paired female	2

[a] Same occasion.

two females from a neighbouring polygynous group. Three polygynous groups with two females changed to pairs when one female moved her range, one female died, and a male newcomer intruded, respectively. In two polygynous groups with three females, one female from each group moved her range away from the group.

Sexual behaviour between territory holders

Intersexual encounters of territory holders usually occurred between mates of the same social mating units because their home ranges almost completely coincided. However, non-mates (opposite sex members from different social units) occasionally encountered each other at the peripheries of the individual home ranges, and females were generally tolerant of non-mates as well as mates. In Akita, of 178 encounters, 170 (95.5%) were between mates and the other eight (4.5%) were between non-mates (Kishimoto & Kawamichi, 1996). Sexual behaviours such as naso-genital sniffing, flehming, foreleg kicking, butting, short chases, and/or mounting were observed in 57% (13% of which included mounting) of the encounters between mates and 50% of the encounters (25% of which included mounting) between non-mates. Based on these observations, there was no particular behaviour indicative of pair bonding. In Kusoudomari, Ochiai (1992) also reported a small number of encounters between non-mates in which behaviour was the same as that between mates, although mounting was not observed during the non-mate encounters. In a number of incidents observed in Akita, several non-territorial males retreated about 30–50 m from either a territorial female or a young female, as well as from territorial males, usually before the females recognized the non-territorial males. This suggests that there were no mating opportunities for the non-territorial males. At neither of the study sites did behavioural observations suggest any intersexual territoriality.

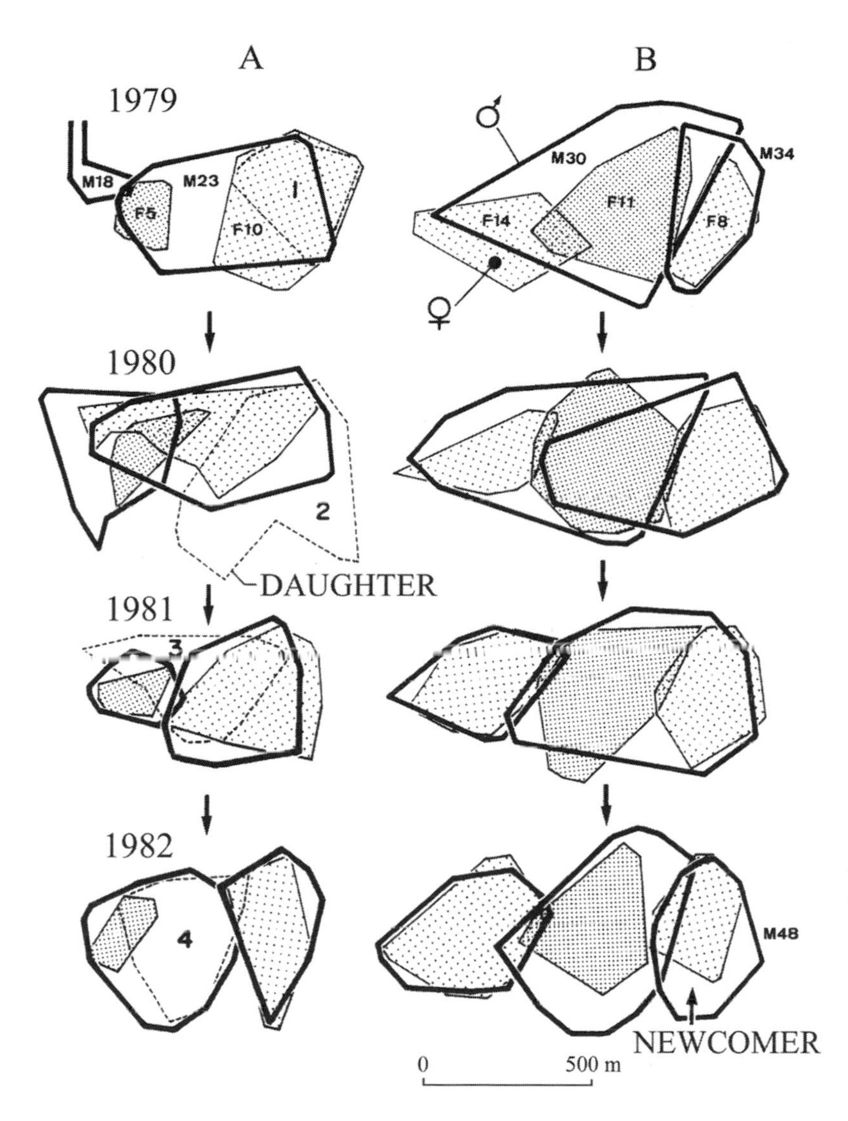

Figure 10.2. Two examples (A and B) of successive changes of home ranges and mates in four or five territorial males (M), and five or six territorial females (F) from 1979 to 1982 in Akita (from Kishimoto, 1989*a*). Females (shaded ranges) occupied almost the same areas throughout the four years. Broken lines show ranges of a daughter of female F10. Arabic numerals 1–4 indicate the daughter's age in each year. A: In autumn of 1980, non-territorial male M18 intruded into the range of socially polygynous male M23, and intercepted female F5, resulting in the formation of two social pairs in 1981. In 1982, the daughter took over part of her mother's range in order to establish her own territory, and M18 occupied this daughter's range, becoming socially polygynous. B: In autumn of 1980, socially paired male M34 intruded into the neighbouring range of socially polygynous male M30 and intercepted female F11, becoming socially polygynous. In 1982, male newcomer M48 intercepted female F8 of M34's socially polygynous group, and two social pairs were re-formed.

The reproductive rates of paired females and polygynous females were analysed in Akita (Kishimoto & Kawamichi, 1996). Each year, there were 9–11 pairs and 1–3 polygynous groups during the rutting season, and observations were conducted to determine whether or not the females gave birth during the following birth season. The reproductive rate was 78.5% for 65 females in pairs, and 72.7% for 22 females in polygynous groups ($\chi^2 = 0.31$, d.f. $= 1$, $P > 0.5$). Although genetic evidence for paternity was not analysed, these similar reproductive rates indicate the likelihood that polygynous males successfully inseminated both mates with nearly the same success rate as did paired males.

Social mating units including mature daughter

Offspring accompany their mothers for nearly a year after birth, and then begin to range solitarily in their natal areas until they either disappear or establish their own territories. In Akita, almost all offspring disappeared or dispersed between the ages of one and three years (Kishimoto, 1989*a*), and in Kusoudomari at age two or three (Ochiai, 1993). At both study sites, this disappearance or dispersal occasionally occurred at age four years of age. Accordingly, a mature daughter remaining within her mother's territory may be regarded as a member of the social mating unit since some daughters gave birth to an infant before establishing their own territory. If mature daughters are included in mating units, the rates of polygynous groups increase significantly. In Akita, eight of the 39 male–female pairs and two of the nine polygynous groups included a mature daughter. This type of mating unit was maintained for two years at most because all the daughters had either disappeared or established their own territories by the age of four. If mature daughters stayed, father–daughter mating may have resulted. However, this kind of mating, if it ever occurred, could not have continued indefinitely because fathers would eventually be replaced by male newcomers or neighbouring territorial males, although this might take several years (Table 10.1).

Territory establishment of offspring

In Akita, of the 20 male offspring that disappeared or became territorial between the ages of one and four, five established their own territories within the study area. Of these five males, three intruded into territories of males that had either injured their foreleg or had disappeared; the other two each took over one female from one of the polygynous groups. All five of these males formed a pair in the neighbouring area or about 1 km away from their natal areas. Of the 12 female offspring that either disappeared or became territorial between the ages of one and four, six remained in their natal areas to establish their own territories. Of these six, three took over their mothers' ranges after their mothers disappeared, and the other three took over a part of their mothers' ranges. Of the 21 offspring that disappeared in Akita, 15 were males and six were females. In Kusoudomari, two of 12 dispersed male offspring and four of nine dispersed female offspring established their territories on sites either contiguous with the mothers' range or about 500 m away (Ochiai, 1993).

Radiotelemetry studies

In the last years of the seven-year study period at Akita, colleagues and I conducted radiotelemetry studies with three territorial males and two territorial females within the study site of Akita (Furumoto *et al.*, 1986; Kishimoto, 1989*b*). This study clearly supported the results from the direct observation study in terms of social mating units. The average home range size in the snow season and non-snow season was 17 ha ($N = 9$, range 7.3–31.7 ha), and no significant seasonal changes in either size or location of home ranges were found. The ranges of the five territory holders overlapped a little with neighbouring ranges of the same sex, and all the mates of these social mating units were known. One socially paired male, one socially polygynous male, and one socially polygynous female continuously used 40–50% of their home range area in a 24-h period and continuously used over 80% in a 72-h period. The polygynous male took no more than 24 h to walk around the two female ranges.

The results of radiotelemetry studies, which later were also conducted in the other study areas, are noticeably different in the average sizes of home ranges (e.g., 109.2 ha for a male: Tano *et al.*, 1994; 99 ha (90% CP), $N = 16$ males, 87 ha (90% CP), $N = 15$ females: Okumura *et al.*, 1996) from the results of the direct observation studies. Although range sizes in direct observation studies seem to be somewhat underestimated, the noticeable differences in home range sizes detected by the two kinds of studies may be attributable to environmental factors such as high altitude and cooler habitat (Tano *et al.*, 1994) and the quantity and distribution of food resources (Miura *et al.*, 1996). Serows probably have large home ranges and there is considerable intrasexual range overlap in the low-density population because of poor food resources. Some residents without radio collars may not have been observed, and the social status of those collared serows whose ranges overlapped each other to a large degree remains unknown. On the other hand, because direct observation of low-density populations of serows is difficult, the radiotelemetry studies are expected substantially to advance the understanding of the spatial organization of this species.

DISCUSSION

Why is the solitary ungulate socially monogamous?

In the Japanese serow, social monogamy comprises three essential elements: (i) solitary living; (ii) intrasexual territoriality in both sexes; and (iii) coincidence between one male territory and one female territory. I will first describe these elements, and then consider why the Japanese serow usually exhibits social monogamy.

Solitary living

African antelopes inhabiting close habitats such as forests or thickets are solitary or form small groups (Estes, 1974; Jarman, 1974; Leuthold, 1977). Jarman

(1974) suggested that selective feeding on high quality parts of browse in close habitats promotes solitary living because it is not economical for large groups to feed on these parts, which are uniformly scattered rather than patchily concentrated like grass in open areas. Jarman (1974) and Estes (1974) also pointed out that solitary living in close habitats is advantageous for implementing anti-predator strategies, for example, hiding. This is true for the Japanese serow, which inhabits mountain forests, with the only stable groups being mother–kid units throughout the year and male–female pairs during the rutting season.

Male and female intrasexual territoriality

Forest habitats provide high-quality and very productive food resources for ungulates, so that the resource defence territory evolved, in which the year-round requisites of life are defendable in a small area (Geist, 1978, 1987). This certainly holds true for the Japanese serow, which defends almost entire home ranges as intrasexual territories throughout the year. In many species of ungulates, males establish their territories in order to attract females for mating (Gosling, 1986), but female Japanese serows independently defend their own territories because female territories are very stable, even when their mates are replaced. Only females expel female intruders, and females as well as males have the large preorbital glands that are used for scent marking, probably as a part of territory defence. Japanese serows are assumed to defend entire ranges in order to safeguard not just food resources but those other necessities for physical maintenance, for example, hiding areas, resting sites and, for breeding, birth sites (see Kishimoto, 1989*b*). Takatsuki *et al.* (1996) reported that the actual population density of Japanese serow is only 3.0–32.3% of the maximum carrying capacity, which is presumed from the mass of winter food, so that Takatsuki (1996) supposed that territory size might be determined not only from food supply but also other factors such as sociological features. For example, male serows may occupy large ranges so that they can guard multiple females.

Coincidence between one male territory and one female territory

In the Japanese serow, the socially monogamous pair is formed by the coincidence of range overlap between one male and one female. Kawamichi and Kawamichi (1979) defined this pairing pattern as a 'solitary-ranging pair' in the tree shrew (*Tupaia glis*). This pairing pattern is rare or has not been studied among solitary ungulates, but is otherwise observed in smaller mammals such as the tree shrew and the elephant shrews (*Rhynchocyon chrysopygus* and *Elephantulus rufescens*: Rathbun, 1979). In the Japanese serow, female territories are comparatively stable, while changes of social mating units usually result from replacement of territorial males or transformation of male home ranges, suggesting that males make their home ranges coincide with particular female ranges. If multiple territorial males divided one female range, competition among the males for approaching the female would inevitably result in actual fighting, and mating with the female would not be ensured for either male, regardless of his willingness and/or ability to pay such a high cost. On the other hand, if a male home range that is defended as a territory coincides with a female range, the male could defend not only all the resources necessary for physical maintenance, but at least one mate as well.

Male and female perspectives on social monogamy

From the male perspective, why does the Japanese serow generally exhibit social monogamy? According to Barlow (1988), polygyny is a possible strategy even when some males cannot obtain more than one female. This seems to apply to the Japanese serow. It may be difficult for males to obtain additional females because of the strict territorial boundaries of neighbouring males, but some males did successfully achieve social polygyny, usually by absorbing neighbouring females. In fact, 19% (Kishimoto, 1989*a*) and 36% (Ochiai, 1993) of territorial males were socially polygynous with two or three territorial females, and these rates were higher when a mature daughter remained within her mother's range. However, occupying the ranges of two or more dispersed females could be costly. Although polygynous males had to occupy the ranges of dispersed females, the radiotelemetry study in Akita indicated that both polygynous and paired males ranged over 80% of their home ranges during a continuous 72-h period, probably because they had to patrol and scent-mark their entire territories every few days in order to defend them. Furthermore, when a polygynous male followed one of his mates during the rutting season, his temporary absence from the range of the other female may have allowed other males to intrude. Therefore, it may be difficult for territorial males to maintain polygynous groups for long periods, although Ochiai (personal communication) observed that some polygynous groups with two females remained intact for six to seven years. Thus, although the male's strategy is polygyny, male serows may be constrained from occupying more than one female range because of the strict territoriality and high cost of guarding additional female ranges. When females are as dispersed as they are for the Japanese serow, monogamy develops. Conversely, when females are patchily distributed, there will be high potential for polygyny (Emlen & Oring, 1977; Wittenberger & Tilson, 1980). If that is the case, the female-biased philopatry in the Japanese serow may imply the next step in the social evolution of polygyny for matrilineal groups.

On the female side, why do females accept single, territorial males who usually exhibit social monogamy? Females may mainly benefit by acquiring good genes for their offspring from successful territorial males, but not from non-territorial males. Non-territorial males, even if they might be acceptable to females, do not seem to participate in mating at all, although we do not know why. In this sense, the intersexual coincidence of territories in the Japanese serow already guarantees females the availability of successful territorial males. Nevertheless, females also seem to accept neighbouring territorial males that appear by chance around the peripheries of their territories, even if such intersexual encounters are rare. To maximize their lifetime reproductive success, female solitary ungulates, especially those that can produce only single young annually, should be inseminated every year. Thus, the strategy of the female Japanese serow may be to accept any territorial males. However, strict intrasexual territoriality ties females to their own territories, and males, by acquiring territories overlapping those of territorial females, preclude females from mating with multiple males. This may be why females are limited to accepting single, territorial males.

This type of monogamy in the Japanese serow has not been confirmed in other solitary ungulates. The bushbuck (*Tragelaphus scriptus*) has a different social organization – two or more same-sex adults sometimes form groups without territoriality being a factor (Waser, 1975). In the common duiker (*Sylvicapra grimmia*), only males seem to hold intrasexual territories, and the territory does not coincide with the home ranges of females (Dunbar & Dunbar, 1979). Even in the Japanese serow, between-population variations in the mating system is assumed to be due both to the large home ranges and the overlaps detected by radiotelemetry studies. More detailed field investigations are necessary in order to understand factors such as food resources and population density, and to increase our knowledge of inter- and intraspecific variations in mating systems.

References

Akasaka, T. & Maruyama, N. (1977). Social organization and habitat use of Japanese serow in Kasabori. *Journal of the Mammalogical Society of Japan*, **7**, 87–102.

Baba, M., Doi, A., Iwamoto, T. & Manabe, T. (1996). *Report of Studies on Wildlife Management Technique for a Counterplan of Pest by a Special Natural Monument, the Japanese Serow.* Oita: The Board of Education of Oita Prefecture (in Japanese).

Barlow, G. W. (1988). Monogamy in relation to resources. In *The Ecology of Social Behavior*, ed. C. N. Slobodchikoff, pp. 55–79. San Diego: Academic Press.

Brotherton, P. N. M. & Manser, M. B. (1997). Female dispersion and the evolution of monogamy in the dik-dik. *Animal Behaviour*, **54**, 1413–24.

Brotherton, P. N. M. & Rhodes, A. (1996). Monogamy without biparental care in a dwarf antelope. *Proceedings of the Royal Society of London, Series B*, **263**, 23–9.

Brotherton, P. N. M., Pemberton, J. M., Komers, P. E. & Malarky, G. (1997). Genetic and behavioural evidence of monogamy in a mammal, Kirk's dik-dik (*Madoqua kirkii*). *Proceedings of the Royal Society of London, Series B*, **264**, 675–81.

Dunbar, R. I. M. & Dunbar, E. P. (1979). Observations on the social organization of common duiker in Ethiopia. *African Journal of Ecology*, **17**, 249–52.

(1980). The pairbond in klipspringer. *Animal Behaviour*, **28**, 219–29.

Emlen, S. T. & Oring, L. W. (1977). Ecology, sexual selection and the evolution of mating systems. *Science*, **197**, 215–33.

Estes, R. D. (1974). Social organization of the African Bovidae. *IUCN New Series*, **24**, 166–205.

Furumoto, H., Kishimoto, R., Kawamichi, T. & Maita, K. (1986). Activity and habitat use of the Japanese serow by radiotelemetry. *Abstracts of the 33rd Annual Meeting of the Ecological Society of Japan*, p. 164 (in Japanese).

Geist, V. (1974). On the relationship of social evolution and ecology in ungulates. *American Zoologist*, **14**, 205–20.

(1978). *Life Strategies, Human Evolution, Environmental Design.* New York: Springer.

(1987). On the evolution of the Caprinae. In *The Biology and Management of Capricornis and Related Mountain Antelopes*, ed. H. Soma, pp. 3–40. London: Croom Helm.

Gosling, L. M. (1986). The evolution of the mating strategies in male antelopes. In *Ecological Aspects of Social Evolution*, ed. D. I. Rubenstein & R. W. Wrangham, pp. 244–81. Princeton, New Jersey: Princeton University Press.

Hanawa, S., Maruyama, N., Nakama, S. & Mori, O. (1980). Ecological survey of Japanese serow *Capricornis crispus* in Wakinosawa Village. *Journal of the Mammalogical Society of Japan*, **8**, 70–7 (in Japanese with English abstract).

Haneda, K. & Furihata, T. (1978). Annual changes of family members and behaviour of the Japanese serow in Mt. Nozoko, Simoina District. In *Report of Urgent Research on Ecology of the Japanese Serow in Nagano*, Number 3, pp. 41–60. Nagano: Shinshu Group for Ecological Research on the Japanese Serow (in Japanese).

Haneda, K., Kira, T., Yoda, K., Muya, A. & Hashido, K. (1979). Report of ecological study of the Japanese serow. Nagano: Nagano Pref. Regional Forest Office (in Japanese).

Hendrichs, H. (1975). Changes in a population of dikdik, *Madoqua (Rhynchotragus) kirki* (Günther, 1880). *Zeitschrift für Tierpsychologie*, **38**, 55–69.

Hendrichs, H. & Hendrichs, U. (1971). Freilanduntersuchungen zur Ökologie und Ethologie der Zwerg-Antilope *Madoqua (Rhynchotragus) kirki* (Günther, 1880). In *Dikdik und Elefanten*, ed. H. Hendrichs & U. Hendrichs, pp. 9–75. Munich: R. Piper & Co. Verlag.

Ito, T. (1971). On the oestrous cycle and gestation period of the Japanese serow, *Capricornis crispus*. *Journal of the Mammalogical Society of Japan*, **5**, 104–8 (in Japanese with English abstract).

Jarman, P. J. (1974). The social organization of antelope in relation to their ecology. *Behaviour*, **48**, 215–67.

Kawamichi, T. & Kawamichi, M. (1979). Spatial organization and territory of tree shrews (*Tupaia glis*). *Animal Behaviour*, **27**, 381–93.

Kishimoto, R. (1986). Territoriality and scent-marking in the Japanese serow. *Abstracts of the 33rd Annual Meeting of the Ecological Society of Japan*, p. 163 (in Japanese).

(1987). Family break-up in Japanese serow, *Capricornis crispus*. In *The Biology and Management of Capricornis and Related Mountain Antelopes*, ed. H. Soma, pp. 104–9. London: Croom Helm.

(1988). Age and sex determination of the Japanese serow, *Capricornis crispus*. *Journal of the Mammalogical Society of Japan*, **13**, 51–8.

(1989*a*). Social Organization of a Solitary Ungulate, Japanese Serow, *Capricornis crispus*. Ph.D. thesis, Osaka City University, Osaka.

(1989*b*). Early mother and kid behavior of a typical "follower", Japanese serow, *Capricornis crispus*. *Mammalia*, **53**, 165–76.

Kishimoto, R. & Kawamichi, T. (1996). Territoriality and monogamous pairs in a solitary ungulate, the Japanese serow, *Capricornis crispus*. *Animal Behaviour*, **52**, 673–82.

Kita, I., Sugimura, M., Suzuki, Y., Tiba, T. & Miura, S. (1987). Reproduction of female Japanese serow based on the morphology of ovaries and fetuses. In *The Biology and Management of Capricornis and Related Mountain Antelopes*, ed. H. Soma, pp. 321–31. London: Croom Helm.

Kiuchi, M., Kudo, H., Kato, S., Yoshida, M., Miyasaka, M., Hoshino, M. & Yamazaki, K. (1978). On Japanese serow in Asahi mountain ranges. In *Conservation Report of Japanese Serow, Special Natural Monuments*, pp. 27–93. Tokyo: Nature Conservation Society of Japan (in Japanese).

Kiuchi, M., Kudo, H., Yosida, M., Miyasaka, M., Hoshino, M., Yamazaki, K., Kato, S. & Umezu, C. (1979). Social organization and habitat use of the Japanese serow in Asahi mountain ranges. In *Conservation Report of Japanese Serow, Special Natural Monuments*, Volume 2, pp. 5–72. Tokyo: Nature Conservation Society of Japan (in Japanese).

Kiuchi, M., Yoshida, M., Yoshida, K., Okasaka, Y., Kazamaki, E., Furihata, S., Kaga, M., Ishimoto, A. & Yaita, M. (1986). *Ecology of the Japanese Serow in Asahi Mountain Ranges*, Volume 3. Tokyo: Nature Conservation Society of Japan (in Japanese).

Kodera, S., Suzuki, Y. & Sugimura, M. (1982). Postnatal development and histology of the infraorbital glands in the Japanese serow, *Capricornis crispus*. *Japanese Journal of Veterinary Science*, **44**, 839–43.

Komers, P. E. (1996). Obligate monogamy without paternal care in Kirk's dikdik. *Animal Behaviour*, **51**, 131–40.

Komori, A. (1975). Survey on the breeding of Japanese serows, *Capricornis crispus*, in captivity. *Journal of the Japanese Association of Zoological Gardens and Aquaria*, **7**, 53–61 (in Japanese).

Leuthold, W. (1977). *African Ungulates. Zoophysiology and Ecology*, Volume 8. Berlin: Springer-Verlag.

Miura, S. (1986). Body and horn growth patterns in the Japanese serow, *Capricornis crispus*. *Journal of the Mammalogical Society of Japan*, **11**, 1–13.

Miura, S., Ito, T. & Okumura, H. (1996). Dispersal behavior of the Japanese serow. In *The Japanese Serow in Nishi-Zao: Report of Studies on Management Technique in the Conservation Area for a Special Natural Monument, the Japanese Serow*, pp. 91–100. Yamagata: The Board of Education of Yamagata Prefecture & Committee of Studies on Management Technique in the Conservation Area for a Special Natural Monument, the Japanese Serow (in Japanese).

Ochiai, K. (1983*a*). Pair-bond and mother–offspring relationships of Japanese serow in Kusoudomari, Wakinosawa

Village. *Journal of the Mammalogical Society of Japan*, **9**, 192–203 (in Japanese with English abstract).

(1983*b*). Territorial behavior of the Japanese serow in Kusoudomari, Wakinosawa Village. *Journal of the Mammalogical Society of Japan*, **9**, 253–9 (in Japanese with English abstract).

(1992). *Life History of the Japanese Serow*. Tokyo: Dobutsusha (in Japanese).

(1993). Dynamics of Population Density and Social Interrelation in the Japanese Serow, *Capricornis crispus*. Ph.D. thesis, Kyushu University, Fukuoka.

(1999). Diet of the Japanese serow (*Capricornis crispus*) on the Shimokita Peninsula, northern Japan, in reference to variations with a 16-year interval. *Mammal Study*, **24**, 91–102.

Okumura, H., Ito, T. & Miura, S. (1996). Home ranges and dispersal of the Japanese serow in Takiyama district, Yamagata City: behavioral analysis by radiotelemetry. In *The Japanese Serow in Nishi-Zao: Report of Studies on Management Technique in the Conservation Area for a Special Natural Monument, the Japanese Serow*, pp. 19–55. Yamagata: The Board of Education of Yamagata Prefecture & Committee of Studies on Management Technique in the Conservation Area for a Special Natural Monument, the Japanese Serow (in Japanese).

Ono, Y., Doi, A. & Nagayama, Y. (1984). *Ecological Studies on Japanese Serow in Sobo-Katamuki Range of Central Kyushu*. Oita: The Board of Education of Oita Prefecture (in Japanese).

Rathbun, G. B. (1979). The social structure and ecology of elephant-shrews. *Zeitschrift für Tierpsychologie*, **20** (supplement), 1–77.

Sakurai, M. (1981). Socio-ecological study of the Japanese serow, *Capricornis crispus* (Temminck) (Mammalia; Bovidae) with special reference to the flexibility of its social structure. *Physiology and Ecology Japan*, **18**, 163–212.

Sugimura, M., Suzuki, Y., Kamiya, S. & Fujita, T. (1981). Reproduction and prenatal growth in the wild Japanese serow, *Capricornis crispus*. *Japanese Journal of Veterinary Science*, **43**, 553–5.

Sugimura, M., Suzuki, Y., Kita, I., Kodera, S. & Yoshizawa, M. (1983). Prenatal development of Japanese serows, *Capricornis crispus*, and reproduction in females. *Journal of Mammalogy*, **64**, 302–4.

Suzuki, K. & Takatsuki, S. (1986). Winter food habits and sexual monomorphism in Japanese serow. *Proceedings of the Biennial Symposium of the Northern Wild Sheep & Goat Council*, **5**, 396–402.

Takatsuki, S. (1996). The relation between food habitats and territoriality of the Japanese serow. In *The Japanese Serow in Nishi-Zao: Report of Studies on Management Technique in the Conservation Area for a Special Natural Monument, the Japanese Serow*, pp. 101–6. Yamagata: The Board of Education of Yamagata Prefecture & Committee of Studies on Management Technique in the Conservation Area for a Special Natural Monument, the Japanese Serow (in Japanese).

Takatsuki, S. & Suzuki, K. (1984). Status and food habits of Japanese serow. *Proceedings of the Biennial Symposium of the Northern Wild Sheep & Goat Council*, **4**, 231–40.

Takatsuki, S., Osugi, N. & Ito, T. (1988). A note on the food habits of the Japanese serow at the western foothill of Mt. Zao, northern Japan. *Journal of the Mammalogical Society of Japan*, **13**, 139–42.

Takatsuki, S., Takemura, K., Abe, M. & Horie, T. (1996). Food supply in the Japanese serow's habitat in Yamagata. In *The Japanese Serow in Nishi-Zao: Report of Studies on Management Technique in the Conservation Area for a Special Natural Monument, the Japanese Serow*, pp. 69–80. Yamagata: The Board of Education of Yamagata Prefecture & Committee of Studies on Management Technique in the Conservation Area for a Special Natural Monument, the Japanese Serow (in Japanese).

Tano, N., Mochizuki, T., Furubayashi, K. & Kitahara, M. (1994). Ecological study of the Japanese serow (*Capricornis crispus*) in sub-alpine zone (I): home ranges. *Transactions of the Japanese Forestry Society*, **105**, 543–6 (in Japanese).

Tiba, T., Sato, M., Hirano, T., Kita, I., Sugimura, M. & Suzuki, Y. (1988). An annual rhythm in reproductive activities and sexual maturation in male Japanese serows (*Capricornis crispus*). *Zeitschrift für Säugetierkunde*, **53**, 178–87.

Tilson, R. L. & Tilson, J. W. (1986). Population turnover in a monogamous antelope (*Madoqua kirki*) in Namibia. *Journal of Mammlogy*, **67**, 610–13.

Waser, P. M. (1975). Spatial associations and social interactions in a "solitary" ungulate: the bushbuck, *Tragelaphus scriptus* (Pallas). *Zeitschrift für Tierpsychologie*, **37**, 24–36.

Wittenberger, J. F. & Tilson, R. L. (1980). The evolution of monogamy: hypotheses and evidence. *Annual Review of Ecology and Systematics*, **11**, 197–232.

PART III

Reproductive strategies of human and non-human primates

CHAPTER 11

Ecological and social complexities in human monogamy

Bobbi S. Low

INTRODUCTION

The term 'monogamy' can mean a number of things. Genetic monogamy is unlikely unless it raises (male) reproductive success enough to compensate for the loss of reproductive success (RS) that would have come from additional mating efforts. We expect to see male-parental polygyny under these conditions: (i) whenever a female can raise offspring successfully alone, and (ii) when male care that sufficiently enhances offspring survivorship and competitive success can be 'generalizable' – no more expensive for several offspring than for one (e.g., a nest or den that can serve several clutches or litters). Consider red-winged blackbirds, in which males watch and warn at the approach of potential predators; a male can do this effectively for several nests, at the same cost as for one.

For a system to be genetically monogamous, then, it is important that the male care be a non-generalizable, true parental investment such as feeding (Trivers, 1974), rather than a more general parental effort (such as a nesting den, which can function for numerous offspring: Low, 1978). Other routes to monogamy also exist: in some species, both mate guarding (that eliminates other mating chances), and the ecology of female dispersion may make males unable to monopolize more than one female, and therefore lead to monogamy (Jarman, 1974; Emlen & Oring, 1977). But when females are controllable and/or must be guarded constantly, or when male effort can be generalized, we see either open polygyny or social monogamy without true genetic monogamy.

Human mating systems are particularly interesting. The extent of within-species variation is extraordinary. Most of the mating systems known in other species occur within the single species *Homo sapiens*, making comparisons telling. There are some complications, however: actual mating data may be more difficult to come by for humans than for many other species; and anthropological definitions of 'monogamy' are actually about marriage rules rather than mating patterns (see Ecological and cultural definitions of monogamy, below).

Paternal care spent on other men's offspring is costly. Because human infants are dependent for a long period and care is expensive, both male-parental investment and mate guarding are especially likely to be important correlates of monogamy in human societies. Nonetheless, human males in many societies appear to have solved several problems that might lead to genetic monogamy without relinquishing polygyny. A man need not guard his wife if he has family (or eunuchs) to do so. Mate guarding is clearly present and effective in many polygynous societies. Human residence patterns mean that men are able to monopolize multiple females under most conditions. Most societies are patrilocal, so that a man lives among his kin and a woman comes to live among her husband's kin. He need not travel to her; and groups of related males can effectively defend resources, including wives.

Some patterns in human mating and marriage systems (below) reflect (i) intensification of human sociality and intergroup conflict; (ii) elaboration of non-parental (mostly nepotistic) care and associated shifts in human life history; and (3) extraordinary elaboration of non-kin 'third-party' influences in human social patterns.

Effective male-parental care might lead to monogamy when resources are limited. Harsh and unproductive habitats may mean that men can do better reproductively by helping to raise a child with true parental investment (Trivers, 1974), rather than continuing their mating efforts in male-parental polygyny. We expect density and sociality in such environments to be limited, and we expect to see relatively little variation in the resources controlled by individuals. Nonetheless, reproductive variance may be high (although not differing between the sexes: Clutton-Brock, 1983) because failure is likely in harsh environments (Dyson-Hudson & Smith, 1978).

Alexander *et al.* (1979) called this kind of monogamy 'ecologically-imposed monogamy'; it is also known as genetic monogamy and as sexual monogamy. In some monogamous species, for example Adélie penguins, ecological circumstances may render any defection costly. As a result, these systems will also be genetically, as well as socially, monogamous. Birds are more likely to be 'ecologically' (genetically) monogamous, but even many supposedly ecologically monogamous birds turn out to be only socially monogamous, with extra-pair paternity in many cases. In humans, too, even in some extreme environments, reproductively profitable extra-pair copulations are likely (see, e.g., Birkhead & Møller, 1998; Birkhead, 2000). The behaviour of men in many societies reflects these conditions. The relatively high level of men's parental investment means that mate guarding to preclude copulation outside marriage is very common, and many societies have social penalties (third-party interventions) against extra-pair copulations. Clearly the costs and benefits differ for men (who may be forced to contribute to their non-marital children) and women (who may lose paternal investment from their spouse).

In whose interest?

Third-party interests complicate matters. Third parties, particularly kin, can influence social patterns in many species; so, too, with humans. Parents may desire particular marriages for their children in order to form alliances with wealthy and powerful families; marriageable children may have other priorities. In some societies, children are betrothed well before puberty; in a few, even before birth.

Humans have further expanded the influence of third-party actors beyond kin influences to include enforcement of social rules by 'disinterested' parties. Social and cultural inventions such as marriage, divorce, and inheritance involve the interests of other individuals – sometimes individuals with no apparent interest in the particular decision (caliphs, ministers, shamans, etc.). Such human-invented social conventions, like physical resource conditions, affect the costs and benefits of monogamy for individuals. Thus, the occurrence and prevalence of monogamy in humans may well follow complex rules, as ecological and sociocultural complexities interact.

Conflict of interest

Reproductive conflicts of interest between the two sexes add further complexity and variation within humans. Women seldom fully share men's reproductive interests, especially in polygynous systems. Further, each partner has kin who may act to shift husbands' and wives' reproductive costs and benefits. Males appear to strive for polygyny, whether as social polygyny or social monogamy with extra-pair copulations (EPCs), whenever possible. Social polygyny is even more predominant when females can be successful, independent units with their offspring.

In genetically monogamous systems, the parents' reproductive interests are identical. In socially monogamous systems generally, the reproductive interests of parents converge to a degree. Yet in humans this does not always protect women's reproductive interests; the complexities of human systems mean that women's interests do not always converge completely with their husbands' even in socially monogamous situations. Consider socially monogamous modern Thailand. Surveys of men's and women's attitudes about sexual matters (Knodel *et al.*, 1997) highlight this sexual conflict of interests. A long tradition exists of men, married as well as single, visiting sex workers in the context of men's social evenings out. In addition, some men have 'minor wives' whom they support at least partially. Thai wives have historically had little ability to constrain these behaviours, and have held a long-standing preference that their husbands, if sexually active outside marriage, visit sex workers because the transaction is brief and represents only a minor diversion of resources from the wife and children, in contrast to a minor wife, who can represent a real drain of resources. Interestingly, with the spread of HIV virus, an increasing proportion of women are responding to this sexual conflict of interests by preferring the husband to have a minor wife: the resource conflict is heightened, but the health risks are reduced.

ECOLOGICAL AND CULTURAL DEFINITIONS OF MONOGAMY

When we investigate monogamy in humans, there is an additional empirical difficulty: we have data on *marriage* systems rather than *mating* systems, and these are not identical (e.g., Betzig, 1997). Human marriage is a social institution involving not just mating patterns but societal rules about allowed number of spouses at any one time, allowable consanguinity in pair bonds, allowable ages of mates, and so on.

'Monogamy' therefore has different meanings for biologists and social anthropologists (e.g., Wickler & Seibt, 1981). Marriage rules, like many other cultural institutions, concern third-party interests (see below); they are not simply about mate choice and parental care.

One central biological aspect of mating system definitions is the relative *variance in reproductive success* between the sexes, clearly a genetic concept. In mammals, especially, since females are equipped to feed dependent offspring, there is a bias toward polygyny, in which variance in reproductive success of males exceeds that of females (review by Low, 2000, Chapters 3, 4). In polygynous mating systems, because only a few males ever acquire a mate, great expenditure and risk taking may be worth while. As a result, in polygynous mating systems more males die than females at most ages. Being a male is a high-risk, high-gain strategy, and for a female, producing a son is also a costly high-risk, high-gain strategy.

Behavioural ecologists use the terms 'monogamous' and 'polygynous' to focus on the genetic impact of sexual selection. Unfortunately, the focus of social anthropological definitions ensures that we have less ability to understand human mating patterns in detail than we have for other species. Indeed, many attempts at categorization (e.g., Murdock, 1967, 1981; Murdock & White, 1969) use labels like 'monogamous/mildly polygynous'. Further, while in some other species we can watch copulations, this is forbidden in most (perhaps all?) human societies. We suspect that there will be extra-pair paternity in humans, and that few if any societies will be genetically monogamous. Not surprisingly, modern data suggest that in humans, extra-pair copulations can result in a mismatch between biological paternity and social paternity (e.g., Baker & Bellis, 1995; Morrell, 1998; Baker, 2000).

Social anthropological definitions of monogamy have thus only a loose and ill-defined relationship to behavioural ecological definitions. This difficulty is exacerbated, because for the variance-related definitions, we need to know about the 'zero success class' – those who fail to reproduce (Conover, 1980) – in order to calculate variance; anthropologists seldom report this category. In addition, others besides the potential mates have interests in, and can act to influence, outcomes, contributing more complications. Others besides the woman may choose her mate for her. Martin King Whyte (1978) found that in the odd-numbered societies of the Standard Cross-Cultural Sample, older relatives, men and women, as well as the bride and groom, had a say in arranging marriages. In the majority of 93 societies examined by Whyte, either men had more say than women (32/93 for the older generation's influence on the younger generation's marriages; 27/93 for bride's vs. groom's influence on their own marriage), or the two sexes had roughly equal say (29/93 for older generation; 46/93 for bride vs. groom). Thus, interested third parties, such as parents and other older relatives of the potential mates, influence marriage patterns.

Many socially monogamous societies – societies with one-spouse-at-a-time rules – would be polygynous in a biological or genetic definition: more men than women fail to marry, and more men than women remarry after death or divorce, producing families in these later unions. The most reproductively successful men have many more children than the most fertile women. All of these phenomena increase the variance of men's reproductive success compared with women's. However, we seldom have the information to calculate the biologically important relationships from anthropological data.

When men's sources of power are unpredictable, and women have sufficient resources to be independent, men cannot always control women. Resource distribution, coalitions for mate competition, inbreeding avoidance, and nepotism (familial coalitions) are all important in marriage patterns. In such societies, 'serial monogamy' (really serial polygyny) results, just as in other polygynous systems, in relatively high variance in men's reproductive success. This is the case, for example, in South America among the Ache, as well as the Cuna Indians, and in the contemporary USA (Nordenskiold, 1949; Hill & Hurtado, 1996). Ache men and women have perhaps ten spouses in a lifetime, the Cuna four or five. The origins of socially-imposed monogamy appear to arise from the pressures of group living, interacting with historical particulars (e.g., Alexander, 1979). As societies increase in size, under many conditions it becomes more difficult for powerful (and thus polygynous) men to maintain reproductive advantages openly.

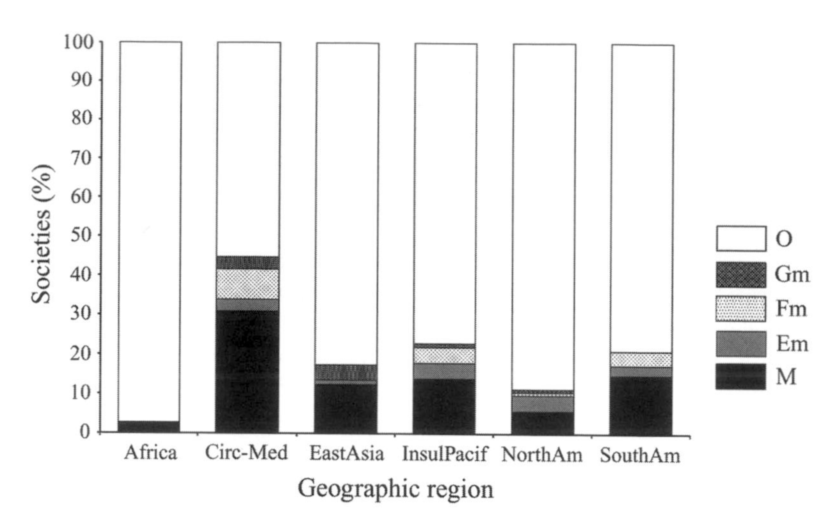

Figure 11.1. Distribution of socially monogamous marriage and family patterns around the world (data from Murdock, 1981). M, independent nuclear monogamy; Gm, minimal extended or stem families; Fm, small extended families; Em, large extended families; O, other (e.g., polygyny, polyandry).

Because we expect cultural diffusion of social patterns, it is not surprising that geographic biases exist in marital organization. For example, although 'independent monogamous families' (the couple lives in their own household) appear in all six of Murdock's geographic categories (Africa, circum-Mediterranean, European, Insular Pacific, North and South America), they are unevenly distributed (Figure 11.1).

Here I will first examine the geographic distribution and ecological correlates of monogamy in human societies, then the social associates and correlates of monogamy. Because most data come from anthropological investigations, and anthropologists focus on cultural aspects, it can be difficult to distinguish anthropological (socially-imposed) 'monogamy' from ecologically-imposed monogamy (e.g., Low, 1988*a*), although evolutionary anthropologists (e.g., Hill & Hurtado, 1996) do include measures of reproductive success. When possible, I will focus on variance-related definitions of monogamy; I will try always to indicate clearly which definition applies.

SOURCES FOR EXAMINING CROSS-CULTURAL VARIATION IN HUMAN SYSTEMS

Our information about mating and marriage patterns comes from several sources. Cross-cultural summaries exist in the Human Relations Area Files (*c.* 1400 societies); this comprises microfiche copies of ethnographies, both in the original form, and cross-referenced by hundreds of topics. Thus, if one asks about marriage rules, one can look up the 'marriage' code for each society of interest, and every relevant page within each ethnography is copied there. There is a minimal effort at quality assessment: the occupation of the reporter, the dates and length of time s/he spent with the society, and the date of publication are given.

Two publications, the *Ethnographic Atlas* (Murdock, 1967) and the *Atlas of World Cultures* (Murdock, 1981) summarize some information for 862 and 563 societies, respectively. The *Standard Cross-Cultural Sample* (Murdock & White, 1969) covers 186 societies, and is restricted to societies studied by qualified ethnographers living with the society for a substantial period of time. To minimize distortions that might arise because of practices spread by cultural diffusion, this sample is stratified by geographic region, and within region, by language group. It is organized in a geographic order, so if one works with a half sample (93 societies), either even- or odd-numbered societies are used.

Most detailed, of course, is research by evolutionary anthropologists, who ask questions in a manner parallel to that of behavioural ecologists (e.g., Blurton Jones, 1986, 1987, 1989, 1997; Borgerhoff Mulder, 1988*a*, *b*, 1990, 1991, 1992, 1997; Hawkes, 1991, 1993; Hill & Hurtado, 1996; Hawkes *et al.*, 1997*a*, *b*, 1998; Strassmann, 1997, 2000).

All of these studies trade detail, precision, and accuracy against sample size. I am using data here primarily from the *Standard Cross-Cultural Sample* and the *Atlas of*

World Cultures, with information added where possible from evolutionary anthropologists' ethnographies.

THE CROSS-CULTURAL DISTRIBUTION OF SOCIAL MONOGAMY IN HUMAN SOCIETIES

Of the 563 societies in the *Atlas of World Cultures*, 100 (17%) are coded as having some form of monogamy; 66 (11.7%) as independent monogamous nuclear families (husband, wife, and children live in own household); 14 (2.5%) as large extended monogamous families; 13 (2.3%) as small extended monogamous families (both these present various combinations of additional non-descendant relatives cohabiting); and seven (1.2%) as monogamous stem families (several generations of descendants cohabit). In all cases, the anthropological definition of (social) monogamy prevails, and these cases are unevenly distributed geographically (Figure 11.1). However, because there are only 100 cases and 24 possible conditions (type × location), no statistical treatment is meaningful. Almost 45% of living arrangements in 'circum-Mediterranean' societies are coded as some form of monogamy, and four forms are described. In contrast, only 2.7% of African societies are considered monogamous, and these three societies are all independent nuclear families.

Monogamy arises in many species when ecological conditions sufficiently modify the costs and benefits for males of paternal care versus mating effort. Human sociality has done much to change the impact of these ecological conditions, both generally and under specific socioecological conditions. The list of clear ecological correlates of human monogamy is shorter than the list of socioecological and cultural–historical interactions.

ECOLOGICAL CORRELATES OF SOCIALLY MONOGAMOUS SYSTEMS

Marriage systems and environmental richness

Many ethnographies suggest that social monogamy is more common in relatively harsh ecological conditions (e.g., very dry, very cold, low plant productivity), which perhaps might make monogamous, two-parent care reproductively advantageous. Although the examples are suggestive, there are no clear statistical patterns. In part, this arises because (i) there are few 'monogamous' societies in the *Standard Cross-Cultural Sample*, and (ii) the definitions of marriage systems for the societies in the *Sample* are anthropological: 'monogamous' means monogamous to mildly polygynous. ('Mildly polygynous' means that polygynous families live openly; it does not include EPCs.)

In the *Standard Cross-Cultural Sample*, highly polygynous societies are found in areas of the world with significantly higher plant productivity (one measure of environmental richness) than others; there is no difference between the plant productivity in 'monogamous' and mildly polygynous societies (Low, 1990*a*; review in Low, 2000). Because we cannot distinguish between the anthropological and the ecological definitions of monogamy, it is hardly surprising that 'monogamous' (anthropological definition, i.e., roughly socially monogamous) societies show no difference from mildly polygynous ones. Reading the ethnographic descriptions of societies deemed 'monogamous' leads one to conclude that even when a few men manage (through skill in hunting, or tapping into novel sources of income) to be polygynously married when most men remain monogamous, it occurs in habitats with a poor resource base, that is, insufficient for many men to manage to gain more than a single wife. Nonetheless, the statistical results arise from the distribution of the highly polygynous societies, not the monogamous ones.

Other measures of ecological stress that might be expected to show a pattern with social monogamy, if human systems responded in ways similar to others species, show little pattern (Low, 1990*a*). Marriage (reduced to 'monogamous,' $<20\%$ polygyny, and $\geq 20\%$ polygyny) shows no associations with heat, cold, wet, or dry extremes (Low, 1990*a*), with hot–wet, hot–dry, cold–wet, cold–dry extremes, or with the range of variation in temperature and rainfall combinations.

Measures of the predictability of rainfall (Colwell, 1974; Low, 1990*a*) do vary with the marriage system. Although there is no pattern with overall predictability of rainfall (Low, 1990*a*), monogamy is more likely as constancy (C) of rainfall increases, and marginally less likely as contingency (M), or seasonality, increases (ANOVA P: d.f. $= 2{,}54$, $F = 0.195$, $P = 0.82$; C: d.f. $= 2{,}54$, $F = 4.11$, $P = 0.02$; M: d.f. $= 2{,}54$, $F = 2.73$, $P = 0.07$). It is difficult to interpret this relationship.

Using cross-cultural measures of the relative contribution of various resources to subsistence (Murdock, 1967, 1981), there are few relationships between subsistence and marriage. Monogamy is associated with the contribution of fishing to subsistence: the more important is fishing, the more likely is monogamy to occur ($N = 179$, $\chi^2 = 11.92$, d.f. $= 2$, $P = 0.03$). No other subsistence contribution (gathering, hunting, animal husbandry, agriculture) showed any pattern with marriage. Measures of perceived stress also showed no pattern: risk of starvation, protein deficiency, general resource scarcity (as reported by subjects in the *Standard Cross-Cultural Sample* to ethnographers), and perceived chronic shortages. In summary, it is difficult to find measures of ecological stress that reflect any pattern, even with social monogamy.

Within monogamous societies, it could be rewarding to explore the strength of ecological influences on family organization, which might be more responsive to ecological constraints than monogamy itself. Social monogamy in humans includes many forms. Some of these seem likely to have quite different biological impacts. Of the 563 societies in the *Atlas of World Cultures*, only 66 (just under 12%) are coded as independent nuclear families. Note, however, that this designation does not exclude societies with sex biases in remarriage, or additional parenthood from EPCs. Other socially monogamous familial arrangements include large extended families (coded *Em* in Murdock's work), small extended families (coded *Fm*), and minimal extended or 'stem' families (coded *Gm*). Currently, we have only descriptive information; are there ecological correlates of family formations, rather than the marriage system itself? Do larger versus smaller, extended families, for example, associate with different subsistence forms?

This exploration raises the question of how we can see an association of monogamy and 'harsh' environments, yet not be able to discern clear patterns of monogamy and any specific environmental variable. I think three factors contribute to this phenomenon. First, social monogamy is not all that common. Second, conditions requiring two-parent care, or promoting intensive male mate guarding (e.g., because of female distribution), are unlikely to be singular (e.g., all rainfall or temperature only). Finally, it is clear that in a primate both as social and as intelligent as humans, there are many routes to solving the ecological dilemmas that in other species restrict males to monogamy. For example, if food-garnering by more than the mother is really profitable, one might expect monogamy – but, in fact, what we often find is that other adults (often the father's female relatives) contribute considerable amounts of food. In sum, men have found many routes to avoid monogamy, resulting in diverse socioecological patterns.

Monogamy and pathogen stress

One ecological variable with a variety of social and sexual correlates is pathogen stress. W. D. Hamilton and colleagues (Hamilton, 1980; Hamilton *et al.*, 1981; Hamilton & Zuk, 1982; Ebert & Hamilton, 1996) suggested that pathogen stress could be involved in the evolution of sexual reproduction itself. Low (1988*b*, 1990*b*) suggested that pathogen stress could be one influence in sexual selection within human systems. Monogamy is absent in areas of high pathogen stress. Polygyny, no matter how it is measured, increases with pathogen stress. Level of pathogen stress alone accounts for 28% of the variation in the degree of polygyny around the world, independent of geographic region or any other factor.

The *type* of polygyny (non-sororal vs. sororal) is further correlated with pathogen stress. Within polygyny, offspring of sisters are more alike than offspring of otherwise related women; thus this pattern is consistent with the 'production of variable offspring in pathogen-laden environments' hypothesis.[1] The degree of polygyny is really a threshold pattern, rather than a linear relationship. Plotting the percentage of men polygynously married shows data points apparently randomly scattered, except for one corner of the graph where data would fall for societies in which few to no men were polygynous but in which pathogen stress was high. Quite simply, in very high pathogen-stress areas, monogamy is absent.[2]

These are intriguing speculations, but a caveat is important. Correlation and association say nothing about causality. All these patterns, for example, do not eliminate another hypothesis: that social systems in which density is high and interindividual contact is frequent, pathogens will provide an ecological challenge. And we cannot yet distinguish the direction of causality.

Ecological–cultural interactions

As noted above, in most species, when monogamy is associated with biparental feeding, care, and protection, males profit by trading off mating opportunities for enhanced offspring survival and success. But human cultural innovations render this association less strong; the presence of kin and others means that multi-parent care can exist in highly polygynous systems. Fathers may hunt, for example, but much meat is shared across families. And others besides the parents contribute to the care, feeding, and protection of a child, including older siblings (Turke, 1988), grandmothers (Hawkes *et al.*, 1997*a*, 1998), and reciprocating adults who may, or may not, be relatives.

For a primate, human life history patterns are odd (Hill & Kaplan, 1999; Low, 2000, pp. 92–112). For example, offspring are weaned much earlier (age three or four rather than age seven) than would be predicted from human size, while fertility is delayed (about age 15 rather than age eight). Kaplan *et al.* (2000) suggest that this multi-person care, involving the acquisition of large game and other foods that require great skill, may even contribute to the evolution of such oddities. At any rate, multi-adult care frees men for polygyny, even when a female and her offspring do not, in fact, operate as an independent ecological–economic unit.

SOCIAL PATTERNS ASSOCIATED CROSS-CULTURALLY WITH MONOGAMOUS SYSTEMS

Although monogamy is relatively rare, and ecological correlates of marriage patterns are weak, the marriage system itself has multiple social impacts, from inheritance to sexual signalling.

Monogamy, polygyny, and inheritance rules

Hartung (1982, 1983, 1997) found that inheritance tends to be male-biased in polygynous systems; in such systems, resources influence male reproductive success more than female reproductive success. In bride price systems, for example, young, high-reproductive-value women may cost more than older women (Borgerhoff Mulder, 1988*b*), with the result that wealthier men can have more, and younger, wives than other men. Cowlishaw and Mace (1996), using a phylogenetic approach, confirmed this pattern, noting that marriage patterns and inheritance 'evolve together in a way that is adaptive' so that strongly polygynous systems have male-biased inheritance and monogamous ones are more characterized by an absence of sex-biased inheritance.

The absence of sex-biased inheritance does not mean an absence of male striving and competition. Resource heritability should be inversely related to striving behaviour. When heritability of resources is high, then the opportunity for gain within strata is likely to be low and competitive behaviour muted.[3] The extent to which resources (status, wealth, etc.) are inherited varies widely. For a society in which the variance in resource control among males is high and resources or status are heritable (e.g., strong patrilines), resource control can create stratification, and thus may well influence the utility of striving and achievement – and risk taking.

Does this mean that men in 'monogamous' systems strive less than men in polygynous systems? Not really. First, as Clutton-Brock (1983) noted, variance in reproductive success (RS) may be quite high in monogamous systems (even biologically defined monogamous systems); only the relative variance between males and females is of interest. Second, most 'monogamous' systems are socially-imposed systems of serial polygyny.

Monogamy and dowry

Dowry, a system in which a bride's family transfers goods or money to the groom or the groom's family, is relatively rare (8 of 186 societies in the *Standard Cross-Cultural Sample*). Dowry reflects an interesting twist on the sexual utility of resources, and the association of monogamy with (i) difficult resource environments, and (ii) intense per capita resource investment in children.

Dowry is 50 times more common in monogamous, stratified societies than in polygynous or non-stratified ones (Gaulin & Boster, 1990; see also Gaulin & Boster, 1997). In such monogamous and stratified societies, males typically vary greatly in their status and wealth, and women married to wealthy, high status men (and their children) benefit reproductively. So it may pay would-be brides (usually, the brides' fathers) to compete, bargaining for wealthier men as mates. In many of these societies, poorer women's families must pay more dowry than wealthy women's families; dowry becomes a form of mate competition among women (Gaulin & Boster, 1990, 1997). In so far as poorer families are unable to pay the price, stratification is intensified.

Such competition can have associated costs. Consider dowry in modern rural India. Since about 1950, demographic shifts have resulted in a decline in potential grooms for potential brides of marriageable ages – and dowries have risen steadily. By 1990, a dowry was likely to constitute over 50% of a household's assets. Wives from poor families, able to pay less in dowry, are less likely to marry; if they do marry, they have a high risk of being abused by their husbands. Further, domestic violence and spousal abuse have correlated with increases in dowry worth; 'insufficient dowry' is an important recorded cause of spousal abuse (Rao, 1993*a*, *b*, 1997).

Monogamy and infanticide

Monogamy has two relationships to patterns of infanticide. First, one major proposed route to monogamy is male protection of their infants from infanticide by other males. In humans, however, we have too few empirical data to infer that monogamy protects infants from infanticide.

A second relationship between monogamy and infanticide arises in part because men's and women's reproductive interests do not necessarily converge completely in socially monogamous human marriage systems. In fact, high divorce rates, with sequential polygyny, mean that parental and step-parental interests often conflict. Whenever monogamy is social rather than genetic, especially because human males are so parental, remarriage exacerbates conflicts of interests between social parents, and can have implications for infant survival. Social monogamy does not preclude remarriage and step-parenthood, which are common in many socially monogamous societies. Human paternal investment is relatively expensive (men give food, shelter, and protection, for example), and when paternal investment is expensive, men may prefer not to invest – even to the extent of infanticide (or coerced infanticide). In other species, such as lions (Packer & Pusey, 1983, 1984), langurs (Hrdy, 1974, 1979, 1999), and gorillas (Watts, 1989), males taking over harems routinely commit infanticide; the result is that females come into oestrus quickly, and males do not spend paternal care on infants not their own.

In modern US populations, stepfathers are often less involved and invest less than biological fathers (e.g., Anderson, 1999; Lancaster & Kaplan, 2000); for example, they are less likely to pay for schooling. A more extreme example, as Daly and Wilson (1984, 1985, 1987, 1988) have shown, is in the greatly increased risks of child abuse and infanticide in some societies when children live with a step-parent. In contrast, in some traditional societies (e.g., Hewlett, 1992; Hewlett *et al.*, 2000), step-parents (especially stepfathers) appear to invest as much in their stepchildren as biological parents do in their biological offspring.

Data exist for too few societies to allow a statistical comparison, but a reading of ethnographies suggests that any patterns that emerge will arise from the kind, and expense, of investment. In societies such as the Aka pygmies, in which men invest time and teaching, men appear to invest equally in biological and stepchildren. In societies in which potential inheritance of great wealth coexists with socially-imposed monogamy and remarriage (such as many modern developed nations), infanticide and child abuse appear with step-parenthood.

Marriage and information in signals

Marital systems have some correlates with the information we communicate to each other. Men's wealth or power status is shown by ornaments in 87 of 138 societies studied (Low, 1979, 1990*a*, *b*). Only four societies distinguish men's marital status by ornament, and two of these are 'ecologically monogamous', living in very poor environments where men have trouble becoming successfully polygynous. In contrast, women's marital status is signalled in all but three of the 138 societies. Many anthropologists argue that marriage has the function of building alliances between families; if this is true, women might be expected to signal family wealth, although not necessarily any separate wealth of their own. Indeed, women's wealth is shown by ornaments in 49 of the societies in the study (Low, 1979, 1990*a*, *b*). Is it their own wealth, or their family's (father's before marriage, husband's after)? Although men usually have greater control of resources, there are societies in which women control significant resources or have considerable influence over resource distribution. But the societies in which women have power and influence are *not* those in which women wear ornaments of status and power.

Women's status advertisement, whether of power and wealth or of marital status, is less likely to be effective when directed towards close kin or household members with whom they interact daily – *those* people cannot be fooled. Across most traditional cultures, women's signals of status largely reflect their husband's or male kins' wealth or standing, consistent with the prediction. Such signals also may represent a conflict of reproductive interest between the man and woman since male resources are used to acquire mates, and signals of 'excess' resources, even if worn by a man's wife, can constitute his sexual advertisement, or mating efforts (Whyte, 1979; Low, 1979, 1990*a*). Patterns in men's and women's signals reflect the observation that males seek resources as a part of mating efforts, competing against other males to whom they are variously, often not, related, and interacting with individuals they know less well; females, on the other hand, seek resources as a form of parental effort, working at or near home with sisters or co-wives.

The one significant cross-cultural relationship between female ornamentation and power is a negative one. The societies in which women can hold political posts are societies in which women do *not* wear ornaments of power or status (Low, 1990*a*, 2000: Chapter 5). The question then remains: in traditional societies, when women do operate independently in the extra-familial community sphere, why do they not signal position and power in the same way, and to the same extent, as men?

Ecologically monogamous societies show distinct signalling patterns compared with other societies (Figure 11.2). For example, it is true that women signal marital status far more than men. The societies in which at least some men signal marital status (and thus 'unavailability') are ecologically monogamous or (socially-imposed monogamous) large nation-states. Ecological constraints mean that, in any case, men cannot profit reproductively from polygyny. Men in 20% of 'ecologically' monogamous societies signal marital status; men in 1.5% of other societies do so. Pubertal and/or age-group status is not discernible for either men or women in ecologically monogamous societies, while women in 12% and men in 67% of other societies signal these states. Finally, men signal wealth and power in 67% of socially (non-ecologically) monogamous societies – those in which such signals might be potent advertisements of their ability to take on additional mates; men do so in only 10% of ecologically monogamous societies. Women in ecologically monogamous societies do not signal wealth or power; women in 38% of other societies do so, though this is not a woman's own wealth or power but rather a reflection of her male relatives' status. When women signal the wealth of men, there is potential for a great conflict of interest.

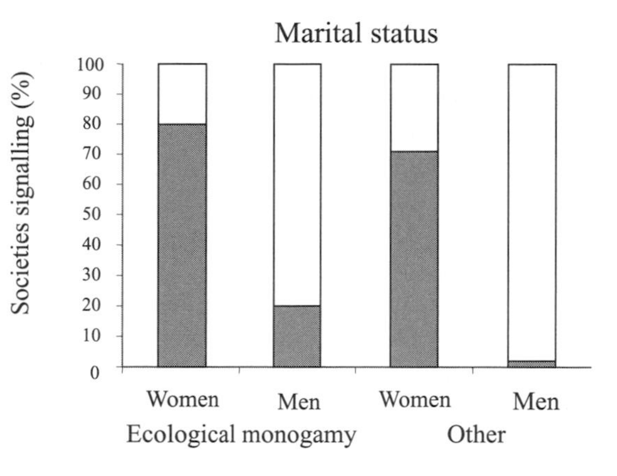

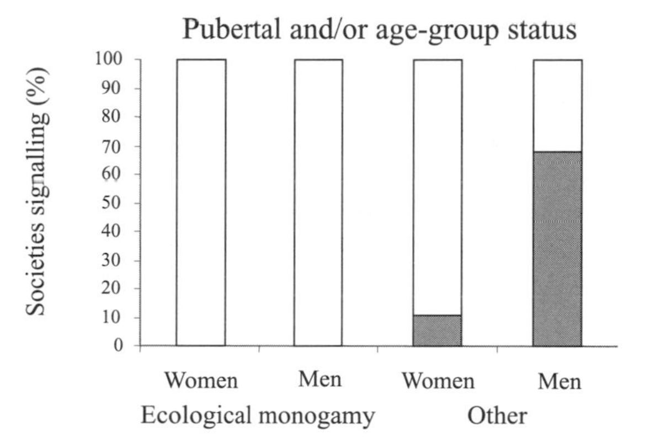

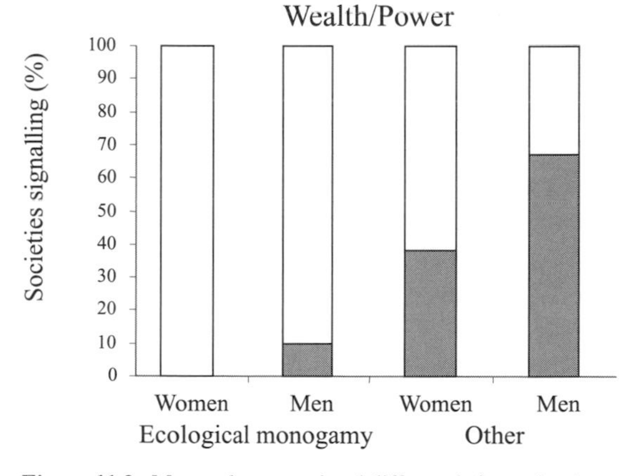

Figure 11.2. Men and women signal different information in dress and ornament in ecologically-imposed, monogamous (marriage) societies ($N = 10$) compared with polygynous, socially-imposed monogamous and polyandrous societies ($N = 128$; data from Low, 1979).

MONOGAMY AND ITS IMPACTS IN HISTORICAL AND MODERN SOCIETIES

One hypothesis about socially-imposed monogamy is that as the size of a society increases and the proportion of men denied full sexual access increases, leaders have more trouble imposing and enforcing uneven distributions of reproductive rewards for risk taking. The result is social monogamy. Modern western developed nations are large, and while 'one-spouse-at-a-time' rules simply make polygyny serial and do not necessarily decrease variance in male RS, they may have considerable modulating influence.

The question of why monogamy-requiring religions or other cultural practices become popular and accepted is an intriguing one. An obvious prediction is that such acceptance is more likely under conditions (ecological or sociocultural) that make male-parental investment reproductively more profitable than mating efforts. Here, groups and coalitions can exert important influences. Certainly, monogamy-promoting religions tend to support in-group members who cooperate, and punish (and perhaps exclude) those who do not. If a group with such characteristics is successful as a group within a population, acceptance of its promotion of social monogamy is likely. This is an area that needs further attention.

One major historical shift may have significantly enhanced the reproductive profits of social monogamy. During the nineteenth century in Europe and North America, resource acquisition and control shifted from reliance on agricultural work and small-scale cottage industries to reliance on major industrial enterprises. Concurrently, fertility fell significantly (e.g., Coale & Watkins, 1986; Coleman & Schofield, 1986). This has not tested out as a clear causal relationship, however.[4]

A more general suggestion from demographers and economists (e.g., Becker & Lewis, 1974; Easterlin, 1978; Becker & Barro, 1988; Becker, 1991) is that any factor that changes the competitive environment faced by one's children might lead to shifts in per capita investment; certainly market forces and industrialization could have such impacts. Could the utility of non-generalizable resource in true parental investment, and competitiveness over resources, represent important influences in the spread of social monogamy in humans?

In societies undergoing demographic transition, as well as in traditional societies, resources affected both men's and women's fertility. There are hints of this pattern even in aggregate data (marriage rates and fertility rates in England and Sweden rose and fell with crop and grain prices; e.g., Wrigley & Schofield, 1981; Low & Clarke, 1992; review by Clarke & Low, 2001), although it is difficult to analyse aggregate data to answer the resource–fertility question.

Analyses of lineage, rather than aggregate, data in Sweden strongly suggest that marriage decisions and fertility of both men and women were influenced by access to resources and investment in children during the demographic transition (e.g., Low & Clarke, 1992; Clarke & Low, 2001). In nineteenth-century Sweden, multi-parish comparisons of a sample of 613 men married for the first time between 1824 and 1840, and their 6811 lineal descendants, found that throughout the demographic transition (which was local and reversible in Sweden), whenever and wherever variation existed in wealth, wealth mattered to family formation and fertility (e.g., Low & Clarke, 1992; Clarke & Low, 1992, 2001). While some wealthy individuals might not marry, very poor individuals often could not marry, and poorer individuals who married did so at an older age – trading reproductive value for resource value as the betrothed scrambled to accumulate goods for a successful marriage. Wealthier families were able to give each child more for marrying, establishing, and raising families. The effects of resources carried across generations (reviewed in Low, 2000).[5]

Controlling for variation in resource type, geography, and temporal patterns of marriage, wealth influenced both men's and women's lives throughout the demographic transition (data from 1824 to 1896: e.g., Low & Clarke, 1992). Children of both sexes born to poorer parents were more likely than richer children to leave their natal parish before reaching maturity. Poorer men, and women whose fathers were poor, were less likely to marry in the parish than others, largely as a result of differential migration; with luck, they might establish and marry in another parish. Men who inherited land had greater fertility than their non-landowning brothers.

In nineteenth century Sweden, the ability to marry was the biggest predictor of fertility – and this is of particular interest in this society, in which perhaps one half of all adults did not marry, and wealth did not necessarily render one likelier to marry. The investment patterns differed: children in wealthy families, if they married, began their marital careers with far more investment than poorer children. Marriage was, even in this egalitarian society, almost exclusively within class: marrying 'up' or 'down' was rare.

Of all adults of both sexes who remained in their home parish and thus generated complete lifetime records, richer individuals had greater lifetime fertility, and more children alive at age ten, than others. Marriages were concentrated in good economic times. Men's age at marriage was not affected by wealth, but daughters of wealthy men married earlier, and had their first child earlier, than poorer women. Finally, wealthier men remarried more than others, becoming effectively polygynous, with high fertility; when women remarried it did not affect their lifetime fertility.

SOCIAL MONOGAMY IN POST-TRANSITIONAL SOCIETIES

Modern societies are curious with regard to social monogamy; how does it function in today's industrialized environments? I believe there is something to demographers' suggestions that the suite of traits we see in demographic transitions – reduced mortality, reduced fertility, and fewer, better-invested children – arises because there is, at least initially, a lineage advantage. Within biology, MacArthur and Wilson (1967) proposed a parallel pattern across species: *when insuring success for offspring requires enormous per capita investment, fertility will fall.* Both sexes will profit from such investment: social and genetic monogamy become more likely and family size will decrease unless resource throughput can be significantly increased (see also Low *et al.*, 1992; Low, 1993). When 'fewer, more-successful children' is a better competitive strategy than 'more numerous, less-successful children', the importance of women's reproductive value (number of likely children) declines, and women might shift to a combined reproductive value/resource value strategy. The result then is relatively small, monogamous families, with intense parental investment by both parents. Of course, this does not preclude extra-pair copulations (and fertilizations), but any resulting children are likely to command less investment than children born within socially recognized unions.

However, as we look at fertility and mortality patterns today, there is a catch. Any initial benefits associated with monogamy, lowered fertility, and intense parental investment, as seen in the demographic transition, may no longer exist in post-industrial societies (Low *et al.*, 2002, 2003; Low, 2003). As women seek more education and greater market participation, their fertility is decreased and delayed, and their lineage success decreases. Even if shifts by women to more resource acquisition, education, and delayed fertility began as adaptive responses, they are no longer seen as such. The conditions under which heightened per capita investment in children can compensate for the effects of women's delayed fertility are extremely restricted, and perhaps even non-existent (Low *et al.*, 2002, 2003; see also Borgerhoff Mulder, 1998).

Nonetheless, people still seek the *proximate* excorrelates of fertility. For example, men may simply seek wealth and status for their own sakes, and these may impact men's sexual access (Betzig & Weber, 1993). And women may still show some preferences for successful men, although the novelty of birth control means that women's preferences no longer mean more children for successful men. Consider Pérusse's findings (1993, 1994) among a sample of Canadian men. A subset of men were monogamously paired by choice. Of those who chose polygynous sexual access, wealthier men had no more children (women used contraception) than poorer men, although they did have greater sexual access. A study of Albuquerque men similarly found that over about a three-generation period, the possible fertility advantages of wealth for men had disappeared (Kaplan *et al.*, 1995; Kaplan & Lancaster, 2003).

Modern conditions in developed nations are evolutionarily novel, with results that significantly affect men, women, and family structure. Divorce and extra-pair copulations appear to be common, rendering us serially polygynous. Successfully polygynous men, and wealthy men, may have no more children than others. Successful career women have high education, relatively high income, and quite low fertility.

CONCLUSION

Genetic monogamy appears to be extremely rare in humans, as in most mammals. Social monogamy is not common (about 17% of societies in the *Ethnographic Atlas*), and often reduces to serial polygyny in a biological sense – with all the attendant sexual conflicts of interest inherent in polygyny. Humans are like other species, although perhaps more extreme, in that socially-monogamous systems do not ensure the absence of extra-pair copulations or 'outside' paternity.

Patterns in human mating and marriage systems are complicated by the elaboration of human intelligence and sociality. Monogamy appears to be less ecologically influenced in humans than in other species. In large part this is because extensive, culturally invented practices solve some of the parental care dilemmas (that drive monogamy in other species) in other ways.

Monogamy has a number of impacts and social correlates, including female contribution to subsistence, female control of the fruits of labour, and information signalled by males and by females. Finally, monogamous *marriage* systems are not necessarily monogamous *mating* systems. All these things mean that as a biological phenomenon genetic monogamy, with its attendant constraints on males, is rare in human societies, although socially-imposed monogamous marriage systems are relatively more common.

Notes

1 Polygynous men, of course, have not only more variable offspring, but more offspring than monogamous men, so we must look further before claiming that offspring variability might have a functional role. Powerful men will promote polygyny whenever they can, whether or not pathogen stress is present. Consistent with the variable-offspring genotype argument are the facts that: (i) polygynous men in pathogen-laden parts of the world are more likely to marry exogamously (outside their group), especially through capture of women from other societies; and (ii) sororal polygyny, in which a man marries sisters (and his children would be less variable genetically), is rare in areas of pathogen stress. Both patterns result in more variable children for men. From a woman's (or her family's) point of view, being the second wife of a healthy man may be preferable to being the sole wife of a parasitized man (cf. Orians, 1969); thus, women may prefer polygyny in highly parasitized or disease-ridden areas, and men's and women's interests (typically more divergent in polygyny than in monogamy) may converge. Polygyny is much more common in Africa and South America than, for example, in Europe, so it is important to ask if there is simply some sort of covariance of pathogens and socially determined patterns of polygyny. Within the high-pathogen but socially diverse tropics (Africa, Eurasia, South and North America, and the Mediterranean) pathogen stress and polygyny covary. Thus neither simple geography nor cultural diffusion of polygynous practices within high pathogen regions is likely to be the source of the patterns we see. In the Pacific, there have been no societies in the 'high' pathogen stress category, and because the relationship is a threshold relationship, no statistical relationship is apparent within that region.

2 Ecological patterns of parasite risk, rainfall seasonality, irrigation, and hunting explain 46% of the observed patterns in human polygyny (Low, 1988*b*, 1990*b*). White and Burton (1988; see critique by Low, 1990*b*) suggested that polygyny is 'favored by homogeneous and high-quality environments', perhaps reasoning that rich environments are easier to exploit. Unfortunately the real relationships are difficult to determine because White and Burton lumped together very different climate zones, as if, e.g., tropical rainforests were comparable with seasonal, high-rainfall areas. It is far too simplistic to assert that there is a single measure of environmental quality, one that decreases with cold or aridity, so that dry polar regions are lowest on environmental quality and moist tropical regions highest. This not only confounds extremeness, range of variation, and predictability (see Low, 1990*a*), but ignores, for example, the fact that pathogen stress, hardly a contributor to 'high environmental quality' (but a factor promoting polygyny) is highest in moist, tropical regions, and that protein availability, probably a requirement for 'high' environmental quality, can be quite high in cold and dry regions.

3 As Low (2000, p. 274) noted, our ability to study behavioural dimorphisms in the context of sexual selection would be more precise if we first partitioned variances in male and female reproductive successes into a non-behavioural component, including morphological characters and inherited resources, and a second component that would predict behaviour. Informally, biologists Steve Frank and James Crow have suggested a simple method for quantifying the *opportunity for reproductive gain* through striving and risky competitive behaviour, in which competitive success has both a stratum-related component independent of competitive behaviour, and a variable component in which success depends on the intensity of striving. Their method suggests that:

   1 When opportunities for gain through striving differ among classes or strata, different behavioural patterns are expected: *high-variance strata will contain the most competitive and risk-taking individuals.*
   2 Within-class variance is most important in strata that are on average most successful.
   3 Among species in which status explains a similar proportion of variance, if two species differ in the amount of total variance explained by heritable rank, then the two species are expected to differ correspondingly in the levels of aggression over status.

   Using this model, an example can be placed in the wider context of partitioning variance into behavioural and non-behavioural components.

4 A reasonable hypothesis, the focus of much demographic work for much of the past 30 years, was that in some (largely unspecified) way, industrialization was at least a proximate cause of fertility decline. Children, once perceived as contributing to the household wealth, came to be seen as costly in an industrial society (whether children ever were actually profitable is debatable: Low *et al.*, 1992). The results have

been disappointing to many demographers (e.g., Coleman & Schofield, 1986), and new data from the developing world also suggest that 'industrialization' is not a 'cause', and in fact is not necessarily strongly associated with fertility declines.

5 In the Swedish study, the areas differed in economic conditions. Locknevi, a southern agricultural parish with some mining and forestry, was limited geographically by surrounding areas unsuitable for agriculture. Early in the nineteenth century in Locknevi, great wealth differentials existed, and corresponding fertility differentials were observed; during the study, wealthy families moved out, landholdings were divided, and fertility became more even. Gullholmen, a small island fishing parish, had high population density, low migration, and very late marriage related to the uncertain returns of fishing (Low & Clarke, 1990). Tuna parish, in mid-Sweden, showed both resource and reproductive differentials throughout the study. In Nedertorneå, in the far north of Sweden, wealth and fertility differentials existed, but were complicated by high infant mortality rates; infants were fed cow's milk (often spoiled) (Low & Clarke, 1990).

References

Alexander, R. D. (1979). *Darwinism and Human Affairs.* Seattle: University of Washington Press.

Alexander, R. D., Hoogland, J. L., Howard, R. D., Noonan, K. M. & Sherman, P. W. (1979). Sexual dimorphism and breeding systems in pinnipeds, ungulates, primates, and humans. In *Evolutionary Biology and Human Social Behavior: An Anthropological Perspective*, ed. N. A. Chagnon & W. Irons, pp. 402–35. North Scituate, Massachusetts: Duxbury Press.

Anderson, K. G. (1999). Parental care by genetic and stepfathers. I: Reports from Albuquerque men. *Evolution and Human Behavior*, **20**, 405–32.

Baker, R. I. M. (2000). *Sperm Wars*, 2nd Edition. London: Pan.

Baker, R. I. M. & Bellis, M. A. (1995). *Human Sperm Competition: Copulation, Masturbation and Infidelity.* London: Chapman & Hall.

Becker, G. (1991). *A Treatise on the Family*, Enlarged Edition. Cambridge, Massachusetts: Harvard University Press.

Becker, G. & Barro, R. J. (1988). Reformulating the economic theory of fertility. *The Quarterly Journal of Economics*, **103**, 1–25.

Becker, G. & Lewis, H. G. (1974). Interaction between quantity and quality of children. In *Economics of the Family: Marriage, Children and Human Capital*, ed. T. W. Schultz, pp. 81–90. Chicago: University of Chicago Press.

Betzig, L. (1997). *Human Nature: A Critical Reader.* Oxford: Oxford University Press.

Betzig, L. & Weber, S. (1993). Polygyny in American politics. *Politics and the Life Sciences*, **12**, 45–52.

Birkhead, T. R. (2000). *Promiscuity: An Evolutionary History of Sperm Competition.* Cambridge, Massachusetts: Harvard University Press.

Birkhead, T. R. & Møller, A. P. (1998). *Sperm Competition and Sexual Selection.* London: Academic Press.

Blurton Jones, N. (1986). Bushman birth spacing: a test for optimal interbirth intervals. *Ethology and Sociobiology*, **7**, 91–105.

(1987). Bushman birth spacing: direct tests of some simple predictions. *Ethology and Sociobiology*, **8**, 183–203.

(1989). The costs of children and the adaptive scheduling of births: towards a sociobiological perspective on demography. In *Sexual and Reproductive Strategies*, ed. A. Rasa, C. Vogel & E. Voland, pp. 265–82. Kent, UK: Croom Helm.

(1997). Too good to be true? Is there really a tradeoff between number and care of offspring in human reproduction? In *Human Nature: A Critical Reader*, ed. L. L. Betzig, pp. 83–6. Oxford: Oxford University Press.

Borgerhoff Mulder, M. (1988*a*). Reproductive success in three Kipsigis cohorts. In *Reproductive Success*, ed. T. H. Clutton-Brock, pp. 419–35. Chicago: University of Chicago Press.

(1988*b*). Kipsigis bridewealth payments. In *Human Reproductive Behaviour: A Darwinian Perspective*, ed. L. Betzig, M. Borgerhoff Mulder & P. Turke, pp. 65–82. Cambridge: Cambridge University Press.

(1990). Kipsigis women's preferences for wealthy men: Evidence for female choice in mammals? *Behavioral Ecology and Sociobiology*, **27**, 255–64.

(1991). Human behavioral ecology. In *Behavioural Ecology*, 3rd Edition, ed. J. R. Krebs & N. B. Davies, pp. 69–98. London: Blackwell.

(1992). Reproductive decisions. In *Evolutionary Ecology and Human Behavior*, ed. E. A. Smith & B. Winterhalder, pp. 339–74. New York: Aldine de Gruyter.

(1997). Marrying a married man: a postscript. In *Human Nature: A Critical Reader*, ed. L. Betzig, pp. 115–17. Oxford: Oxford University Press.

(1998). The demographic transition: are we any closer to an evolutionary explanation? *Trends in Ecology and Evolution*, **13**, 266–70.

Clarke, A. L. & Low, B. (1992). Ecological correlates of human dispersal in 19th century Sweden. *Animal Behaviour*, **44**, 677–93.

(2001). Testing evolutionary hypotheses with demographic data: recent progress. *Population and Development Review*, **27**, 633–60.

Clutton-Brock, T. H. (1983). Selection in relation to sex. In *From Molecules to Men*, ed. D. S. Bendall, pp. 457–81. Cambridge: Cambridge University Press.

Coale, A. J. & Watkins, S. C. (1986). *The Decline of Fertility in Europe*. Princeton, New Jersey: Princeton University Press.

Coleman, D. & Schofield, R. (ed.) (1986). *The State of Population Theory*. Oxford: Basil Blackwell.

Colwell, R. K. (1974). Predictability, constancy, and contingency of periodic phenomena. *Ecology*, **55**, 1148–53.

Conover, W. J. (1980). *Practical Non-Parametric Statistics*, 2nd Edition. New York: John Wiley.

Cowlishaw, G. & Mace, R. (1996). Cross-cultural patterns of marriage and inheritance: a phylogenetic approach. *Ethology and Sociobiology*, **17**, 87–97.

Daly, M. & Wilson, M. (1984). A sociobiological analysis of human infanticide. In *Infanticide: Comparative and Evolutionary Perspectives*, ed. G. Hausfater & S. B. Hrdy, pp. 487–502. New York: Aldine de Gruyter.

(1985). Child abuse and other risks of not living with both parents. *Ethology and Sociobiology*, **6**, 197–210.

(1987). Children as homicide victims. In *Child Abuse and Neglect: Biosocial Dimensions*, ed. R. J. Gelles & J. B. Lancaster, pp. 201–14. New York: Aldine de Gruyter.

(1988). *Homicide*. New York: Aldine de Gruyter.

Dyson-Hudson, R. & Smith, E. A. (1978). Human territoriality. *American Anthropologist*, **80**, 21–42.

Easterlin, R. (1978). The economics and sociology of fertility: a synthesis. In *Historical Studies of Changing Fertility*, ed. C. Tilly, pp. 57–134. Princeton, New Jersey: Princeton University Press.

Ebert, D. & Hamilton, W. D. (1996). Sex against virulence: the coevolution of parasitic diseases. *Trends in Ecology and Evolution*, **11**, 79–82.

Emlen, S. T. & Oring, L. W. (1977). Ecology, sexual selection, and the evolution of mating systems. *Science*, **197**, 215–23.

Gaulin, S. J. C. & Boster, J. S. (1990). Dowry as female competition. *American Anthropologist*, **92**, 994–1005.

(1997). When are husbands worth fighting for? In *Human Nature: A Critical Reader*, ed. L. Betzig, pp. 372–4. Oxford: Oxford University Press.

Hamilton, W. D. (1980). Sex versus non-sex versus parasite. *Oikos*, **35**, 282–90.

Hamilton, W. D. & Zuk, M. (1982). Heritable true fitness and bright birds: a role for parasites? *Science*, **218**, 384–7.

Hamilton, W. D., Henderson, P. A. & Moran, N. (1981). Fluctuation of environment and coevolved antagonist polymorphism as factors in the maintenance of sex. In *Natural Selection and Social Behavior: Recent Research and Theory*, ed. R. D. Alexander & D. W. Tinkle, pp. 363–82. New York: Chiron Press.

Hartung, J. (1982). Polygyny and the inheritance of wealth. *Current Anthropology*, **23**, 1–12.

(1983). In defense of Murdock: a reply to Dickemann. *Current Anthropology*, **24**, 125–6.

(1997). If I had it to do over. In *Human Nature: A Critical Reader*, ed. L. Betzig, pp. 344–8. Oxford: Oxford University Press.

Hawkes, K. (1991). Showing off: tests of a hypothesis about men's foraging goals. *Ethology and Sociobiology*, **12**, 29–54.

(1993). Why hunter-gatherers work: an ancient version of the problem of public goods. *Current Anthropology*, **34**, 341–61.

Hawkes, K., O'Connell, J. F. & Blurton Jones, N. G. (1997*a*). Hadza women's time allocation, offspring provisioning, and the evolution of long postmenopausal life spans. *Current Anthropology*, **18**, 551–77.

Hawkes, K., O'Connell, J. F. & Rogers, L. (1997*b*). The behavioral ecology of modern hunter-gatherers, and human evolution. *Trends in Ecology and Evolution*, **12**, 29–31.

Hawkes, K., O'Connell, J. F., Blurton Jones, N. G., Alvarez, H. & Charnov, E. L. (1998). Grandmothering, menopause, and the evolution of human life histories. *Proceedings of the National Academy of Sciences of the USA*, **95**, 1336–9.

Hewlett, B. S. (1992). Husband–wife reciprocity and the father–infant relationship among Aka pygmies. In *Father–Child Relations: Cultural and Biosocial Contexts*, ed. B. S. Hewlett, pp. 153–75. New York: Aldine de Gruyter.

Hewlett, B. S., Lamb, M. E., Leyendecker, B. & Scholmerich A. (2000). Parental strategies among Aka foragers, Ngandu farmers, and Euro-American Urban-industrialists. In *Adaptation and Human Behavior: An Anthropological Perspective*, ed. L. Cronk, N. Chagnon & W. Irons, pp. 155–78. New York: Aldine de Gruyter.

Hill, K. & Hurtado, A. M. (1996). *Ache Life History: The Ecology and Demography of a Foraging People.* New York: Aldine de Gruyter.

Hill, K. & Kaplan, H. (1999). Life history traits in humans: Theory and empirical studies. *Annual Review of Anthropology*, **28**, 397–430.

Hrdy, S. B. (1974). Male–male competition and infanticide among the lemurs (*Presbytis entellus*) of Abu Rajasthan. *Folia Primatologia*, **22**, 19–58.

(1979). Infanticide among animals: a review, classification, and implications for the reproductive strategies of females. *Ethology and Sociobiology*, **1**, 13–40.

(1999). *Mother Nature.* New York: Pantheon Books.

Jarman, P. J. (1974). The social organization of antelope in relation to their ecology. *Behaviour*, **48**, 215–67.

Kaplan, H. S. & Lancaster, J. (2003). The life histories of men in Albuquerque: An evolutionary–economic analysis of parental investment and fertility in modern society. *American Journal of Human Biology* (in press).

Kaplan, H., Hill, K., Lancaster, J. & Hurtado, A. M. (2000). A theory of human life history evolution: diet, intelligence, and longevity. *Evolutionary Anthropology*, **9**, 149–86.

Kaplan, H. S., Lancaster, J., Johnson, S. E. & Bock, J. A. (1995). Does observed fertility maximize fitness among New Mexican men? *Human Nature*, **6**, 325–60.

Knodel, J., Saengtienchai, C., Low, B. & Lucas, R. (1997). An evolutionary perspective on Thai sexual attitudes. *Journal of Sex Research*, **34**, 292–303.

Lancaster, J. B. & Kaplan, H. (2000). Parenting other men's children: costs, benefits, and consequences. In *Adaptation and Human Behavior: An Anthropological Perspective*, ed. L. Cronk, N. Chagnon & W. Irons, pp. 179–202. New York: Aldine de Gruyter.

Low, B. S. (1978). Environmental uncertainty and the parental strategies of marsupials and placentals. *American Naturalist*, **112**, 197–213.

(1979). Sexual selection and human ornamentation. In *Evolutionary Biology and Human Social Behavior*, ed. N. Chagnon & W. Irons, pp. 462–86. North Scituate, Massachusetts: Duxbury Press.

(1988*a*). Measures of polygyny in humans. *Current Anthropology*, **29**, 189–94.

(1988*b*). Pathogen stress and polygyny in humans. In *Human Reproductive Behaviour: A Darwinian Perspective*, ed. L. Betzig, M. Borgerhoff Mulder & P. Turke, pp. 115–28. Cambridge: Cambridge University Press.

(1990*a*). Human responses to environmental extremeness and uncertainty: a cross-cultural perspective. In *Risk and Uncertainty in Tribal and Peasant Economies*, ed. E. Cashdan, pp. 229–55. Boulder, Colorado: Westview Press.

(1990*b*). Marriage systems and pathogen stress in human societies. *American Zoologist*, **30**, 325–39.

(1993). Ecological demography: a synthetic focus in evolutionary anthropology. *Evolutionary Anthropology*, **1993**, 106–12.

(2000). *Why Sex Matters*. Princeton, New Jersey: Princeton University Press.

(2003). Families: An evolutionary anthropological perspective. In *Families in Global Perspective*, ed. J. L. Roopnarine & U. Gielen. Boston, Massachusetts: Allyn & Bacon (in press).

Low, B. S. & Clarke, A. L. (1990). Family patterns in nineteenth-century Sweden: Impact of occupational status and landownership. *Journal of Family History*, **16**, 117–38.

(1992). Resources and the life course: patterns in the demographic transition. *Ethology and Sociobiology*, **13**, 463–94.

Low, B. S., Clarke, A. L. & Lockridge, K. (1992). Toward an ecological demography. *Population and Development Review*, **18**, 1–31.

Low, B. S., Simon, C. P. & Anderson, K. G. (2002). An evolutionary perspective on demographic transitions: modeling multiple currencies. *American Journal of Human Biology*, **14**, 149–67.

(2003). The biodemography of modern women: tradeoffs when resources become limiting. In *The Biodemography of Human Fertility and Reproduction*, ed. J. L. Rodgers & H.-P. Kohler. Amsterdam Kluwer Press (in press).

MacArthur, R. H. & Wilson, E. O. (1967). *The Theory of Island Biogeography*. Princeton, New Jersey: Princeton University Press.

Morell, V. (1998). A new look at monogamy. *Science*, **281**, 1982–3.

Murdock, G. P. (1967). *Ethnographic Atlas*. Pittsburgh: University of Pittsburgh Press.

(1981). *Atlas of World Cultures*. Pittsburgh: University of Pittsburgh Press.

Murdock, G. P. & White, D. (1969). Standard cross-cultural sample. *Ethnology*, **8**, 329–69.

Nordenskiold, E. (1949). The Cuna. In *Handbook of South American Indians*, Volume 4, ed. J. Steward. Washington, DC: US Government Printing Office.

Orians, G. H. (1969). On the evolution of mating systems in birds and mammals. *American Naturalist*, **103**, 589–603.

Packer, C. & Pusey, A. E. (1983). Adaptations of female lions to infanticide by incoming males. *American Naturalist*, **121**, 716–28.

(1984). Infanticide in carnivores. In *Infanticide: Comparative and Evolutionary Perspectives*, ed. G. Hausfater & S. B. Hrdy, pp. 31–42. New York: Aldine de Gruyter.

Pérusse, D. (1993). Cultural and reproductive success in industrial societies: testing the relationship at proximate and ultimate levels. *Behavioral and Brain Sciences*, **16**, 267–322.

(1994). Mate choice in modern societies: testing evolutionary hypotheses with behavioral data. *Human Nature*, **5**, 255–78.

Rao, V. (1993*a*). Dowry "inflation" in rural India: a statistical investigation. *Population Studies*, **47**, 283–93.

(1993*b*). The rising price of husbands: a hedonic analysis of dowry increases in rural India. *Journal of Political Economy*, **101**, 666–77.

(1997). Wife-beating in rural south India: a qualitative and econometric analysis. *Social Science & Medicine*, **44**, 1169–80.

Strassmann, B. I. (1997). Polygyny as a risk factor for child mortality among the Dogon. *Current Anthropology*, **38**, 688–95.

(2000). Polygyny, family structure, and child mortality: a prospective study among the Dogon of Mali. In *Adaptation and Human Behavior: an Anthropological Perspective*, ed. L. Cronk, N. Chagnon & W. Irons, pp. 45–63. Hawthorne, New York: Aldine de Gruyter.

Trivers, R. L. (1974). Parent–offspring conflict. *American Zoologist*, **14**, 249–64.

Turke, P. W. (1988). Helpers at the nest: Childcare networks in Ifaluk. In *Human Reproductive Behaviour: A Darwinian Perspective*, ed. L. Betzig, M. Borgerhoff Mulder & P. Turke, pp. 173–88. New York: Cambridge University Press.

Watts, D. P. (1989). Infanticide in mountain gorillas: new cases and a reconsideration of the evidence. *Ethology*, **81**, 1–18.

White, D. R. & Burton, M. L. (1988). Causes of polygyny: ecology, economy, kinship, and warfare. *American Anthropologist*, **90**, 871–87.

Whyte, M. K. (1978). Cross-cultural codes dealing with the relative status of women. *Ethnology*, **17**, 211–37.

(1979). *The Status of Women in Pre-industrial Society.* Princeton, New Jersey: Princeton University Press.

Wickler, W. & Seibt, U. (1981). Monogamy in Crustacea and man. *Zeitschrift für Tierpsychologie*, **57**, 215–34.

Wrigley, E. A. & Schofield, R. (1981). *The Population History of England, 1541–1871.* Cambridge, Massachusetts: Harvard University Press.

CHAPTER 12

Social monogamy in a human society: marriage and reproductive success among the Dogon

Beverly I. Strassmann

INTRODUCTION

In studying monogamy versus polygyny in humans, one can consider the diversity of marital and mating arrangements within societies or between societies. This chapter will focus on the diversity that occurs within a society, by asking two empirical questions: (i) what predicts who is monogamously or polygynously married for both males and females? and (ii) what are the consequences of both marriage types for reproductive success? Data are not available on extra-pair copulations (EPCs) or on genetic paternity in the study population. Therefore, throughout this chapter, the terms 'monogamy' and 'polygyny' refer to the number of concurrent spouses. This usage for the term 'monogamy' is akin to 'social monogamy', or 'pair living' as described in species for which actual mating behaviour has not been quantified.

The study of socially monogamous partnerships in a population that has both monogamous and polygynous unions helps to control for the cultural and historical complexity that obscures cross-cultural comparisons. In essence, a society with both monogamy and polygyny provides a natural experiment since these two types of union can be contrasted while holding constant a wide array of confounding variables. As demonstrated in this volume, much has already been learned about the occurrence of monogamy from parallel examples of intraspecific variation in other species. Ideally, this variation permits monogamy to be linked to relevant socioecological causes while holding phylogeny constant.

Attempts to explain the evolution of monogamy are often predicated on the assumption that monogamy is a trait that has been favourably selected. However, in humans and many other species, monogamy and polygyny are expressions of phenotypic plasticity (Vehrencamp & Bradbury, 1984). A diversity of mating arrangements arises within the species reaction norm, such that the same individual can be monogamously mated in one year and polygynously or polyandrously mated in another. One implication is that in seeking to understand monogamy, neither the origin nor the maintenance of a trait is at issue. Instead, the critical questions have to do with whose fitness interests are served by alternative outcomes (monogamy, polygyny, polyandry), and when interests conflict, who wins, who loses, and what are the compromises?

Human behavioural ecologists have reached no consensus on the costs and benefits of alternative mating systems for males and females. One school accepts the fundamental argument of the polygyny threshold model (PTM), which is that polygyny results from female choice (non-humans: Verner, 1964; Verner & Willson, 1966; Orians, 1969; humans: Borgerhoff Mulder, 1988, 1990; Josephson, 1993; Winterhalder & Smith, 2000; and see Hames, 1996). Another school questions the utility of the PTM, given the confusion about what it actually predicts and whether its assumptions are reasonable. In the parent field of animal behaviour, the PTM could be described as a morass (see Altmann *et al.*, 1977; Vehrencamp & Bradbury, 1984; Davies, 1989).

Perhaps the foremost problem is that the PTM ignores the potential for conflicts of interest between members of the two sexes (non-humans: Downhower & Armitage, 1971; Irons, 1983; Davies, 1989; humans: Chisholm & Burbank, 1991; Strassmann, 1997*b*, 2000; Sellen *et al.*, 2000). A useful theory of human mating systems should address the possibility that a male who marries polygynously may thereby gain higher fitness, even if each of his wives achieves lower fitness than she would under monogamy. If monogamy is advantageous for female fitness, it is unlikely that polygyny can be satisfactorily explained by female choice. Rather than focusing only on differences in male mate value, a useful theory should also examine the numerous constraints on female options, such as differences in female quality (see Altmann *et al.*, 1977).

In 1986 I initiated a longitudinal investigation of the evolutionary ecology of the Dogon of Mali, West Africa. A primary motivation for this research was the expectation that a focus on conflicts of interest between the sexes would clarify the potential costs of polygyny (and advantages of monogamy) for female fitness. Prior to fieldwork, my goal was to test the following hypotheses.[1]

1 Males potentially gain higher fitness under polygyny than under monogamy (Trivers, 1972).
2 In a male-dominated society, husbands are able to pursue their own fitness interests at the expense of those of wives. Thus, polygynous males are wealthier in terms of total resources, but poorer on a per capita basis. Due to the dilution of wealth under polygyny, females in polygynous marriages experience lower fertility and offspring survivorship.
3 On account of the reproductive costs of polygyny for members of their sex, females constrained by their own low mate value (especially older, less fecund, or less well nourished women) are obliged to accept polygynous marriage.

SITE, TIME, AND DOGON SOCIOLOGY

The geographic origins of the Dogon are unknown, but since the fifteenth or sixteenth centuries they have lived along the Bandiagara Cliff, a sandstone fault in a region of rocks and thorn savanna 250 km south of Timbuktu (Pern, 1982). This escarpment, 260 km long by about 500 m high, provided a safe refuge from raids by neighbouring pastoralists, the Fulani and Mossi. The first French arrived at the end of the nineteenth century and the Dogon of Tabi took a last stand against the colonial government in the revolt of 1920 (Cazes, 1993). During the period of French occupation, which lasted until Malian Independence in 1960, intertribal conflict was suppressed. Dogon settlements then spread away from the cliff to undefended sites on the plateau and out onto the sandy plains below.

The Dogon are organized into clans that are comprised of men from one or more villages who bear a common 'surname'. The Dolo clan lives in the 11 villages of Sangha, situated just above the cliff face at the edge of the plateau. The study village, Sangha Sangui (14° 29′ N, 3° 19′ W), is further subdivided into four patrilineages, each of which traces its ancestry to a single male founder within the Dolo clan.

Due to the expansion of the population and the poor quality of the habitat, arable land is a scarce resource. It is also exclusively in male ownership. A man and his married sons and all of their wives and children form an economic group. I call these groups of economically interdependent individuals work-eat groups (WEGs) since they work together in the same millet fields and eat together from the harvest. After the death of the oldest male, extended families used to remain together in WEGs that could exceed 100 individuals, but in recent times, the patriarch's sons by different wives usually stop working together and even sons who are full siblings usually separate some years after their father's death. The WEGs in Sangui at the time of this study ranged in size from one to 41 individuals and included both nuclear and extended families.

Arable land, especially millet fields, is the major resource critical for survival and reproduction. In an eight-year prospective study, the wealth rank of a child's WEG was a major predictor of survivorship adjusting for other significant predictors such as monogamy and a child's sex and age (Strassmann, 1997*b*, 2000). A woman has access to millet for herself and her offspring only through males. As a girl she is dependent on her father, then her husband, and ultimately her nephew or other male members of her natal patrilineage. After the masked dances that honour her husband after his death, an elderly widow usually returns to her natal patrilineage to work a small parcel of land either by herself or in the company of a grandchild. Widows who supported themselves in this manner headed the 13 poorest WEGs in Sangui. Across all ages women spent 21% more time working than men, and men spent 29% more time resting (Strassmann, 1996).

Co-wives are never sisters, and related women, such as first cousins, are forbidden from marrying into the same patrilineage. The result is male solidarity based on kinship and weakened alliances among female kin. Co-wives are rivals but equals. Custom dictates that a husband should sleep with his wives on alternate nights, but this rule is open to ambiguity and disputes. He owes each of his wives an equal share of poor quality millet to store in her granary; the first wife cuts the pile and the second wife is first to choose. The good quality or 'male' millet, he keeps under lock and key in his own granaries (Strassmann, 1997*b*).

The Dogon have a system of arranged marriages that operates alongside a more flexible system of spousal choice. Parents or grandparents choose a girl's first husband, but after divorce she exerts her own choice. As is typical in West Africa, most divorce is female initiated and not greatly stigmatized (Strassmann, 1997*b*, 2000). Boys have at least one, and in rare cases up to three spouses chosen by their relatives. To acquire an additional wife they must coax away the wife of another; thus, a non-arranged wife is referred to as 'cut-off' (*ya kezu*). Bride price is not practised, but the young fiancé of a nubile girl helps out his future parents-in-law in the fields and offers small gifts such as firewood and chickens. Because of this obligation, an arranged wife is called a 'work wife' (*ya bire*). A levirate wife (*ya pani*) is a woman married to her deceased husband's brother. The Dogon practise patrilocal residence, but it is considered preferable for a young woman to take up residence with her husband's patrilineage only after the birth of two offspring. If either infant survives, he or she will be raised by the maternal grandparents (Paulme, 1940; Strassmann, 1997*b*, 2000). A wife becomes *tanga* once she resides with her husband's family full-time; beforehand, as a *tanganu* wife, she resides with her natal family.

Abortion and infanticide are extremely rare among the Dogon, but not unheard of in cases of illegitimacy (Paulme, 1940; Strassmann, 1992, 1996[1]). Dogon males attest to an abhorrence of cuckoldry and all little girls in the study village undergo clitoridectomy. No man in the data set had ever formally adopted a genetically unrelated offspring, but in a couple of instances a new wife was allowed to bring with her a daughter from a previous marriage. This temporary arrangement was accepted as daughters ultimately marry outside the lineage and do not threaten the key concern, which is to prevent unrelated males from gaining access to the land owned by the patrilineage. When one man is cuckolded, his entire lineage is cuckolded because descendants of the 'imposter' will inherit resources that otherwise would have gone to the descendants of the original lineage (Strassmann, 1992, 1996[1]). Thus, Dogon patrilineages can be seen as coalitions of related males who are organized into socially and spatially cohesive units for the purpose of resource defence.

In humans ovulation is concealed and there are no reliable cues of paternity (Alexander, 1979; Alexander & Noonan, 1979; Strassmann, 1981); nonetheless, knowledge of the timing of menstruation can be helpful in paternity assessments. Dogon women are therefore required (via supernatural threats and social reprisals) to advertise the timing of menstruation to their husband's entire lineage (Strassmann, 1992, 1996[1]). This advertisement takes place at a menstrual hut and there is usually one hut per patrilineage.

A Dogon woman aged 25–30 typically experiences nine months of pregnancy, some 20 months of lactational amenorrhoea, and has a monthly probability of conception of 0.2 after the resumption of cycling (Strassmann & Warner, 1998). Thus, she has only about five months of cycling between births. If the duration of her marriage is 0.5 standard deviations shorter than the mean for her age, or if her husband is less than 35 years of age, or her past fecundity was 0.5 standard deviations higher than the mean, then her monthly probability of conception is about 0.3 (Strassmann & Warner, 1998). Women in this category have only about three months of cycling between births. It is the comparative rarity of menstruation in the absence of contraception that makes it so informative, even in the absence of precise knowledge of the timing of ovulation within a given cycle.

Husbands and other patrilineage members perpetually monitor female reproductive status and know that a wife who is cycling is potentially fecundable, whereas a wife who is pregnant or in lactational amenorrhoea is not. They keep a watchful eye on their fecundable women, who are less likely to successfully deceive them about paternity. Menstrual taboos might also help to deter EPCs, as male offspring of ambiguous paternity become outcasts, while their mothers lose social status (further details in Strassmann, 1992, 1996[1], 1997*b*).

The optimal fertility level can be defined as the number of livebirths that maximizes reproductive success. Among the women of Sangui, the predicted maximum reproductive success of 4.1 surviving offspring was attained at a fertility of 10.5 livebirths, which was similar to the modal fertility of ten births. Eighty-three per cent of women had seven to 13 births. Their predicted reproductive success was within the confidence limits (3.4–4.8) for reproductive success at the maximum. Thus, the fertility behaviour of Dogon women appears to make adaptive sense (Strassmann & Gillespie, 2002). Among females, variation in offspring mortality rather than fertility was the primary determinant of lifetime reproductive success (Strassmann & Gillespie, 2002).

METHODS

Study population

I gathered the data for this study during 35 months of fieldwork in Sangui between 1986 and 1998. In January 1988 the total population of the village was 460 inhabitants and by June 1998 it had increased to over 600. The data set is the entire population of one village, rather than a random sample of multiple villages. This approach enhanced data quality by making it possible to know all members of the community personally and to gain as much knowledge as possible by direct observation and measurement rather than interviews.

The analyses included men from age 25 to 74 years (the oldest male) ($N = 72$ in 1988, $N = 82$ in 1998). Rarely, a man younger than 25 years had a *tanganu* wife or two, but on account of the pronounced age difference among spouses none of these young women had given birth. The modal age at first birth for Dogon women is 19 years due to the late onset of menarche at about age 16 years and prolonged adolescent subfecundity (see Strassmann & Warner, 1998). Thus, for the men of my data set, any paternity before age 25 years would have happened in the context of extra-pair copulations (EPCs) and was not quantified. The sample size and age range for women varied by year (1988, 1998, or 1988–98) and whether the particular analysis included all married women (*tanga* and *tanganu*), *tanga* women only, cycling women only, women who last gave birth ten years prior to the interview, and so forth (details below).

Cooperation was essentially 100% as no one declined to participate. Subjects were remunerated through small cash payments (US $1.00 for measuring a family's cereal fields with their assistance, US $0.20 for women's marital and reproductive interviews). I also provided habitual first aid to the village and emergency transportation to the hospital in Mopti. Through the American Embassy in Bamako, I obtained financing for a small development project requested by the village: a dam and dyke for retaining rainwater for the onion gardens.

Wealth

I calculated five wealth variables: hectarage in cereal crops (millet, sorghum, rice, fonio), hectarage in onion gardens, estimated value of cereal harvest, value of livestock (mostly sheep and goats), and estimated revenues from commerce. I ranked each WEG with respect to the others across these five variables and then averaged the five ranks. In previous research, I developed seven economic indices that differed in how they weighted the various wealth variables (Strassmann & Warner, 1998). However, none captured the relative wealth of different families better than the unweighted average rank. The wealthiest family had a rank of 59 and the poorest a rank of one. I then excluded from the sample the 13 poorest WEGs because they were headed by widows and did not have any married adults, which left a total of 46 WEGs. I also excluded a solitary, unmarried male, who was the only person in his WEG, because he could not be accurately classified as either monogamous or polygynous.

Fields were measured with a compass and metre tape and areas calculated from a trigonometric program for a hand calculator. The estimated value of the harvest was obtained by counting all baskets of grain as they were brought back to the village. The fields are on the opposite side of a stream and foot traffic had to funnel across a bridge. I posted an observer at this bridge throughout the 1987 harvest season, then weighed representative baskets and bundles of each size. Livestock were counted and revenues from commerce, which make up only 5% of the village economy, were estimated by a Dogon man familiar with the various merchant activities (such as butcher, fish seller, onion middle-man). The total estimated value of the economic resources of Sangui in 1987 was about US $50 000. Further details regarding the procedures for measuring WEG wealth are found in Strassmann and Warner (1998).

Interviews

I conducted the interviews for this study in the local language (Sangha-Sò, a dialect of Toro-Sò) after I had lived in the village for more than two years. Few Dogon know their birth dates, so age had to be estimated from interviews. Fortunately, the Dogon belong to age classes that span only one birth interval ($2\frac{1}{2}$ years) and individuals know their own birth order within their age class; thus, relative age could be established with precision. An exception is a resident manual labourer who had a missing value for age because he came from outside Sangha.

I asked each married man to name his age-mates in order of age. I then used the known birth dates for a few individuals, who were born at the time of a datable event, to convert relative ages into chronological ages. To determine female ages, I asked each woman to name her closest male and female age-mates in the village. Female ages were then established by reference to the male chronology.

Other data obtained from interviews included marital and reproductive histories, including the putative reproductive success of both males and females based on self-reports. I knew most of these reported offspring because personal interactions, formal censuses, and quantitative behavioural scans made me familiar with the family composition of each WEG. However, putative offspring whom I did not observe directly included daughters who had married outside the village, sons who had left for the city, and offspring born to women before they married a man of Sangui.

Female reproductive success is defined as the number of self-reported offspring who survived to age ten years. Ten is an appropriate age cut-off because 20% of offspring did not survive to age 12 months and 46% did not survive to age five years (Strassmann, 1992), but few deaths occurred between age ten and sexual maturity. Male reproductive success is defined as the number of putative offspring who were still alive at the time of the interview, adjusted for the father's age. It would have been preferable to count offspring who survived to age ten years (Strassmann & Gillespie, 2002), but males professed little knowledge of offspring's ages at death. Deceased mothers and mothers who had divorced out of the village were not interviewed, so mothers' assessments regarding age at death could not be substituted. When both parents were still resident in the village, it was possible to compare the self-reports of mothers and fathers. There were no discrepancies, which is probably not altogether surprising. I interacted with the families on a daily basis and knew the names of their resident children, which meant that such offspring could not be invented or forgotten without my noticing it. Nonetheless, given the lack of genetic data, it is appropriate to refer to reproductive success in this study as 'putative'.

I also asked each individual about religious preference (indigenous religion, Islam, Catholicism, Protestantism, or none), number of years of schooling if any, and number of months of urban work. I collected data on each man's patrilineage from interviews, but in most cases these affiliations were also evident from the spatial location of compounds.

Monogamy versus polygyny

I defined a WEG as monogamous if the ratio of married women to men was 1.0 and polygynous if this ratio was greater than one. For the marital status of males, I compared both the total number of concurrent wives (which was public knowledge) and the total cumulative number of wives ever impregnated (based on self-reports). Unless otherwise indicated, the data on males refer to 1988. I assigned females the marital status that applied to them in two of three interviews for the ten-year period from 1988 to 1998.

Statistics

I used linear regression when the dependent variable was quantitative and logistic regression (Hosmer & Lemeshow, 1989) when it was binary. I also calculated the Pearson partial correlation coefficient between wealth and number of wives, with age controlled. When the relationship between age and a dependent variable was non-linear, I included both age and age squared in regression models. Analyses were carried out in the statistical software program SPSS 10.0.

Body mass

I measured each woman's height and weight to compute her body mass index (wt [kg]/height squared [m]) as described previously (Strassmann & Warner, 1998).

RESULTS

Marriage system

Variance in the putative reproductive success of males was greater than that of females (Figure 12.1). Such a pattern has been found in most mammals and was formerly used as the criterion for defining a mating system as polygynous (e.g., Daly & Wilson, 1983, pp. 83, 151–2). If the data set were from a closed breeding system, the mean reproductive success of males should have equalled that of females, rather than being twofold higher. However, Sangui has superior streamside land and more Sangui men have multiple wives than do men in most surrounding villages. Thus, the higher reproductive success of Sangui males may be balanced by the lower reproductive success of males from other villages. Moreover, Figure 12.1 includes only individuals who are age 42 years and older, which corresponds to the age at last birth for females. Males who survived

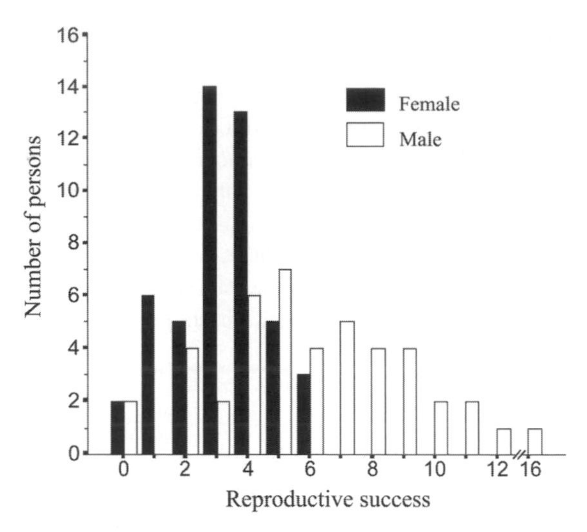

Figure 12.1. Variance in putative reproductive success for males ($N = 44$) and females ($N = 48$) in 1988. Reproductive success refers to the number of offspring alive at the time of the interview in 1988 for both males and females 42 years and older. The mean (± SD) number of offspring alive for males was 6.14 (± 3.27) and the median was 6.00. For females, respective values were 3.19 (± 1.50) and 3.00.

to this age will have enjoyed a female-biased sex ratio resulting from excess male mortality. They are also eight years older than their wives, on average. The difference in age at marriage combined with the population age pyramid further skews the operational sex ratio towards females (Dorjahn, 1959; Pison, 1985; Chisholm & Burbank, 1991; Strassmann, 1997b, 2000).

Although the marriage system could be described as polygynous, about half of the married males were in socially monogamous unions (54% in 1988, 56% in 1998). Data are not available on extra-pair paternity, so these percentages do not refer to genetic monogamy. In Sangui in 1988, 8% of the men age 25 years and older had no wives, 49% had one wife, 31% had two, and 10% had three wives concurrently ($N = 72$ men). Ten years later, the number of wives per man age 25 years and older was almost unchanged. Eight per cent had no wife, 51% had one, 31% had two, 8% had three and 1% had four wives ($N = 82$ men). Only 36% of married women (*tanga* and *tanganu* wives combined) were in socially monogamous marriages in both 1988 and 1998 (Table 12.1). An even smaller percentage of women (21%) were in WEGs in which no man had more than one wife.

Table 12.1. *Number (and percentage) of married women by marital status*

Wife	1988	1998
Sole	36 (36.0)	42 (35.9)
First	30 (30.0)	33 (28.2)
Second	28 (28.0)	33 (28.2)
Third	6 (6.0)	8 (6.8)
Fourth	0 (0.0)	1 (0.9)
Total	100 (100.0)	117 (100.0)

Table 12.2. *Predictors of WEG polygyny: 0, monogamous; 1, polygynous (N = 46 WEGs)*

Predictor	Coefficient	Odds ratio	*P*
Indigenous religion (0, no; 1, yes)	2.18	8.82	0.028
Wealth rank (46 = highest)	0.24	1.27	0.002
Intercept	−8.00	–	–

Goodness-of-fit statistic: 32.2. −2 log likelihood = 34.7.

Work-Eat Groups (WEGs)

If the WEG boss practised the indigenous Dogon religion instead of Islam or Christianity, the odds that the WEG was polygynous instead of monogamous increased 8.8-fold ($P = 0.028$) (Table 12.2). After adjusting for religion, the odds that the WEG was polygynous also increased with wealth. Specifically, as WEG wealth increased by one rank (out of 46), the odds that the WEG was polygynous increased by 27% ($P = 0.002$) (Table 12.2). WEG religion (traditional or other) and wealth rank were not significantly correlated (Spearman's rho = 0.32, $P = 0.138$, $N = 46$). After WEG wealth was standardized by the energy requirements of the WEG (based on its age and sex composition), the finding that wealthier WEGs were more likely to be polygynous continued to hold (Table 12.3). Thus, monogamous WEGs were more likely to have a boss who identified himself as a Muslim or Christian and monogamous WEGs were poorer both absolutely and on a per capita basis.

Table 12.3. *Predictors of WEG polygyny: 0, monogamous; 1, polygynous (N = 46 WEGs). In this model, wealth rank is standardized by the energy requirements of the WEG as described in the text*

Predictor	Coefficient	Odds ratio	*P*
Indigenous religion (0, no; 1, yes)	2.62	13.64	0.028
Standardized wealth rank	0.22	1.25	0.001
Intercept	−8.22	–	–

Goodness-of-fit statistic: 25.4. −2 log likelihood = 26.6.

Table 12.4. *Predictors of polygyny defined as the number of wives a man was married to concurrently (N = 71 men, R^2 = 0.28)*

Predictor	Coefficient	SE	*P*
Age (years)	0.109	0.193	0.026
Age squared	−0.001	0.002	0.012
Wealth rank (46 = highest)	0.024	0.007	0.001
Intercept	−1.493	–	–

Males

The analyses of males included both the number of wives a man was married to simultaneously, which was factual public information, and the cumulative number of wives a man said he had impregnated, which was non-verifiable. After adjusting for a man's age and age squared, as the wealth rank of a man's WEG increased by ten units, he acquired 0.24 extra concurrent wives. Together, the age and wealth variables explain 28% of the variance (Table 12.4). The percentage of variance explained decreased to 22% when the dependent variable was the total cumulative number of wives supposedly impregnated. In the latter model (not shown) the coefficient for wealth rank is nearly identical (0.023), but the *P*-value decreases from 0.001 to 0.06.

The Pearson partial correlation coefficient between the number of concurrent wives and wealth rank after controlling for a man's age and age squared was 0.38 ($P = 0.001$). For the reported cumulative number of wives the partial correlation coefficient was 0.23 ($P = 0.06$).

Table 12.5. *Predictors of the putative reproductive success of males (N = 71 men, R^2 refers to adjusted R^2)*

Predictor	β^a	SE	*P*
Model 1 ($R^2 = 0.31$)			
Age (years)	0.607	0.200	0.003
Age squared	−0.005	0.002	0.014
Number of wives (concurrent)	1.766	0.478	<0.001
Intercept	−13.372	–	–
Model 2 ($R^2 = 0.35$)			
Age (years)	0.889	0.187	<0.001
Age squared	−0.008	0.002	<0.001
Wealth rank (46 = highest)	0.107	0.029	<0.001
Intercept	−20.741	–	–
Model 3 ($R^2 = 0.35$)			
Age (years)	0.777	0.204	<0.001
Age squared	−0.007	0.002	0.002
Number of wives[b,c] (concurrent)	1.246	0.502	0.016
Wealth rank[b] (46 = highest)	0.008	0.031	0.013
Intercept	−19.595	–	–

[a] All coefficients are for an increase of one unit in the independent variable.

[b] Correlation between number of wives and wealth rank: Spearman's rho = 0.37, *P* (2-tailed) = 0.001.

[c] If the cumulative number of wives (based on self-reports) is included instead of concurrent wives, then the coefficient becomes 1.136 and adjusted R^2 becomes 0.47.

After adjusting for age and age squared, a man got 1.3 extra offspring for each additional wife that he said he had married cumulatively (not shown) and 1.8 extra offspring per concurrent wife present at the time of fieldwork (Table 12.5). For an increase in wealth of ten ranks out of 46, he got 1.1 extra offspring (Table 12.5). Sangui has four different lineages, the oldest of which possesses most of the high quality riparian land that is immediately adjacent to the village. Nonetheless, when lineage membership was included in the model as a categorical variable, it did not predict a man's putative reproductive success. Nor did his religion, schooling, or whether he had worked in the city.

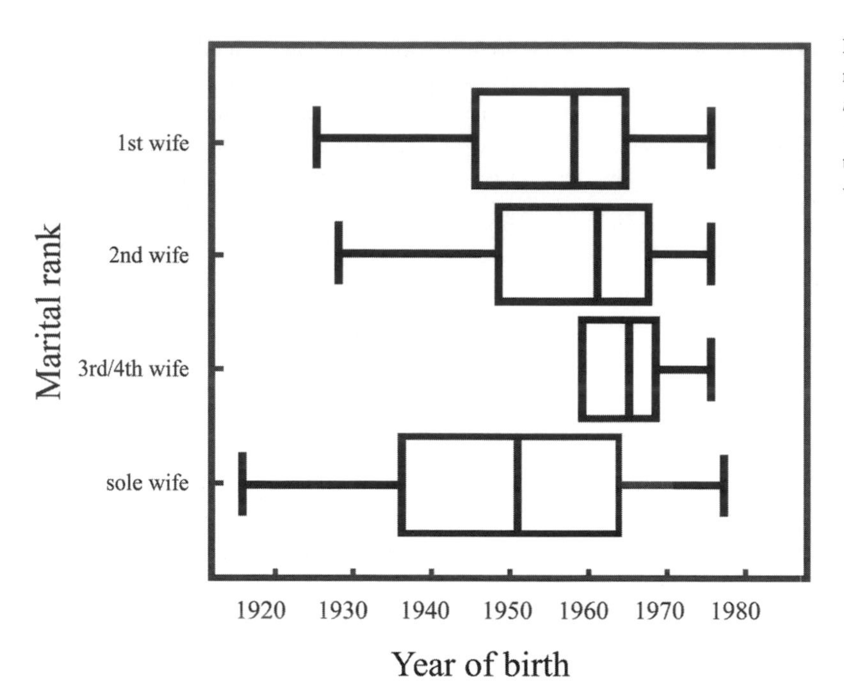

Figure 12.2. Boxplots of female marital status by age ($N = 143$ *tanga* women). Data are for 1988, 1994, and 1998 combined, and refer to the marital status in which a woman spent the most years.

Females

Monogamously married women were slightly older than polygynously married women (Figure 12.2). This result was the opposite from that predicted. Using a *post hoc* test for multiple comparisons (Bonferroni), the mean ($\pm$ SEM) difference in year of birth for sole wives and first wives was 5.01 ± 3.13 years ($P = 0.67$). The difference between sole wives and second wives was 8.61 ± 3.03 years ($P = 0.03$), and the difference between sole wives and third or fourth wives was 13.33 ± 5.06 years ($P = 0.06$). At $P = 0.05$, only the difference between sole wives and second wives was statistically significant, but the trend was for wives of successively higher order to be progressively younger (Figure 12.2).

The body mass (weight over height squared) of monogamously and polygynously married women was indistinguishable (Table 12.6). However, after controlling for whether a woman was cycling (as opposed to pregnant or in lactational amenorrhoea), the sample size was small ($N = 24$). If age and body mass are indicative of mate value, then I found no evidence that mate value was higher for women in monogamous unions. They did, however, have higher reproductive success. After adjusting for other significant covariates, women who spent most of the decade between 1988 and 1998 in monogamous marriages had approximately one more offspring who survived to age ten years than did women who had been in polygynous marriages (Table 12.7).

Table 12.6. *The relationship between female marital status and body mass, defined as weight (kg) over height squared* (m^2) *(N = 24)*[a]

Predictor	Coefficient	SE	*P*
Age (years)	0.0	0.0	0.91
1st wife[b]	−0.6	1.0	0.52
2nd/3rd wife[b]	−1.0	0.9	0.29
Intercept	21.6	–	–

[a] Data are for women who were not pregnant or in lactational amenorrhoea at the time of measurement in 1987 (based on urinary steroid hormone levels).
[b] Reference category: sole wife.

To evaluate the effect size of one extra offspring, it is helpful to consider the mean ($\pm$ SD) number of offspring who survived to age ten years for all women who last gave birth at least ten years prior to their last interview. Lifetime reproductive success calculated in this way is 3.58 ± 1.79 offspring. Thus, one extra surviving offspring for women in monogamous marriages, after

Table 12.7. *Predictors of female reproductive success (offspring survivorship to age ten years) (N = 104 women, $R^2 = 0.14$)*[a]

Predictor	Coefficient	SE	P
Mother's year of birth	0.011	0.013	0.391
1st wife[b]	−1.071	0.409	0.010
2nd, 3rd, 4th wife[b]	−1.215	0.408	0.004
Arranged wife[c]	−1.780	0.769	0.023
Non-arranged wife[c]	−1.670	0.759	0.030
Intercept	−15.874	–	–

[a] This analysis uses the Kaplan Meier method as described in Strassmann and Gillespie (2002).

[b] Reference category is sole wife. First to fourth wives did not differ significantly from each other.

[c] Reference category is levirate wife. Arranged and non-arranged wives did not differ significantly from each other.

adjusting for covariates, represents a substantial gain in reproductive success over that of women in polygynous unions.

The reproductive success of women in arranged marriages did not differ significantly from that of women in non-arranged marriages, but levirate wives had 1.8 more offspring.

DISCUSSION

The human mating system examined here echoes that of polygynous species in general in that variance in reproductive success was greater among males than females. The data support the hypothesis that polygynous WEGs were wealthier than their monogamous counterparts, but there was no evidence for the dilution of wealth under polygyny. *Assuming that all else is equal*, females should prefer wealthier males, but nearly three-quarters of the variance in the number of wives cannot be attributed to wealth differences. Therefore, resource defence provides a partial but not sufficient explanation for Dogon polygyny.

Controlling for age, the Pearson partial correlation coefficient between the wealth rank of a man's WEG and the number of wives he was married to concurrently was 0.38 ($P = 0.001$); cumulatively the coefficient was 0.23 ($P = 0.06$). In a meta analysis, W. Irons (unpublished data), rated the quality of research design for studies of the Kipsigis, the Mukogodo, and the Mormons as particularly high. These studies are also among the few that report effect sizes for the association between wealth and polygyny and therefore provide a useful basis for comparison. In the Kipsigis, who are agropastoralists of Kenya, the Pearson correlation coefficient between wealth and polygyny in six cohorts of men ranged from 0.13 (n.s.) to 0.91 ($P < 0.001$) per cohort with a mean ± SD of 0.48 ± 0.26 (Borgerhoff Mulder, 1987). In the Mukogodo pastoralists of Kenya, the partial correlation between livestock wealth and number of wives for all men of six cohorts was 0.41 ($P < 0.001$) and the correlation for individual cohorts ranged from 0.07 (n.s.) to 0.57 ($P < 0.05$) with age controlled (Cronk, 1991). In nineteenth century Mormons of Utah, the partial correlation between wealth and number of wives was 0.17 ($P < 0.006$) to 0.27 ($P < 0.001$) controlling for a man's age and rank (Mealey, 1985). Thus, the partial correlation coefficient of 0.38 between wealth and concurrent wives for the Dogon (controlling for age) is similar to the coefficients for these other three populations.

Returning to the linear regression results, age and wealth explain 28% of the variance in number of wives in the study village. But what explains the remaining 72% of the variance? The answer may be other indicators of male mate value. In a survey of 37 societies, females in every sample ranked 'kind and understanding' and 'intelligent' as more important than earning power (Buss, 1989). Part of the emphasis on wealth in human behavioural ecology may stem from the fact that wealth is more quantifiable.

In the Dogon, it would be interesting to test the role of masked dancing as a male display comparable in function to the peacock's tail or the bowerbird's bower. The house of eight storeys mask could break the dancer's neck and may be a costly signal of male quality. Even more dangerous is the stilt-walker mask, in which the man's legs are lashed to tall sticks. Masked dancing is a performance that is intently watched by the women, who can identify the individual dancers. It is tempting to speculate that the weight and difficulty ratings for individual masks might correlate with the dancer's number of wives or offspring. Doubtless many other male qualities also contribute, perhaps even the indigenous explanation that if a man is to retain multiple wives he will require 'unusual patience'.

The data support the hypothesis that polygynous males had higher reproductive success than monogamous males. For each extra concurrent wife, a man gained 1.8 extra offspring (after adjusting for age). In contradiction to the notion of equal female reproductive success under monogamy and polygyny (first critiqued by Altmann *et al.*, 1977), females in polygynous unions appeared to incur a cost. This cost was approximately one fewer surviving offspring out of a mean of 3.6 surviving offspring per woman. Consistent with the absence of a hierarchy among co-wives by order of marriage, the cost was approximately the same for first wives and higher order wives. It is possible that the apparent cost of polygyny merely reflects a tendency for lower quality women to enter polygynous marriages, but in that case one would not expect first wives to bear the same fitness cost as second and subsequent wives. Moreover, contrary to my original expectations, I could detect no superiority in the mate quality of women in monogamous versus polygynous unions.

The mean year of birth for monogamously married women was 1950 and that of third and fourth wives was 1964. The finding that polygynously married women were generally younger suggests two hypotheses for future research. One is that the prevalence of polygyny has increased over time, despite the lack of evidence for this between 1988 and 1998. Another hypothesis is that older women were more likely to have lost their former co-wives via death or divorce and their husbands had not succeeded in replacing them.

The ideology that males should marry females younger than themselves, coupled with the population age pyramid, assures that the number of women exceeds the number of men on the marriage market. The result is demographic pressure for polygyny as a male strategy throughout West Africa, despite essentially universal marriage and no major perturbation of the sex ratio (such as warfare) (Dorjahn, 1959; Pison, 1985; Chisholm & Burbank, 1991; Strassmann, 1997*b*, 2000). This male strategy is tempered by a female strategy of combating polygyny through divorce. Among the Dogon, female–female competition is rife in polygynous marriages. When a woman loses a child or fails to conceive, she and her husband frequently blame her co-wife. Distrust of a co-wife was cited by women as the main precipitating factor in 10% of all divorces ($N = 88$ divorces) (Strassmann, 1997*b*, 2000). Thus, female–female antagonism contributes to Dogon monogamy.

Two previous prospective studies help shed light on the reproductive cost of polygyny. The first of these used the monthly probability of conception as the dependent variable. After controlling for age of each spouse, marital duration, parity, and nursing status, whether a woman was monogamously or polygynously married did not affect her monthly probability of conception (Strassmann & Warner, 1998). Only one woman in the village had primary sterility, and contrary to my expectation, polygynously married women were not less fecund. The second study examined the odds of death for 176 children who were prospectively followed for up to eight years (Strassmann, 1997*b*, 2000). Controlling for age, sex, economic status, and the ratio of children to adults in the family, the odds of death were 4.6-fold higher as the ratio of married women to men in the WEG increased by one extra woman per man ($P < 0.005$). The new analyses presented here suggest that the high child mortality under polygyny translates into lower reproductive success for mothers. I cannot, however, exclude the possibility that women in polygynous unions recoup this fitness cost in the grandoffspring generation through ‘sexy sons’ (Weatherhead & Robertson, 1979; Hartung, 1982; Josephson, 1993).

Rather than being of high genetic quality, as envisioned for birds (Weatherhead & Robertson, 1979), the sons of polygynists might command more material resources. However, sons are in competition for the same limited inheritance of fields, which may explain why polygynous fathers are less likely than monogamous fathers to invest in the health of sons, but are equally likely to invest in daughters (author’s unpublished data). Polygynous males produce a surplus of sons relative to the available resources, whereas daughters marry out of the lineage and therefore do not compete over land. Further data are required before a conclusion can be reached about fitness effects in the grandoffspring generation; nonetheless, many sons of polygynists may be obliged to accept monogamy.

Levirate wives in this study had higher reproductive success than did wives in arranged or non-arranged marriages. A plausible explanation is that widows who have many surviving offspring marry their husband’s brother so as to continue raising them, whereas widows with fewer children are more likely to remarry elsewhere (Strassmann & Gillespie, 2002). Further data are needed, however, on the causes of variance in reproductive success among women.

Several of the factors that have been mentioned in connection with monogamy in other species also characterize the Dogon. These include extensive male paternal investment, especially through patrilineal inheritance of millet fields, the critical resource for survival and reproduction (Strassmann, 1992, 1997*b*). Females are dependent on male investment since they do not control critical resources of their own. Females are vulnerable to male coercion, as demonstrated, for example, by female compliance with menstrual taboos imposed by males (Strassmann, 1992, 1996[1]). Mate guarding among the Dogon is extreme, and includes excision of the clitoris, whereas in other primates genital mutilation is found only in the context of aggression. However, the above observations characterize the Dogon as an ethnic group. They apply to individuals in both socially monogamous and socially polygynous unions, and thus did not prove helpful for teasing apart these two kinds of marriage in this human society.

CONCLUSION

1. Among the Dogon of Sangui, about half of all males and a third of all females were in socially monogamous marriages. As is true for polygynous species, the variance in putative reproductive success was greater among males than females.
2. Controlling for age, monogamously married men were significantly poorer, but wealth differences explained only 28% of the variance in the number of concurrent wives. These results agree with other studies of mate choice in humans, which demonstrate that wealth is but one criterion for a desirable spouse. Among the Dogon, it is unclear what aspects of male mate value may account for the remaining 72% of the variance in number of wives. The partial correlation coefficient between wealth and number of wives (0.38, $P = 0.001$) was similar to that of other polygynous societies. Resource defence clearly contributes to polygyny among the Dogon, but further research is needed on other causes.
3. Females in monogamous and polygynous marriages did not differ significantly with respect to body mass or fecundity. Thus I found few correlates of monogamous marriage for females. Instead, polygyny is facilitated by the 8-year difference in age at marriage between the sexes. Countervailing pressures for monogamy include female–female antagonism and the inability of some males to support more than one wife and her offspring.
4. For each extra wife that a man married, he gained 1.8 extra offspring, controlling for age. Thus, monogamously married men had lower reproductive success. Wealthier males had higher reproductive success above and beyond the additional wives that they married.
5. Monogamously married women had higher reproductive success as measured by offspring survival to age ten years. Compared with women in polygynous marriages, they raised one extra child each. Monogamy is advantageous for females even after controlling for wealth, which may be due to the dilution of paternal investment under polygyny.
6. The polygyny threshold model is consistent with the common sense notion that when all else is equal women prefer wealthier men, but the argument that polygyny results from female choice disregards the demonstrable costs of polygyny for female fitness among the Dogon. Moreover, nearly half of women in polygynous marriages were first wives and I learned of no instance of a woman encouraging her husband to marry again. The ideology that age at marriage should be markedly lower for women serves the interests of wealthier, more desirable males because it constrains the options of females, channelling many into polygynous unions. Thus, Dogon polygyny provides an interesting example of how normative beliefs are compatible with the reproductive interests of high status males.
7. Conflict exists both within and between the sexes over the optimal mating system; identification of the causes of this conflict is a useful theoretical starting point for understanding monogamy versus polygyny in humans.

Note

1 Strassmann, B. I. (1985), The Ecology of Polygyny among the Dogon, Grant Proposal to the L. S. B. Leakey Foundation, and (1986), The Ecology of Polygyny among the Dogon, US National Science Foundation Grant (BNS-8612291).

References

Alexander, R. D. (1979). *Darwinism and Human Affairs*. Seattle: University of Washington Press.

Alexander, R. D. & Noonan, K. M. (1979). Concealment of ovulation, parental care, and human social evolution. In *Evolutionary Biology and Human Social Behavior: An Anthropological Perspective*, ed. N. A. Chagnon & W. G. Irons, pp. 436–53. North Scituate, Massachusetts: Duxbury Press.

Altmann, S. A., Wagner, S. S. & Lemington, S. (1977). Two models for the evolution of polygyny. *Behavioral Ecology and Sociobiology*, **2**, 397–410.

Borgerhoff Mulder, M. (1987). On cultural and reproductive success: Kipsigis evidence. *American Anthropologist*, **89**, 617–34.

(1988). The relevance of the polygyny threshold model to humans. In *Human Mating Patterns*, ed. C. G. N. Mascie-Taylor & A. J. Boyce, pp. 209–30. Cambridge: Cambridge University Press.

(1990). Kipsigis women's preferences for wealthy men: evidence for female choice in mammals? *Behavioral Ecology and Sociobiology*, **27**, 255–64.

Buss, D. (1989). Sex differences in human mate preferences. *Behavioral and Brain Science*, **12**, 1–14.

Cazes, M.-H. (1993). *Les Dogon de Boni*. Paris: Institut National d'Etudes Démographiques.

Chisholm, J. S. & Burbank, V. K. (1991). Monogamy and polygyny in southeast Arnhem Land: male coercion and female choice. *Ethology and Sociobiology*, **12**, 291–313.

Cronk, L. (1991). Wealth, status, and reproductive success among the Mukogodo of Kenya. *American Anthropologist*, **93**, 345–60.

Daly, M. & Wilson, M. (1983). *Sex, Evolution, and Behavior*, 2nd Edition. Belmont, California: Wadsworth Publishing Co.

Davies, N. B. (1989). Sexual conflict and the polygamy threshold. *Animal Behaviour*, **38**, 226–34.

Dorjahn, V. R. (1959). The factor of polygyny in African demography. In *Continuity and Change in African Cultures*, ed. W. R. Bascom & M. J. Herskovits, pp. 87–112. Chicago: University of Chicago Press.

Downhower, J. F. & Armitage, K. B. (1971). The yellow-bellied marmot and the evolution of polygamy. *American Naturalist*, **105**, 35–70.

Hames, R. (1996). Costs and benefits of monogamy and polygyny for Yanomamo women. *Ethology and Sociobiology*, **17**, 181–99.

Hartung, J. (1982). Polygyny and inheritance of wealth. *Current Anthropology*, **23**, 1–12.

Hosmer, D. W. & Lemeshow, S. (1989). *Applied Logistic Regression*. New York: John Wiley.

Irons, W. (1983). Human female reproductive strategies. In *Social Behavior of Female Vertebrates*, ed. S. Wasser, pp. 169–213. New York: Academic Press.

Irons, W. (2003). Cultural and reproductive success in traditional societies. Paper presented at the Evolution and Human Behavior Society annual meeting.

Josephson, S. C. (1993). Status, reproductive success, and marrying polygynously. *Ethology and Sociobiology*, **14**, 391–96.

Mealey, L. (1985). The relationship between social status and biological success: a case study of the Mormon religious hierarchy. *Ethology and Sociobiology*, **6**, 249–57.

Orians, G. H. (1969). On the evolution of mating systems in birds and mammals. *American Naturalist*, **103**, 589–603.

Paulme, D. (1940). *Organisation Sociale des Dogon*. Paris: Les Editions Domat-Montchrestien.

Pern, S. (1982). *Masked Dancers of West Africa: The Dogon*. Amsterdam: Time-Life Books.

Pison, G. (1985). La démographie de la polygamie. *La Recherche*, **168**, 894–901.

Sellen, D. W., Borgerhoff Mulder, M. & Sieff, D. F. (2000). Fertility, offspring quality, and wealth in Datoga pastoralists; testing evolutionary models of intersexual selection. In *Adaptation and Human Behavior: An Anthropological Perspective*, ed. L. Cronk, N. Chagnon & W. Irons, pp. 91–114. New York: Aldine de Gruyter.

Strassmann, B. I. (1981). Sexual selection, paternal care, and concealed ovulation in humans. *Ethology and Sociobiology*, **2**, 31–40.

(1992). The function of menstrual taboos among the Dogon: Defense against cuckoldry? *Human Nature*, **3**, 89–131.

(1996). Menstrual hut visits by Dogon women: a hormonal test distinguishes deceit from honest signaling. *Behavioral Ecology*, **7**, 304–15.

(1997*a*). The biology of menstruation in *Homo sapiens*: total lifetime menses, fecundity, and nonsynchrony in a natural fertility population. *Current Anthropology*, **38**, 123–9.

(1997*b*). Polygyny as a risk factor for child mortality among the Dogon. *Current Anthropology*, **38**, 688–95.

(2000). Polygyny, family structure, and child mortality: a prospective study among the Dogon of Mali. In *Adaptation and Human Behavior: An Anthropological Perspective*, ed. L. Cronk, N. Chagnon & W. Irons, pp. 49–67. New York: Aldine de Gruyter.

Strassmann, B. I. & Gillespie, B. (2002). Life history theory, fertility, and reproductive success in humans. *Proceedings of the Royal Society of London, Series B*, **269**, 553–62.

Strassmann, B. I. & Warner, J. (1998). Predictors of fecundability and conception waits among the Dogon of Mali. *American Journal of Physical Anthropology*, **105**, 167–84.

Trivers, R. L. (1972). Parental investment and sexual selection. In *Sexual Selection and the Descent of Man*, ed. B. Campbell, pp. 136–79. Chicago: Aldine.

Vehrencamp, S. L. & Bradbury, J. W. (1984). Mating systems and ecology. In *Behavioural Ecology: An Evolutionary Approach*, ed. J. R. Krebs & N. B. Davies, pp. 251–78. Sunderland, Massachusetts: Sinauer Associates.

Verner, J. (1964). Evolution of polygamy in the long-billed marsh wren. *Evolution*, **18**, 252–61.

Verner, J. & Willson, M. F. (1966). The influence of habitats on mating systems of North American passerine birds. *Ecology*, **47**, 143–7.

Weatherhead, P. J. & Robertson, R. J. (1979). Offspring quality and the polygyny threshold: the "sexy son hypothesis." *American Naturalist*, **128**, 499–512.

Winterhalder, B. & Smith, E. A. (2000). Analyzing adaptive strategies: human behavioral ecology at twenty-five. *Evolutionary Anthropology*, **9**, 51–72.

CHAPTER 13

Social monogamy in gibbons: the male perspective

Ulrich H. Reichard

INTRODUCTION

To date, McCann's (1933) encounter with a male hoolock (*Hylobates hoolock*) is the only documented sighting of a wild gibbon male (other than a siamang, *H. syndactylus*) carrying an infant. McCann shot both the adult male carrying the four-month-old infant, as well as another male, not yet fully grown. No female that could have belonged to this group was heard or seen in the neighbourhood. But, this anecdote aside, there is no other evidence to indicate that males in the genus *Hylobates* – *H. syndactylus* excepted – exhibit any measurable amount of direct paternal care in the form of infant carrying (Chivers, 1974; Fischer & Geissmann, 1990). Unless one assumes the unlikely scenario that with the exception of the siamang, direct paternal care was lost secondarily during hylobatid evolution, selective pressures other than the need for paternal care must be considered as elements in the evolution of social monogamy (cf. Komers & Brotherton, 1997; van Schaik & Kappeler, 1997).

In particular, benefits to or constraints on males need to be identified in order to understand the evolution of social monogamy (Clutton-Brock, 1989). The classical constraints on male reproductive strategies are the temporal distribution of fertile females, the spatial distribution of resources, and relationships between females. When females favour a non-gregarious, widely dispersed lifestyle (cf. Sterck *et al.*, 1997) and live in exclusive ranges, males are left with limited options to maximize their fitness. They may be forced to accept social monogamy as a consequence of their inability to defend areas large enough to encompass several female ranges (Emlen & Oring, 1977; Wrangham, 1980; Rutberg, 1983; van Schaik & van Hooff, 1983). Alternatively, conditions may arise in which the fitness gains of pair living for males exceed the benefits of roving and trying to defend multiple females. Males may either stay with one female to ensure paternity, maintaining a close sociospatial relationship in order to improve the chances of future breeding with a particular female (Brotherton & Komers, chapter 3), or males could provide indirect benefits to offspring survival, which may make pair living a successful, preferred male strategy. Protecting offspring from predation, infanticide, or to secure resources have all been suggested as important for the evolution of social monogamy (Dunbar & Dunbar, 1980; Rutberg, 1983; Brockelman & Srikosamatara, 1984; van Schaik & Dunbar, 1990; van Schaik & Kappeler, chapter 4).

Gibbons (family: *Hylobatidae*) are among the standard-bearers of social monogamy in primates. Since the brief, pioneering study of a wild, white-handed gibbon population (*H. lar*) by Carpenter (1940) in the late 1930s, there has been little doubt that the basic social unit of the genus is the 'monogamous family' (Kleiman, 1981; Leighton, 1987). Interestingly, however, even Carpenter's study (1940) noted an additional, albeit senile, male in each of two study groups (*H. lar*). Recently, long-term documentation of group structures from a natural, white-handed gibbon population at Khao Yai, Thailand, also revealed an unexpected degree of social flexibility (Sommer & Reichard, 2000; Reichard, unpublished data). Table 13.1 summarizes cases where two adult males, two adult females or more than three adults were observed in wild gibbon groups. Most of the anecdotal observations of other configurations in addition to the 'monogamous family' are considered to reflect exceptional situations (Srikosamatara & Brockelman, 1987; cf. Bleisch & Chen, 1991), but the systematic study from Khao Yai, Thailand, suggests that viable alternatives to social monogamy may exist.

This chapter is about male reproductive strategies and gibbon social monogamy primarily based on data from a white-handed gibbon population (*Hylobates lar*) at Khao Yai, Thailand (Figure 13.1). I will show that Khao Yai males are capable of defending areas large enough to encompass several female ranges and I will

Table 13.1. *Variation in group structures of hylobatids*

Species	Site	Group structure (number of groups)			Duration	Remarks	Reference
		2 M – 1 F	2 F – 1 M	>2 full adult size individuals			
Hylobates hoolock	West Bhanugach, Bangladesh	1 gr			≥7 mo	Females floated between staying with male and staying alone; (one) female disappeared later (Islam & Feeroz, 1992)	Ahsan, 1995
	West Banugach, Bangladesh	1 gr	1 gr		–	Survey data; disturbed forest	Siddiqi, 1986
	Barail, India		1 gr		–	Survey data	Choudhury, 1990, 1996
	Lohit, India			1 all male gr (5 ad, 2 juv)	–	Survey data	Mukherjee *et al.*, 1992
H. syndactylus	Kuala Lompat, Malaysia	1 gr			≥6 mo	(One) female disappeared later (Aldrich-Blake & Chivers, 1973)	Chivers & Raemaerkers, 1980
H. klossii	North Pagai, Indonesia			9 full-size ad	4 mo	No details of group composition	Fuentes, 1999
H. concolor	Wuliang, China	3 gr		2 gr (4 ad F + 1 ad M; 3 ad F + 1 ad M)	– –	Survey data; auditory & visual spot observations	Haimoff *et al.*, 1986
	Bawanglin, China	2 gr			–	Population recovering from drastic decline; restricted to 2 isolated forest patches	Liu *et al.*, 1989
H. pileatus	Khao Soi Dao, Thailand	1 gr			≥1.5 y	Both females initially carried young infants	Srikosamatara & Brockelman, 1987
H. lar	Doi Dao, Thailand		2 gr		–	–	Carpenter, 1940
	Khao Yai, Thailand	10.1% (SD ± 6.2)	2.5% (SD ± 3.5)	– –	Variable 1 d – >4 y	–	Reichard, unpublished data

ad, adult; juv, juvenile; F, female(s); M, male(s); y, year(s); mo, month(s); d, day; gr, group(s).

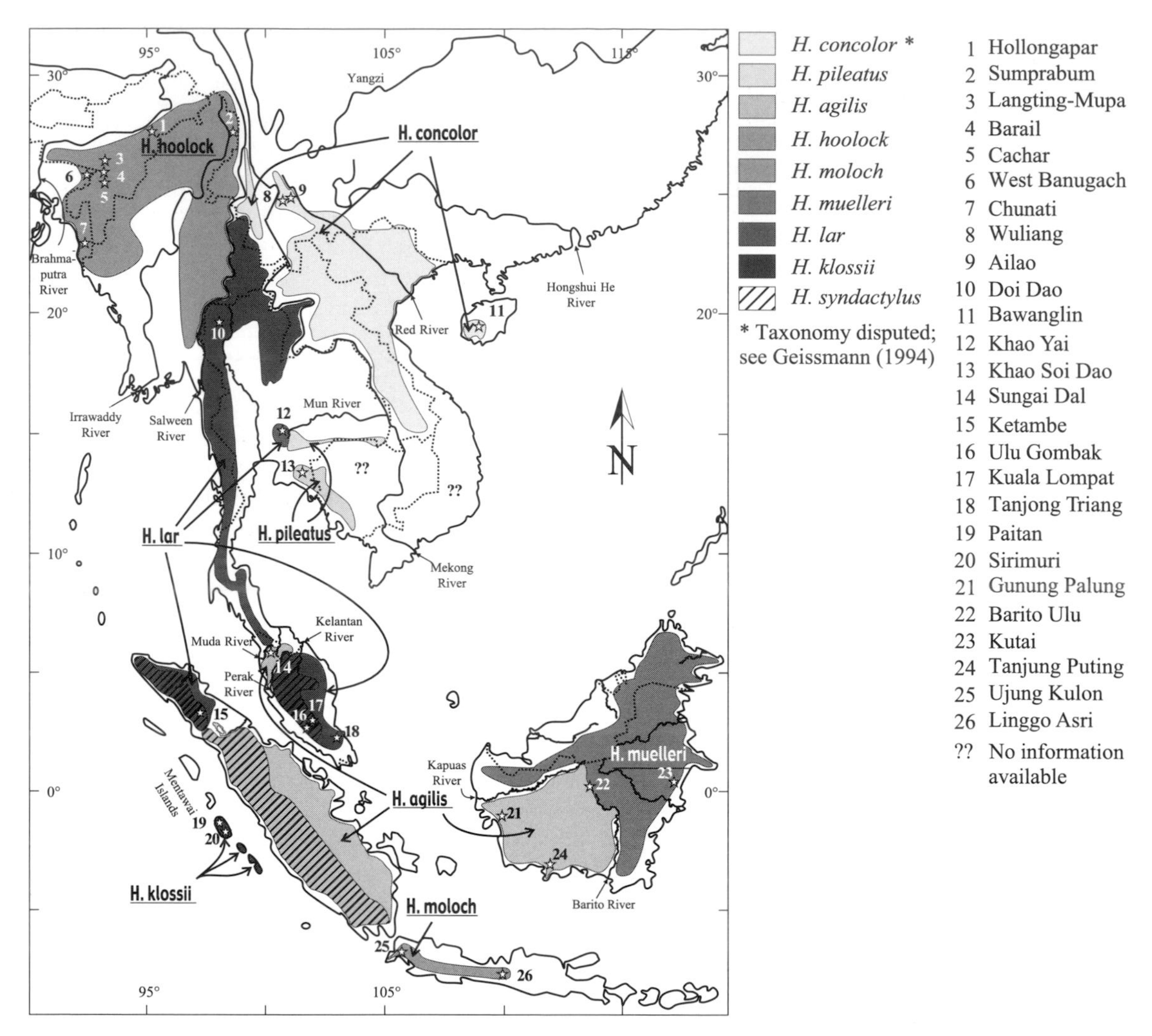

Figure 13.1. Approximate geographic distribution of gibbon species and study sites in Southeast Asia based on Alfred & Sati (1990), Brockelman (1975), Brockelman & Gittins (1984), Chivers (1971, 1977, 1978), Choudhury (1990, 1996), Dallmann & Geissmann (2001), Daltry & Momberg (2000), Dao Van Tien (1983), Dehua & Peikun (1990), Ellefson (1974), Evans *et al.* (2000), Gittins (1978), Gittins & Akonda (1982), Groves (1967), Haimoff *et al.* (1987), Islam & Feeroz (1992), Kappeler (1984), Lan (1993), Lan *et al.* (1990), Liu *et al.* (1987, 1989), Ma & Wang (1986), Ma *et al.* (1988), Marshall (1990), Mather (1992), Mootnick *et al.* (1987), Mukherjee (1982, 1986), Mukherjee *et al.* (1992), Siddiqi (1986), Srikosamatara (1984), Tenaza (1975), and Whitten (1980).

also discuss the potential benefits males may gain from pair living. My focus will be twofold: the majority of the chapter takes a traditional, male-centred perspective in which a male is seen as part of a single male–female unit. At the end, I present a perspective in which a male is seen as part of the local community in/with which he lives. Mating systems can be understood as the visible outcome of compromises of male and female reproductive strategies (Davies, 1991), and where male options are evaluated the female perspective of social monogamy in gibbons is therefore incorporated. The assumption here is that gibbon females live in exclusive ranges primarily to reduce within-group contest feeding competition (cf. Sommer & Reichard, 2000; Uhde & Sommer, 2002). Gibbon anti-predation strategies are probably efficient enough to decrease predation risk to such a level that the benefits of further reducing predation through forming multi-female groups are outweighed by increased within-group feeding competition (Uhde, 1997; Reichard, 1998; Sommer & Reichard, 2000). Currently, there is no comprehensive study on the genetic structure of the Khao Yai or any other gibbon population, which restricts the scope of this chapter to sociosexual male–female relationships and the assumed reproductive benefits of alternative male strategies.

SITE, TIME, AND SPECIES

I studied white-handed gibbons (*Hylobates lar*) at Khao Yai National Park, Thailand (Figure 13.1), for about 2900 hours spread over *c.* 650 days between 1991 and 2002. The site is located 730–890 m above sea level inside a seasonally wet, evergreen rainforest. Long-term demographic and behavioural data were collected next to and west of the park headquarters, an area known as Mo Singto, and at a newly established site northwest of Mo Singto, known as Klong E-Tau. Mo Singto and Klong E-Tau are part of the same continuous forest. Unrestricted migration and dispersal are possible between the areas, distinguished here only to acknowledge local references and descriptions. Gibbons from both areas are considered a single population.

A total of 48 different groups was recognized and studied over the 12-year study period (Mo Singto: 25 groups; Klong E-Tau: 23 groups). Two groups (K and Y) dissolved and disappeared from the site during the study. Few patches of tall secondary growth exist at the site, and where possible, these are used by gibbons for travel and foraging. Except for illegal exploitation of one tree species, *Aquilaria crassna* (family: *Thymeleaceae*), and one exceptional case in which a gibbon was shot, the study population was well protected and believed not to be threatened by poaching or forest destruction.

Gibbons are small-bodied, highly agile, arboreal apes that have long been known to live in socially monogamous pairs (Carpenter, 1940; Chivers, 1974). They inhabit the remaining rainforests of South and Southeast Asia, from the northeastern tip of India to southern China, across Bangladesh and the Malaysian peninsula to Borneo, Java, and Sumatra, including the Mentawai Islands (Figure 13.1). All gibbon species live allopatrically, with the exception of the siamang (*H. syndactylus*), which occurs sympatrically with white-handed gibbons (*H. lar*) in northern Sumatra and agile gibbons (*H. agilis*) on southern Sumatra and the Malaysian mainland. A few small natural hybrid zones exist in Thailand, Malaysia, and in Kalimantan, Indonesia (Brockelman & Gittins, 1984; Mather, 1992).

Age classes were established in order to help correctly identify adults and immature individuals (Table 13.2). Adults are defined as fully grown individuals, independent of a paired or unpaired social status. At Khao Yai, full adult body size was not reached before eight years of age. Age classifications used in this study correspond to those used in other gibbon studies (cf. Leighton, 1987; Mitani, 1990). In the field, age classes were determined by comparing body sizes of all group members. Additionally, physical cues including characteristics of the typical white face-ring, relative strength of the voice, development of genitalia and visible cues of female reproductive stage (ano-genital swellings), as well as behavioural criteria like relative proximity to other group members, singing performance, and within-group foraging dynamics (Table 13.2), were used to identify individuals.

LARGE-RANGE SOCIOREPRODUCTIVE POLYGYNY (LRP) IN KHAO YAI GIBBON MALES?

Males benefit from access to multiple, reproductively active females under most conditions (Trivers, 1972). When females live in exclusive ranges, the potential for socioreproductive polygyny is reduced because it becomes more difficult for a male to defend an area large enough to encompass two or more female ranges than to defend a small group of gregarious females (Emlen & Oring, 1977). However, Komers and Brotherton (1997) found that the space inhabited by gregarious, socially polygynous mammalian females is relatively larger compared with the space inhabited by socially monogamous females. Consequently, if females live in exclusive but relatively small ranges, a male's opportunity for socioreproductive polygyny increases. It therefore seems plausible to expect that a gibbon male would try to maximize his fitness by defending an area large enough for more than one female and reproducing with all the females in his range. This kind of male strategy can be defined as *large-range socioreproductive polygyny* (LRP). And just such a strategy has been observed in some mammals (Green *et al.*, 1998; Fisher & Owens, 2000). The critical question is whether or not a male would be physically capable of realizing the LRP-strategy and defending a range substantially larger than a single female's

Table 13.2. *Age classes with corresponding physical and behavioural development in wild white-handed gibbons, Khao Yai, Thailand*

Age class	Approximate age (years)	Physical features	Behavioural development[a]
Infant	0–2	Small body size	Carried by mother; suckling
Juvenile	2.1–4	Weaned; about $\frac{1}{3}$ of adult size; thick white face ring; high voice	Independent travel; close proximity to mother; prolonged play with siblings; females exceptionally singing great call with mother
Adolescent	4.1–6	About $\frac{1}{2}$ of adult size; high voice	Decreasing proximity to mother
Female		Vulva may become visible	Occasionally singing great call with mother
Male		Testes may become visible	Occasional singing of morning solo song
Subadult	6.1–8	About $\frac{3}{4}$ of adult size; decreasing thickness of white face ring; strong but slightly higher voice than adults	Decreasing proximity to other group members; occasionally exclusion from small food sources by adult pair; decreasing play activity
Female		Vulva visible	Regular singing of great call with mother
Male		Testes visible	Regular singing morning solo song; occasionally encounters neighbouring individuals alone; onset of sexual behaviour
Adult	>8	Full adult body size; strong adult voice	
Female (*mature offspring*)		Onset of sex-skin swelling cyclicity; first parturition possible	Occasional or rare singing of great call with mother; regularly excluded from food sources; frequently peripheral from group (*c*. 20–50 m and further)
Male (*mature offspring*)			Frequent morning solo singing; regularly excluded from food sources; frequently peripheral from group (*c*. 20–50 m and more); occasional sexual behaviour; occasional departure from natal home range for consecutive days
Female		Regular sex-skin swelling cyclicity	Regular duetting with adult male; regular sexual behaviour during fertile period
Male			Regular duetting with adult female; frequent morning solo singing; regular sexual behaviour during female cycle
Old Adult		Thin white face ring; sparse hair around neck, shoulders and chest (females); legs appear 'thin' in both sexes	Decreasing singing activity

[a] Note: Due to individual variation in development, behavioural development does not always tightly correspond to age classes. Not all individuals necessarily show all behaviours listed.

range. For gibbons, as for many other socially monogamous species (cf. Wittenberger & Tilson, 1980), it is commonly assumed that males would benefit from LRP but that they are unable to defend large enough ranges (Ellefson, 1968; Gittins & Raemaekers, 1980).

I tested whether Khao Yai gibbon males were theoretically able to realize an LRP-strategy, and followed van Schaik and Dunbar's (1990) model to calculate the number of females a male could expect within his range if he defended the largest possible home range. First, I calculated the maximal defendable area. The basic assumption for this calculation was that an individual male had to be able to reach the perimeter of his range frequently enough to detect potential intruders. The maximal defendable area was modelled as a circle. It was determined by the relationship between day range length and home range size, which represented a circle's diameter and a circle respectively (cf. Mitani & Rodman, 1979). In a second step, the probability that a male would find the females within his range when those females were fertile was calculated using a 'gas-model' approach, as suggested by Dunbar (1988). This approach is based on the assumption that a male would randomly meet a female within his territory in the way that atoms would collide with each other in a gas cloud. The calculations of the LRP-strategy model yielded two results: a theoretical number of females that could live in the maximal defendable area of a male, and a probability value for the number of females a male could expect to meet by chance at least once while the females are fertile, if he randomly moved around his home range. Parameters and equations used in the following calculations are summarized in Table 13.3, and Table 13.4 summarizes the database for Khao Yai.

LRP-model, step one: maximal defendable area and number of females per male range

The maximum area (A_{max}) a male could successfully defend against intruders based on Mitani and Rodman's (1979) inverted equation for territory defensibility in primates, is given by van Schaik and Dunbar (1990) as

$$A_{max} = 0.25 \cdot \pi \cdot d^2 \tag{13.1}$$

and the number of females f that could live in the range is given by

$$f = \frac{A_{max}}{A_{female+offspring}} \tag{13.2}$$

with $A_{female+offspring}$ representing the size of a home range a female and offspring would need. Based on observed mean day journey lengths d for several gibbon populations, van Schaik and Dunbar (1990) concluded that with few exceptions, males of the tested wild populations should be capable of defending areas large enough to be inhabited by two to five females.

Database I for Khao Yai

Mean day journey length $d = 1.19$ km/day for Khao Yai ($N = 79$ days; pooled data for three groups; March–June & November 1992: Nettelbeck, 1993; Neudenberger, 1993) was only marginally shorter than those reported for other white-handed gibbon populations (e.g., 1490 m/day; $N = 91$ days: Raemaekers, 1979). To calculate the maximum number of females that could reside within a male home range it was necessary to determine the average proportion of a home range a male would need for himself (A_{male}). Following van Schaik and Dunbar (1990), the area needed by a male was assumed to be a function of group size divided by mean home range size because observed home range size was assumed reliably to reflect the area needed by a gibbon group. Offspring were considered to need as much food as adults for unconstrained development, which was a conservative assumption because any bias due to this assumption would lead to an underestimation of the number of females that could reside in a male's home range. Mean group size at Khao Yai was 4.0 ± 0.3 individuals ($N = 170$ social units; pooled data across nine census years, 1992–2001: Table 13.4). Therefore, a male was assumed to use one-quarter or 25% of the home range resources. At worst, this somewhat arbitrary value for the area needed by a male is an overestimate that will lead to a smaller range for females and offspring. Average home range size at Khao Yai was 0.24 km^2 (pooled data for three groups, August–December 1992: Neudenberger, 1993). A male was expected to need on average $A_{male} = \frac{0.24}{4} = 0.06$ $km^2$of the home range for himself. The female and her offspring were then assumed to need the remainder of the home range, or on average $A_{female+offspring} = 0.24 - 0.06$ $km^2 = 0.18$ km^2 for their survival and development. All following calculations of the maximum number of females per male home range implicitly considered that each female would have sufficient space and resources for herself and a number of offspring accompanying her within the home range.

Table 13.3. *Parameters, equations, and values calculated for gibbon male Large Range Socio-reproductive Polygynous (LRP-) strategy*

	Maximum defendable area[a] A_{max} (km^2)	Maximum number of females per male home range f	Fertile females[b, c] E_{fs}	Probability of meeting *non-receptive* female[b] P_0	Expected frequency of locating fertile female[b] m	Female receptivity probability S	Female mating period S_f (days)
Equation	$A_{max} = 0.25 \cdot \pi \cdot d^2$		$E_{fs} = f \cdot (1 - P_0)$	$P_0 = \frac{e^{-m} \cdot m^x}{x!} = e^{-m}$	$m = \frac{2 \cdot r \cdot d \cdot S \cdot B}{A}$	$S = \frac{S_f}{T_f}$	$S_f = \frac{\sum_{i=1}^{n} c}{B}$
Calculated value	1.05	5.6	4.19	2.0×10^{-6}	13.1		
Remarks	Corrected for area needed by male ($A_{male} = 0.06\ km^2$)	$A_{max} = 1\ km^2$	$A_{max} = 1\ km^2$	$A_{max} = 1\ km^2$	$A_{max} = 1\ km^2$		

[a] Equation after Mitani & Rodman (1979) and van Schaik & Dunbar (1990).
[b] Equation after van Schaik & Dunbar (1990).
[c] Expected number of females met at least once during their fertile period by a randomly searching male.

Table 13.4. *Combined databases I & II for Khao Yai gibbon male Large Range Socio-reproductive (LRP–) strategy*

	Day journey length d (km/day)	Group size z	Detection radius r (km)	Female active period T_f (days)	Female fertile period c (days)	Cycles before conception n	Birth interval B (years)	Weaning period w (months)	Gestation period g (days)	Cycling window t (days)	Menstrual cycle length M (days)
Equation										$t = B - (w + g)$	
Mean	1.19	4.0	0.1	365	6.1	9	3.43	23.1	210	270	30
SD		±0.3			±2.4		±0.96	±1.7			
Range		3.6–4.6					2.66–5.91	21–25			
N	79 days	107 social units			16 cycles		11	6			
Remarks	3 groups pooled; Mar–Jun & Nov 1992	Across 9 years	Guesstimate		11 captive females				Captive females		Captive females
Reference	Nettelbeck, 1993; Neudenberger, 1993	Reichard, unpublished data	This study	This study	Dahl & Nadler, 1992	This study	This study	This study; Brockelman *et al.*, 1998[a]	Napier & Napier, 1967; Geissmann, 1991	This study	Carpenter, 1941; Breznock *et al.*, 1977; Hayssen *et al.*, 1993

[a] One birth interval.

Empty cells = no data given or no calculation possible.

Calculation of maximum defendable area (A_{max}) and maximum number of females (f) per home range for Khao Yai males

Inserting the data from Khao Yai into equation (13.1) reveals

$$A_{\max} = 0.25 \cdot \pi \cdot 1.19^2 = 1.11\ \text{km}^2$$

The maximal defendable area for a Khao Yai male was calculated as $A_{\max} = 1.11\ \text{km}^2$. Of this portion the male needed $A_{male} = 0.06\ \text{km}^2$ for himself, which left an area of $A_{\max} = 1.11 - 0.06\ \text{km}^2 = 1.05\ \text{km}^2$ for females and offspring. Using equation (13.2) to calculate the number of females per male home range reveals a maximum number of

$$f = \frac{1.05}{0.18} = 5.8 \text{ females} + \text{offspring}$$

to live within a male's range. Even if maximum defensibility was set at a more conservative value of $A_{\max} = 1\ \text{km}^2$, because Mitani and Rodman (1979) did not find any primate defending an area larger than 1 km^2,

$f = 5.56$ females + offspring could live in a Khao Yai male's home range.

LRP-model, step two: number of females a male could expect to meet while females are fertile

In a second step, van Schaik and Dunbar (1990) tested the probability that a roving male would locate a female within his range at a time when the female is fertile. I applied Dunbar's (1988) 'gas-model' approach and the adjustments suggested by van Schaik and Dunbar (1990), but used a slightly different equation to calculate female fertility period. It is worth noting that the logic of the LRP-model is conservative because it assumes random movements of males and females. In natural habitats, however, the distribution of fruiting trees and hence female movements, for example, are probably not random, which may increase the chances that a male will find a female above what would be predicted by random search. The expected value of finding a female (E_{fs}) at least once during her fertile period for a randomly roving Khao Yai male was calculated as

$$E_{fs} = f(1 - P_0) \tag{13.3}$$

where f is the maximum number of females that could live within the male home range as calculated above, and P_0 is the probability that any of the females within the male's home range will *not* be fertile at the time the male encounters them. A Poisson distribution was used to model P_0 because the probability of finding a fertile female was considered small in relation to the length of a complete reproductive cycle (assuming cycles to be independent of each other),

$$P_0 = \frac{e^{-m} \cdot m^x}{x!} = e^{-m} \tag{13.4}$$

where x is the number of receptive females encountered and the Poisson parameter m is the expected frequency with which a randomly searching male can locate a given female while she is fertile during the course of a reproductive cycle. An entire reproductive cycle (D, days) lasts from one birth to the next, which can be measured as the birth interval B (years). The expected frequency m can be calculated by the following equation:

$$m = \frac{2 \cdot r \cdot d \cdot S \cdot D}{A} = \frac{2 \cdot r \cdot d \cdot S \cdot B \cdot 365}{A} \tag{13.5}$$

where r is the distance at which a male is expected to be able to detect a female (visibility, km) and d is the mean day journey length (km). A is the average size of a male home range (km^2), which here again was set at the conservative value of $A_{max} = 1\ \text{km}^2$, based on Mitani and Rodman's (1979) finding of maximum defendable area in primates.

A female's fertility period (S) was calculated as the female's receptivity probability, which is the time a female was ready to mate (S_f, days/year) as part of her total active time (T_f, days/year), which in the case of gibbons is a complete 365-day year.

$$S = \frac{S_f}{T_f} = \frac{S_f}{365} \tag{13.6}$$

The mating period of the female (S_f) is a function of the number and length of her oestrous cycles before conception.

$$S_f = \frac{\sum_{i=1}^{n} c}{B} \tag{13.7}$$

where c is the duration of receptivity in days, n is the number of oestrous cycles females experience before conception, and B is the birth interval (in years). Substituting S_f in equation (13.6) yields:

$$S = \frac{\sum_{i=1}^{n} c}{B \cdot 365} \tag{13.8}$$

which leads to a final equation for m:

$$m = \frac{2 \cdot r \cdot d \cdot B \cdot 365 \cdot \frac{\sum_{i=1}^{n} c}{B \cdot 365}}{A} = \frac{2 \cdot r \cdot d \cdot \sum_{i=1}^{n} c}{A} \qquad (13.9)$$

Database II for Khao Yai

Visibility at Khao Yai was judged to be 25–30 m for a human observer. Considering that gibbons enjoy a greater field of vision in the canopy owing to the absence of thick undergrowth, detection radius (r) for spotting conspecifics was assumed to be 50 m on either side of a male's travel path ($r = 0.1$ km). Mean day journey length for three groups was $d = 1.19$ km ($N = 79$ days; March–June & November 1992; pooled data for three groups: Nettelbeck, 1993; Neudenberger, 1993), and maximum male home range size was set at $A_{max} = 1.0$ km^2. The period of a female's maximum genital swelling as observed in captivity was used as a fair approximation for the duration of receptivity (c) because the swollen phase in gibbon females reliably signals fertility (Kollias & Kawakami, 1981). Dahl and Nadler (1992) and Nadler and colleagues (1993) showed that for captive gibbon females sex hormone concentrations and genital swellings change in the predicted way during the menstrual cycle, with ovulation being most likely at the end of the swollen period. They recorded significant swelling scores for a mean duration of 6.1 days (SD $\pm$ 2.4; $N = 16$ cycles of 11 captive gibbon females: Dahl & Nadler, 1992). I calculated birth intervals for Khao Yai females based on 20 live births of nine females in my study population. Because surviving infants best reflected the average birth interval important in this context, and birth dates of infants were known to within ± 1 month, consecutive births were considered only when the previous infant survived until the female's subsequent conception period. Mean birth interval was 3.43 years (SD $\pm$ 11.5 months, range 2.66–5.91 years, $N = 11$: Table 13.4). Calculation of the number of oestrous cycles before conception was based on the period between completion of lactation/weaning and subsequent conception. It was assumed that females would not cycle during lactation due to lactational amenorrhoea (see Dempsey, 1940). In the Khao Yai population, I calculated mean weaning age as 690 days after birth ($N = 6$; SD $\pm$ 50.7 days; range 630–750 days: Table 13.4), which corresponded to weaning ages estimated in other white-handed gibbon studies (Treesucon & Raemaekers, 1984; Harvey & Clutton-Brock, 1985; Hayssen *et al.*, 1993). Studies of captive females indicated that gestation in white-handed gibbons generally lasts about 210 days (Napier & Napier, 1967; Geissmann, 1991) and Khao Yai females were assumed to conform to this pattern. When gestation and lactation periods were subtracted from the mean Khao Yai birth interval (3.43 years = 1170 days), females at Khao Yai were assumed to have experienced an average cycling window of 270 days between weaning and subsequent conception. Mean female menstrual cycle length in captivity was found to be 30 days (Carpenter, 1941; Breznock *et al.*, 1977; Hayssen *et al.*, 1993). Accepting this menstrual cycle length for Khao Yai as well revealed that females could be assumed to experience a maximum of nine menstrual cycles until conception.

Calculation of expected frequency (m) of locating a female and expected number of fertile females found (E_{fs}) by a randomly searching male

Entering all parameter values for Khao Yai into equation (13.9) leads to the following result for m:

$$m = \frac{2 \cdot r \cdot d \cdot \sum_{i=1}^{n} c}{A} = \frac{2 \cdot 0.1 \cdot 1.19 \cdot \sum_{i=1}^{9} 6.1}{1}$$

$$m = 13.1$$

From equation (13.3) the expected number of fertile females (E_{fs}) found by a randomly searching male in a 1 km^2 home range inhabited by 5.6 females at least once during their fertile periods can be calculated as

$$E_{fs} = 5.6 \cdot (1 - e^{-13.1}) = 5.6$$

A Khao Yai male living in a 1 km^2 home range is thus expected to meet all females living in his range by chance at least once while they are fertile.

The above results can be interpreted as realistically indicating that a Khao Yai gibbon male could defend an area large enough for several females and would probably profit from such an LRP-strategy because he could expect to reproduce with all females inhabiting his home range. However, one could argue that some of the assumptions made were unrealistic and resulted in unusually long female fertility periods. Responding to this argument, the expected number of females a randomly searching male could find in his home range during their fertile period was modelled under varying conditions (I–III) that produce a range of circumstances making it

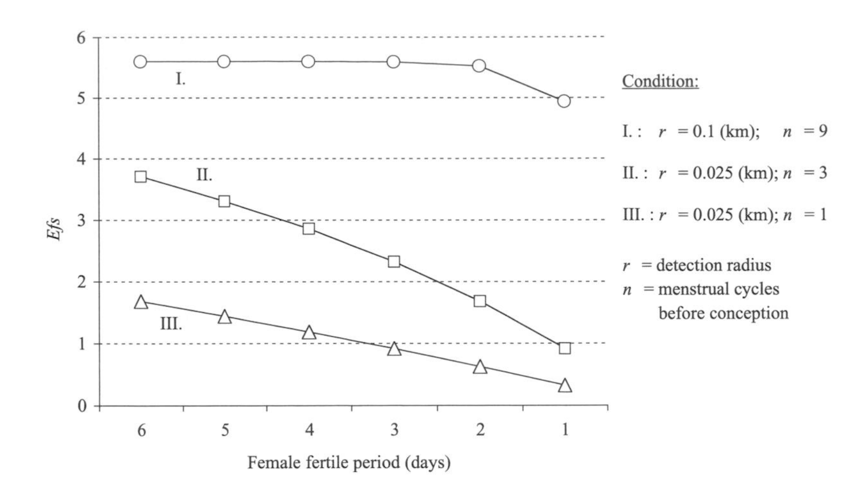

Figure 13.2. Expected number of females found (*Efs*) by a male randomly searching a 1 km^2 home range at least once while the female is fertile.

increasingly difficult for a male to find a female at the right time.

Condition I (Figure 13.2, ○) assumed a detection radius and number of menstrual cycles until conception at their original values ($r = 0.1$ km; $N = 9$), and only the length of the female fertile period (c) was gradually reduced from 6.1 days to 1 day. A fertility period of $c = 1$ day assumed that the male had to meet the female on her day of ovulation.

Under **Condition II** (Figure 13.2, □), detection radius was reduced to $r = 0.025$, the number of cycles before conception was decreased to $N = 3$, and again female fertility period (c) was gradually reduced from 6.1 days to 1 day.

Finally, under **Condition III** (Figure 13.2, Δ), detection radius was set at $r = 0.025$ and only one cycle was allowed until conception ($N = 1$), while female fertility period (c) was gradually reduced from 6.1 days to 1 day.

Resulting graphs were compared with a socioreproductively monogamous male with only one female ($E_{fs} = 1$). Only under a single circumstance of Condition II, i.e., when females were assumed to experience a single cycle before conception, and under a few circumstances of Condition III, i.e., when females were assumed to have three or fewer menstrual cycles before conception, would an LRP-strategy become disadvantageous for a male compared with socioreproductive monogamy because the value for the expected number of females (E_{fs}) would drop below one (1) female. In conclusion, under most circmustances produced by the three conditions, the expected reproductive output for a Khao Yai male following the LRP-strategy would exceed what he could expect to gain from socioreproductive monogamy, which was in accordance with calculations for other gibbon populations (van Schaik & Dunbar, 1990).

Resolving a puzzle

The above results reveal a puzzling phenomenon: on one hand, a high reproductive potential for LRP, but on the other a complete absence of such a male strategy in wild populations. How can this be resolved? First, in addition to a male's physical capacity to achieve exclusive access to dispersed females, female reproductive strategies need to be considered. In the model presented here, female interest was assumed to be neutral and constant, which may not be the case in gibbon females. When social monogamy is beneficial to female reproduction, females may overlap their sexual activities and conceptions to enforce social monogamy on males (cf. Knowlton, 1979; van Schaik & Kappeler, 1993; Nunn, 1999*a*). Potential benefits of LRP diminish rapidly when females temporally overlap in reproductive periods because it becomes increasingly difficult for a male to successfully maintain mating exclusivity (e.g., Chism & Rogers, 1997; Cords, 2000). Such a situation may push average reproductive success for males following LRP below the reproductive output of socially monogamous males. It is speculation whether such a scenario could apply to gibbons because close overlap in female fertile periods primarily characterizes seasonal breeders, whereas data for wild gibbon populations indicate at best weak breeding seasonality (Leighton, 1987; Sommer & Reichard, 2000). On the other hand, the development of visible sexual swellings in females (Dahl &

Nadler, 1992; Nadler *et al.*, 1993) may function to increase the duration and overlap of female mating activity when breeding is non-seasonal (van Schaik *et al.*, 1999; Nunn, 1999*b*; Reichert *et al.*, 2002). Endocrinological studies of ovarian cycles of wild females are needed to test whether neighbouring females are close enough in their sexual activities and conceptions to make LRP less efficient for males than social monogamy. Second, as females within a male's home range experience consecutive reproductive cycles, benefits of LRP may decrease because females will randomly overlap in fertile periods even if they do not benefit from this. It is likely that females vary in reproductive performance, for example, number of menstrual cycles until conception, length of lactation periods, or unexpected infant loss. To minimize reproductive loss from overlapping female cycling, a male will have to switch from roving to a systematic search strategy to find females that are in the right phase of their menstrual cycles, which is presumably a more costly strategy than random searching, even when female movements are not random. This scenario requires a new model to calculate the expected number of females a male could find within his home range while the females are fertile. Nonetheless, it seems questionable that reproductive loss due to female cycle overlap could push reproductive success of LRP below what a socially monogamous male could expect from only one female. Third, costs for LRP may generally exceed the assumed cost of lost mating opportunities associated with social monogamy (Brotherton & Komers, chapter 3). High costs for LRP could arise from fights resulting from an increase in male–male contest competition for access to females, or simply attributable to an increase in encounters as shared borders become longer. Intense male–male competition may lead to short male tenure periods that in turn could lower lifetime reproductive success of males following LRP compared with socially monogamous males. Male–male competition has rarely been considered in gibbons (but see Cowlishaw, 1996), but clearly this neglected aspect of gibbon socioecology is a promising and important field for future research. Fourth, the permanent presence of a male with a female and her offspring may be beneficial for male reproductive success, which is incompatible with LRP. Possible benefits of social monogamy to males are considered in the following section. However, studies are still needed that specifically address the four arguments discussed here for the absence of LRP in gibbons. Until more data become available, it has to be assumed that males choose social monogamy because they gain higher fitness returns from staying with one female than from trying to stay with several females.

BENEFITS OF SOCIAL MONOGAMY TO MALES

Theories about benefits of pair living revolve around defence strategies or services a male can provide to a female and offspring, which in turn increase male fitness above what could be achieved with alternative reproductive strategies. These strategies and potential benefits are discussed as they may apply to gibbons, primarily in light of data from Khao Yai.

Defending a female from other males

This hypothesis suggests that the origin of social monogamy was intense male mate guarding, and assumes that where females live in exclusive ranges, low opportunity and/or high costs of socioreproductive polygyny make permanently guarding one female the best strategy for males. The critical question is, however, why a male needs to stay beyond a female's fertile period. It can be argued that continuous spatial proximity makes it possible to: (i) advertise a female's paired status; (ii) restrict her ability to gain information about other potential partners (cf. Brotherton *et al.*, 1997); and (iii) impede extra-pair copulation, which could facilitate female desertion and pair formation with an EPC partner (cf. Wagner, chapter 6). The male mate-guarding hypothesis identifies loss of paternity as well as loss of social pair relationships as threats to male reproduction, and continuous mate guarding functions to forestall mate loss (i–iii), resulting in males favouring social monogamy over social polygyny. Recent observations of partner changes (Brockelman *et al.*, 1998; Reichard, unpublished data) suggest that paired white-handed gibbon males may indeed face a risk of mate loss, which makes male mate guarding a plausible hypothesis for the evolution of male social monogamy in gibbons. Does evidence exist to support this hypothesis?

Advertise paired status (i)

Gibbons are well known for their vocal duets (Marshall & Marshall, 1976; Haimoff, 1984; Gittins, 1984*a*), which consist of a coordinated sequence of sex-specific components repeated once or several times. The prominent part of a duet is the female's loud 'great call', which is usually followed by a male 'reply'. It has

Table 13.5. *Gibbon sexual activities, Khao Yai 1989–2002*

Female						
Name	Social partner(s)	Extra-pair partner(s)	Extra-pair partner status	In-pair copulation	EPC attempt	EPC
Andromeda	Fearless	Bard	paired	105	1	1
		Cassius II	paired	–	–	7
		Efendi	paired	–	1	1
Brenda	Elias			56	–	–
	Amadeus	Chet	paired	5	1	–
Bridget	Chet			6	–	–
Brit	Actionbaby	Marlon	paired	6	–	1
		Marty	unpaired	–	1	–
Cassandra	Cassius II	Lung	paired	26	2	1
Eclipse	San			2	–	–
Jenna	Frodo	Wjoran[a]	unpaired	–	–	9
Natasha	Claude			1	–	–
Serai	Efendi			4	–	–
Total				211	6	20

[a] Subadult individual.

EPC, extra-pair copulation.

been suggested that duets function as the main agent to advertise a paired status (Raemaekers *et al.*, 1984). If dueting evolved as a male mate-guarding strategy, duets could be expected to be primarily male initiated because they are in the interest of males. There are no data to test this prediction. Similarly, if males duet to advertise their mate's paired status, they would be expected to contribute their part to the performance. Contrary to this expectation, almost one-quarter of the great calls given by white-handed gibbon females at Khao Yai during encounters with neighbouring groups were not followed by the typical male reply (Reichard & Sommer, 1997). Furthermore, a general analysis of gibbon singing behaviour likewise questioned the mate-guarding function of duets. Cowlishaw (1992) inferred from the literature on gibbon song structure that the female part of a duet primarily functions as an intrasexual territorial defence mechanism between females. He further concluded that duetting is unrelated to male mate guarding because no positive correlation between duetting and the density of unpaired males (floaters) was found. Floater males were seen as the main risk for a paired male and, therefore, duetting should be more frequent in populations with a high male-floater density. From the presently available data, there is little support that gibbon duets function mainly as a mate-guarding mechanism.

Restrict female information gathering and sexual flexibility (ii) & (iii)

Like other socially monogamous species, gibbons show little morphological and behavioural sexual dimorphism (Carpenter, 1940; Leighton, 1987). It appears questionable altogether whether gibbon males could interfere with female information gathering, sexual activities or partner choice(s) if such behaviour would benefit female reproduction. In any case, Khao Yai males seemed unable or unwilling to control or restrict female reproductive strategies. Frequent encounters with neighbours presented ample opportunities for females to assess both their present mate and a number of neighbouring males. Group encounters in the Khao Yai population occurred on average every 1.7 days ($N = 98$ days; Reichard & Sommer, 1997). Likewise, female sexual activity appeared largely unrestricted based on the observation of EPCs of a number of females in this population (Reichard, 1995*b*; Table 13.5), and the fact that male partner changes also occurred occasionally (Brockelman *et al.*, 1998; Reichard, unpublished data).

Alternatively, however, the frequency of EPCs and male partner changes might be significantly higher in the absence of male mate guarding.

So far, little evidence is available to support the hypothesis that male mate guarding was at the core of social monogamy evolution in gibbon males. However, this preliminary conclusion must be treated with caution because convincing, specific tests of predictions delineated from the mate-guarding hypothesis are still outstanding. Whether, for example, male presence has a negligible influence on infant survival as predicted by the mate-guarding hypothesis has yet to be investigated. Furthermore, quantitative data from wild populations, for example, about the function(s) and fine structure of gibbon duets or alternative mechanisms of pair-status advertisement have yet to be accumulated. If, for example, gibbon duets mainly serve sex-specific functions, and the female part of the duet evolved as an intrasexual territorial signal (cf. Cowlishaw, 1992), either males or females may be expected to initiate duets. Both sexes then would be necessary to communicate either a female's paired status (males) or occupancy of a territory (females). The occasional failure (25%) of Khao Yai males to respond to their mate's great calls during group encounters also does not necessarily contradict expectations derived from the mate-guarding hypothesis. Close proximity to potential competitors, for example, during group encounters, perhaps decreases the need for a primarily long-range vocal signal like the answer to a great call to demonstrate a female's guarded status. Finally, not all socially monogamous mammals use vocal displays to advertise a paired status (cf. Brotherton & Rhodes, 1996).

Defending resources for a female

This hypothesis assumes that a male defends a territory for a female that contains all of the resources necessary for female reproduction (cf. Rutberg, 1983; Brockelman & Srikosamatara, 1984). Male territorial defence minimizes resource loss to conspecifics and relieves the female of having to engage in territorial behaviour. The female gains time to maximize food intake, which enhances her reproductive rate, and hence also increases male reproductive success. It is mainly an increased female reproductive rate that makes resource-defence social monogamy beneficial to a male. However, the food-defence hypothesis makes the assumption that even the strongest males in a population would be unable to defend an area large enough for several females. Even where females live in exclusive ranges they would not suffer a disadvantage if a single male could defend sufficient resources for all of them. This assumption appears inappropriate for Khao Yai males because a high potential for LRP was found. It seems unlikely that male resource defence was the origin for male social monogamy in gibbons.

Nonetheless, the resource-defence hypothesis may partially be rescued if we assume that pair living initially evolved for other reasons, but that the male food resource-defence became important later for the maintenance of social monogamy. If so, we can predict that males aggressively defend a territory to prevent resource depletion by conspecifics irrespective of an intruder's sex. This is an important prediction because if males defend a territory only against one sex, the resident female loses her resource-access advantage derived from having a constant male partner. However, data from Khao Yai do not support the prediction. Encounters in areas of the home ranges that overlap probably reflect situations of direct resource competition between neighbours, but these encounters were not always dominated by aggression. During 35% of encounters ($N = 162$) neighbouring individuals sat or travelled together or even shared food sources (Reichard & Sommer, 1997). Furthermore, in contrast to an expectation delineated from the resource-defence hypothesis (cf. Brotherton & Rhodes, 1996), aggressive interactions during encounters were sex-specific. Compared with expected frequencies based on the sex ratio (χ^2 goodness-of-fit test: $\chi^2_1 = 47.3$, $N = 126$, $P < 0.001$), all males chased males much more than they chased intruding females, although attacks involving contact aggression were not sex-specific (χ^2 goodness-of-fit test: $\chi^2_1 = 0.75$, $N = 13$, $P < 0.5$). Play-back experiments, which simulated territorial intrusion, likewise prompted primarily intrasexual range defence (Mitani, 1984; Raemakers & Raemaekers, 1985). The hypothesis that defending food for females was important for gibbon males was rejected.

Defending offspring from predators

Male proximity to dependent offspring may be important to protect infants from predation (Dunbar & Dunbar, 1980; Sommer, chapter 7; Fietz, chapter 14). When predation risk for infants is high, socially monogamous males that efficiently protect their progeny could leave more surviving infants than socially polygynous males. This hypothesis assumes that a female alone cannot successfully protect infants from predators because high energetic demands due to reproduction bring food consumption into conflict with predator vigilance. If predator detection and deterrence can be shifted to a male, both sexes profit. The female could maximize her food intake, which positively influences her reproductive rate, and hence further increase the male's reproductive success, in addition to increasing offspring survival.

Only limited information exists regarding the magnitude of predation pressure on gibbons, and evaluating predation risk in gibbons is inconclusive. Potential gibbon predators include tigers (*Neofelis tigris*), leopards (*Panthera pardus*), clouded leopard (*Neofelis nebulosa*), marbled cat (*Pardofelis marmorata*), Asian golden cat (*Profelis temmincki*), leopard cat (*Prionailurus bengalensis*), python (*Python reticulatus*), changeable hawk-eagle (*Spizaetus cirrhatus*), mountain hawk-eagle (*Spizaetus nipalensis*), Asian black eagle (*Ictinaetus malayensis*), and crested serpent eagle (*Spilornis cheela*). Most of these predators are widespread throughout Southeast Asia (von Leyhausen, 1988; von Leyhausen *et al.*, 1988) and occur within the hylobatid geographical range. Even if some raptors or smaller cats may not prey on adult gibbons, infants, juveniles, and adolescents are expected to be vulnerable to predation, an important consideration when the evolutionary significance of predation pressure is taken into account. However, neither predation events nor predator attacks have yet been reported.

Despite an empirical lack of evidence for predation of gibbons, it is possible to develop some predictions to see if available indirect data would support the hypothesis. If males perform predator detection, they are expected to save offspring from being killed (van Schaik & Hörstermann, 1994). This prediction cannot be tested in the absence of empirical data. Instead, exhibiting vigilant behaviour may be used to infer predator detection efforts. If males are the main protectors against predators, we may expect them to perform this task better than other group members (van Schaik & van Noordwijk, 1989) and expect a male-biased vigilance. Additionally males, more frequently than females, should detect, warn, mob, and deter predators.

In general, Khao Yai gibbons gave alarm calls predominantly in response to potentially dangerous animals: tiger, python, and large birds of prey, whereas encounters with large, but apparently harmless, animals including bears (*Helarctos malayensis*, $N = 2$; personal observation) usually did not elicit alarm calls (Uhde, 1997; Uhde & Sommer, 2002). Other anti-predation strategies have also been identified (Uhde, 1997; Reichard, 1998; Uhde & Sommer, 2002). Reichard (1998) described an encounter in which a mountain hawk-eagle (*Spizaetus nipalensis*) landed close by. The male gibbon approached and slowly chased off the bird of prey while the group gave loud alarm calls and slowly followed the male. However, Uhde found no evidence for male-biased vigilance (Uhde & Sommer, 2002) as predicted by the predation-avoidance hypothesis. But overall, it seems that predation risk in gibbons may not be negligible and further observational and experimental studies are needed in order to quantify the impact predation may have on gibbon social evolution. Whether or not females without a male have lower reproductive success because of higher infant mortality from predation, and whether they show increased vigilance behaviour to compensate for the absence of a protector male, should be tested. Until more data become available, the conclusion is that gibbon males did not become socially monogamous by predation avoidance.

Defending against infanticide

The threat of infanticide to infant-carrying primates has been suggested as the primary cause selecting for year-round male–female association in primates, of which social monogamy represents the extreme case of a single female on the continuum of female group size (van Schaik & Kappeler, 1997, chapter 4). In the presence of an infanticide risk, males may favour social monogamy over social polygyny because protection of a female and progeny increases infant survival, and hence male reproductive success beyond fitness gains derived from social polygyny. Male infanticide is predicted to occur when three conditions are met simultaneously: (i) the male accompanying the female and infant has a low probability of being the infant's sire; (ii) the female returns to ovarian cycling more quickly if she loses the infant; and (iii) the infanticidal male has a high probablity of siring the female's next offspring (van Schaik, 2000).

In general, gibbon reproductive biology is consistent with expectations derived from an existing infanticide risk. And male changes in group composition occur more frequently than previously assumed (Brockelman *et al.*, 1998; Sommer & Reichard, 2000; Reichard, unpublished data), occasionally including aggressive male 'take-over' (Treesucon & Raemaekers, 1984). It has been suggested that immigration into existing social groups may even be the most common mode for mature offspring to acquire a breeding position (Brockelman *et al.*, 1998). This generates groups where a female's new social partner is unlikely to be the sire of her dependent offspring and hence increases the risk of infanticide. Gibbon females experience long lactation periods compared with gestation (see Table 13.4), which supports the assumption that a gibbon female that loses an infant during early lactation would return to ovarian cycling faster than a female that completed lactation. Finally, qualitative observations of male immigration at Khao Yai did not contradict the expectation that a male committing infanticide might have a greater chance of siring a female's next infant. Male immigration was documented nine times. In four cases no systematic observation followed, but in five cases females were observed to shift their sociosexual attention from the old resident to the new immigrant, and all immigrants were seen to copulate with the female soon after immigration (author's unpublished observation). The occurrence of EPCs (Table 13.5) is also compatible with the idea of a risk of infanticide. Infanticide decreases female reproductive success, and females vulnerable to male infanticide are therefore expected to evolve strategies to avoid it. Copulating with multiple males may be interpreted as part of such a female infanticide avoidance strategy (cf. van Noordwijk & van Schaik, 2000). This may also function to confuse paternity in case a male change occurs while an infant is vulnerable to infanticide (Reichard & Sommer, 1997; van Schaik *et al.*, 1999, 2000).

Despite indirect hints of potential risks of male infanticide, no plausible anecdotes (Alfred & Sati, 1991) or other empirical evidence yet exist that could confirm that infanticide is part of gibbon male reproductive strategies. Perhaps the exception so far is one case from Khao Yai where an unweaned infant died seven weeks after a male change (Reichard & Bracebridge, unpublished data). The infant's putative father was severely injured by the immigrant and left the group a few days after the immigration event. He was last seen alive in the home range about a month later, appearing thin and still handicapped from the wounds he had incurred. A dead, buff adult gibbon found in the home range a few weeks later was assumed to be the old resident male, who had presumably died from the infections of his wounds. What caused the infant's death was not observed directly and further analyses of the data are necessary to clarify the most likely cause. Preliminary inspection of circumstances surrounding the event, however, agreed with predictions derived from the sexual selection hypothesis for male infanticide. The male immigrated when the female was still carrying the unweaned, one-year-old infant most of the time. It is unlikely that the immigrant was the infant's sire because he was a resident somewhere else at the time the infant was conceived. The female returned to ovarian cycling faster than she would have if her infant had survived because she gave birth again after 27 months, whereas her two previous birth intervals had been 36 months each. Finally, the immigrant male copulated regularly with the female soon after immigration, and therefore had a high probability of having sired the new infant.

Taken together, the hypothesis that the risk of infanticide was the ultimate force for the evolution of social monogamy in gibbons is supported indirectly by a suite of gibbon behavioural and physiological traits compatible with the risk of male infanticide. In contrast, direct evidence for male infanticide is lacking, except perhaps for the one case from Khao Yai (cited above) in which an unweaned infant's death might be attributed to male infanticide. The present lack of direct empirical evidence may, however, not be surprising. Male immigration events that would allow the hypothesis to be tested more directly have rarely been recorded in gibbons. Even in the Khao Yai population, for which probably the best record of group composition changes exists (Reichard, unpublished data), only nine male immigration episodes have been documented so far. Alternatively, the current lack of empirical evidence perhaps reflects the efficiency of a permanent sociospatial male–female association to avoid infanticide (cf. van Schaik & Kappeler, chapter 4). Clearly, more quantitative data are needed to allow more direct testing of evidence for (or against!) an existing infanticide risk in gibbons to be able to evaluate the significance of this threat for the evolution of social monogamy. Specifically, a sufficiently large number of male immigration cases are needed to compare predicted infanticide with infant survival after a male change. Additionally, the role of males as infant protectors, and variable male–infant relationships based on an infant's age and a mother's mating histories with neighbouring males, need to be investigated.

Until more data become available, it appears impossible to either reject or accept the hypothesis that the risk of male infanticide caused the evolution of social monogamy in gibbons.

LOCAL-COMMUNITY MODEL (LCMP) OF GIBBON SOCIAL MONOGAMY EVOLUTION

Until now, I have discussed the alternative male strategies of LRP and socioreproductive monogamy. A simple approach was used to model these strategies. In this approach, *expected* male reproductive success with one female (social monogamy) was compared with *expected* male reproductive success with several females (LRP). Such an approach may, however, not adequately reflect the situation of pair-living male gibbons.

Behavioural and molecular studies have defeated the long held assumption of an essential link between social monogamy and reproductive monogamy (Hughes, 1998). Extra-pair copulation (EPC), i.e., those copulations between individuals who do not maintain a recognizable sociospatial pair relationship, occurs regularly in pair-living species. Most reports of EPC behaviour come from studies of socially monogamous birds (cf. Birkhead & Møller, 1992; Hasselquist & Sherman, 2001; Bennett & Owens, 2002). However, EPCs are not restricted to the class Aves and have been documented for a variety of mammals, including primates (Mason, 1966; Barash, 1981; Foltz, 1981; Hubrecht, 1985; Richardson, 1987; Ågren *et al.*, 1989; Gursky, 2000).

In gibbons, extra-pair copulation behaviour has been observed at two sites. At Ketambe research station (Gunung Leuser Ecosystem, Sumatra, Indonesia), five EPC episodes of a paired siamang female (*H. syndactylus*) with three males of a neighbouring group were witnessed (Palombit, 1994). Tragically, the Ketambe gibbon population became infected by a lethal disease that spread throughout the population (Palombit, 1992), making further analyses problematic. At Khao Yai, EPCs were recorded in a healthy white-handed gibbon (*H. lar*) population (Reichard, 1995*b*). Long-term records revealed that overall, EPCs accounted for 9.5% of observed copulations ($N = 211$). Four out of nine females observed to be sexually active between 1989 and 2002 participated in at least one EPC episode (Table 13.5). All extra-pair partners of females were direct neighbours and six out of eight males that attempted or succeeded in EPCs were already paired with another female. The observations from Khao Yai highlight the potential for paired males to maximize their reproductive success by copulating with neighbouring females in addition to copulating with their social-partner female. That EPCs have so far been documented at only two gibbon research sites is presumably a result of the usually short duration and small number of groups studied. In gibbons, as in most pair-living mammals, it is only possible to observe EPCs if, in addition to a study group, neighbouring individuals are likewise tolerant of a human observer. It is plausible to assume that even at Khao Yai, where the largest number of individuals engaging in EPC behaviour has been documented so far (Table 13.5), the true number of individuals following such a strategy is considerably larger than can currently be verified.

If a majority of socially paired males engage in EPCs, apparent and realized male reproductive success may differ significantly (cf. Wade & Arnold, 1980; Gibbs *et al.*, 1990; Møller, 1998). Reproductive success through extra-pair paternity can be high (cf. Goossens *et al.*, 1998; Fietz, chapter 14), and a more realistic modelling of expected reproductive benefits from social monogamy therefore should incorporate chances of extra-pair mating (EPC) and extra-pair paternity (EPP). If EPP can be confirmed in gibbons, then the relevant fitness perspective for a pair-living male may become the local community, defined by individuals that share and interact in overlapping home range areas, rather than just the female with whom a male shares the home range. In natural habitats, gibbon groups usually live in communities in which single groups occupy home ranges that overlap with each other and where range boundaries are shared with neighbouring individuals (Gittins, 1980, 1984*b*; Leighton, 1987; Reichard, 1995*a*). Encounters between neighbouring groups in overlapping areas are frequent, at least in the Khao Yai population (Reichard & Sommer, 1997), and therefore males may be more accurately seen as part of their local community, comprised of groups that meet in overlapping home ranges. A comprehensive model incorporating this scenario for males would then be defined as: *local-community social-monogamy combined with mating polygyny* (LCMP).

This model presumes that males would live in pairs with females, but strive to mate and reproduce with all neighbouring females accessible to them. Under these conditions, the question of why Khao Yai gibbon males do not follow an LRP-strategy may have to be asked again because expected reproductive success from LRP has to be compared with the expected reproductive success from LCMP. Realized male reproductive success may include the social female partner and some females from the neighbourhood. Even a small average increase in reproductive success from local-community mating polygyny may be sufficient to tip the cost:benefit balance between LRP and social monogamy over to the latter. This kind of scenario would lead to an increase in the variance in male reproductive success, and some males would gain the benefit of reproduction with several females without the cost of having to defend a large home range. In order to explore if and to what degree gibbon males may profit from EPCs, genetic data on the structure of gibbon groups are essential for testing the LCMP model. At the same time it will be important to quantify costs of LRP when compared with a combination of social monogamy and local-community reproductive polygyny.

CONCLUSION

Large-range socioreproductive polygyny (LRP) was identified as the reproductive strategy that theoretically would best maximize gibbon male fitness. Because this kind of strategy has not been observed in wild gibbons, it was assumed that social monogamy provided as yet undetected benefits to males that increased male reproductive success above what could be gained from LRP.

In a subsequent analysis of hypotheses aiming to explain the benefits and hence the evolution of social monogamy in mammals (cf. Dunbar & Dunbar, 1980; Wittenberger & Tilson, 1980; van Schaik & Dunbar, 1990; Brotherton & Rhodes, 1996) only one could be rejected for gibbons. The hypothesis that social monogamy evolved as a strategy where males defend exclusive access to territorial resources vital for a female and her offspring (cf. Rutberg, 1983) was rejected because aggression against territorial intruders was largely sex-specific. Three other commonly advocated hypotheses either received little support from existing gibbon field studies or could not be tested adequately because relevant data are lacking. The hypothesis that intensive male mate guarding was the origin of social monogamy (cf. Brotherton & Rhodes, 1996) appeared unlikely because gibbon duets do not function unequivocally to advertise a female's mated status, nor did Khao Yai males seem able to otherwise control female exploratory and/or sexual behaviour. However, convincing tests of this hypothesis are still outstanding because the function(s) of duets is not yet entirely clear and mechanisms of pair-status advertisement other than duetting have not been investigated. The hypothesis that predation pressure triggered the evolution of social monogamy (cf. Dunbar & Dunbar, 1980) in gibbons received indirect support because potential predators are widespread in gibbon habitats and gibbons correctly identify predators, react specifically to potential predators, and show anti-predation behaviour. On the other hand, empirical evidence of either predation events or predator attacks on gibbons is lacking, and further behavioural and experimental observations are needed to evaluate the importance of predation for gibbon social evolution. The hypothesis that the need for infanticide protection led males to become socially monogamous (cf. van Schaik & Dunbar, 1990) was also indirectly supported because gibbon socioecology and reproductive biology are in agreement with predictions of the risk of infanticide in this species. However, there is no empirical evidence to support the idea, except perhaps for the one Khao Yai case in which an unweaned infant died shortly after a male change. Further testing of the hypothesis is essential before the risk of infanticide as an explanation for the evolution of social monogamy in gibbons can be rejected or accepted.

Finally, none of the commonly advocated explanations for social monogamy in mammals (cf. Kleiman, 1977; Wittenberger & Tilson, 1980; van Schaik & Dunbar, 1990) has yet convincingly demonstrated the predicted causal link to social monogamy in gibbons. It therefore may have to be asked if perhaps the assumption that males became socially monogamous because of benefits associated with either increased female reproductive rates or enhanced infant survival has been misleading. Extra-pair copulations have been observed in gibbons, and instead of assuming that social monogamy is associated with reproductive monogamy a more realistic modelling of reproductive benefits derived from pair living may have to consider potential reproductive benefits for males from copulating with neighbouring females. A local-community mating polygyny model (LCMP) was suggested as yet another alternative to explain the evolution of social monogamy in gibbons. The LCMP model combines pair living with mating and potentially reproductive polygyny in the local neighbourhood, which includes all individuals sharing areas that overlap between their adjacent home ranges. The LCMP model differs from previous attempts to explain the origin of social monogamy in gibbons in so far as it does not initially require additional forces, e.g., a risk of predation or infanticide, to make pair living a preferred male strategy. Instead its basic requirement is a positive cost: benefit equation in favour of pair living compared with large-range polygyny (LRP).

The questions arise, however, of why and how the characteristically close, sociospatial pair relationships of gibbons fit into this picture, since a close sociospatial relationship between the male and female is not required nor predicted by the LCMP model. Two scenarios appear plausible. Perhaps close, sociospatial pair relationships evolved after pair living, allowing more benefits from pair living to materialize. Alternatively, close, sociospatial pair relationships may have to be reinvestigated in light of the LCMP model presented here. A close *spatial* association may be a consequence of optimized female resource exploitation. The function of a close spatial association between a male and a female may be to avoid uncontrolled resource use by the male sharing the female's range. Preliminary data on feeding strategies from Khao Yai are in agreement with such interpretation because all females of three study groups enjoyed priority access to food sources (Reichard & Sommer, 1997), suggesting that male feeding strategies indeed decrease interference with female resource exploitation. And a close *social* pair relationship may in fact reflect efforts to maximize chances for reproduction with additional partners (Gowaty, 1996*a*, *b*). It will be interesting to learn if male singing activities are perhaps better understood as an outcome of male–male competition for extra-pair females rather than the more traditional understanding of gibbon singing activities primarily reflecting the mutual interests of male and female pair partners (Leighton, 1987; Geissmann & Orgeldinger, 2000). Recent observations from Khao Yai indicate that at the very least gibbon pair relationships are less durable than commonly assumed, and that they rarely, if ever, last a lifetime (Brockelman *et al.*, 1998; Sommer & Reichard, 2000; Reichard, unpublished data).

Acknowledgements

I thank Claudia Barelli, Luca Morino, and Tommaso Savini, as well as the Thai field assistant team of the Thai-Gibbon-Project, Samrong Bangjunud, Chanakan Mungpoonklang, Sommai Seeboon, Sombat Sornchaipoom, for their help in collecting census data, and Warren Brockelman, Volker Sommer, Nicola Uhde, and the participants of the workshop on monogamy for stimulating discussions on gibbon social systems. Christophe Boesch, Pete Brotherton, Karen Chambers, Susan Perry, Daniel Stahl, and Carel van Schaik made valuable comments on an earlier draft of this chapter. Thanks are also due to the German Science Foundation (DFG), which sponsored the workshop on monogamy. Research permits to work at Khao Yai National Park were kindly granted by the National Research Council of Thailand (NRCT) and the Royal Forest Department of Thailand (RFD). I thank Christophe Boesch and the Max-Planck-Institute for Evolutionary Anthropology for the interest in and support of my research on the Lesser Apes.

References

Ågren, G., Zhou, Q. & Zhong, W. (1989). Ecology and social behaviour of Mongolian gerbils *Meriones unguiculatus*, at Xiliuhot, Inner Mongolia, China. *Animal Behaviour*, **37**, 11–27.

Ahsan, F. (1995). Fighting between two females for a male in the Hoolock gibbon. *International Journal of Primatology*, **16**, 731–7.

Alfred, J. R. B. & Sati, J. P. (1990). Survey and census of the hoolock gibbon in West Garo Hills, Northwest India. *Primates*, **31**, 299–306.

(1991). On the first record of infanticide in the hoolock gibbon – *Hylobates hoolock* in the wild. *Records of the Zoological Survey of India*, **89**, 319–21.

Aldrich-Blake, F. P. G. & Chivers, D. J. (1973). On the genesis of a group of siamang. *American Journal of Physical Anthropology*, **38**, 631–6.

Barash, D. P. (1981). Mate guarding and gallivanting by male hoary marmots (*Marmota caligata*). *Behavioral Ecology and Sociobiology*, **9**, 187–93.

Bennett, P. M. & Owens, I. O. F. (2002). *Evolutionary Ecology of Birds: Life Histories, Mating Systems, and Extinction*. Oxford: Oxford University Press.

Birkhead, T. R. & Møller, A. P. (1992). *Sperm Competition in Birds: Evolutionary Causes and Consequences*. London: Academic Press.

Bleisch, W. & Chen, N. (1991). Ecology and behavior of wild black-crested gibbons (*Hylobates concolor*) in China with a reconsideration of evidence for polygyny. *Primates*, **32**, 539–48.

Breznock, A. W., Harrold, J. B. & Kawakami, T. G. (1977). Successful breeding of the laboratory-housed gibbon (*Hylobates lar*). *Laboratory Animal Science*, **27**, 222–8.

Brockelman, W. Y. (1975). Gibbon populations and their conservation in Thailand. *Natural History Bulletin of the Siam Society*, **26**, 133–57.

Brockelman, W. Y. & Gittins, S. P. (1984). Natural hybridization in the *Hylobates lar* species group: implications for speciations in gibbons. In *The Lesser Apes: Evolutionary and Behavioural Biology*, ed. H. Preuschoft, D. J. Chivers, W. Y. Brockelman & N. Creel, pp. 498–532. Edinburgh: Edinburgh University Press.

Brockelman, W. Y. & Srikosamatara, S. (1984). Maintenance and evolution of social structure in gibbons. In *The Lesser Apes: Evolutionary and Behavioural Biology*, ed. H. Preuschoft, D. J. Chivers, W. Y. Brockelman & N. Creel, pp. 298–323. Edinburgh: Edinburgh University Press.

Brockelman, W. Y., Reichard, U., Treesucon, U. & Raemaekers, J. J. (1998). Dispersal, pair formation and social structure in gibbons (*Hylobates lar*). *Behavioral Ecology and Sociobiology*, **42**, 329–39.

Brotherton, P. N. M. & Rhodes, A. (1996). Monogamy without biparental care in a dwarf antelope. *Proceedings of the Royal Society of London, Series B*, **263**, 23–9.

Brotherton, P. N. M., Pemberton, J. M., Komers, P. E. & Malarky, G. (1997). Genetic and behavioural evidence of monogamy in a mammal, Kirk's dik-dik (*Madoqua kirkii*). *Proceedings of the Royal Society of London, Series B*, **264**, 675–81.

Carpenter, C. R. (1940). A field study in Siam of the behavior and social relations of the gibbon (*Hylobates lar*). *Comparative Psychology Monographs*, **84**, 1–212.

(1941). The menstrual cycle and body temperatures in two gibbons (*Hylobates lar*). *Anatomical Record*, **79**, 291–6.

Chism, J. B. & Rogers, W. (1997). Male competition, mating success and female choice in a seasonally breeding primate (*Erythrocebus patas*). *Ethology*, **103**, 109–26.

Chivers, D. J. (1971). The malayan siamang. *Malayan Nature Journal*, **24**, 78–86.

(1974). The siamang in Malaya: a field study of a primate in a tropical forest. In *Contributions to Primatology*, Volume 4, ed. H. Kuhn, C. R. Luckett, C. R. Noback, A. H. Schultz, D. Starck & F. S. Szalay, pp. 1–335. Basel: Karger.

(1977). The feeding behavior of Siamang (*Symphalangus syndactylus*). In *Primate Ecology: Studies of Feeding and Ranging Behavior in Lemurs, Monkeys and Apes*, ed. T. H. Clutton-Brock, pp. 335–82. London: Academic Press.

(1978). The gibbons of peninsular Malaysia. *Malayan Nature Journal*, **30**, 565–91.

Chivers, D. J. & Raemaekers, J. J. (1980). Long-term changes in behaviour. In *Malayan Forest Primates: Ten Years' Study in Tropical Rain Forest*, ed. D. J. Chivers, pp. 209–60. New York: Plenum Press.

Choudhury, A. (1990). Population dynamics of hoolock gibbons (*Hylobates hoolock*) in Assam, India. *American Journal of Primatology*, **20**, 37–41.

(1996). A survey of hoolock gibbon (*Hylobates hoolock*) in southern Assam, India. *Primate Report*, **44**, 77–85.

Clutton-Brock, T. H. (1989). Mammalian mating systems. *Proceedings of the Royal Society of London, Series B*, **236**, 339–72.

Cords, M. (2000). The number of males in guenon groups. In *Primate Males. Causes and Consequences of Variation in Group Composition*, ed. P. M. Kappeler, pp. 84–96. Cambridge: Cambridge University Press.

Cowlishaw, G. (1992). Song function in Gibbons. *Behaviour*, **121**, 131–53.

(1996). Sexual selection and information content in gibbon song bouts. *Ethology*, **102**, 272–84.

Dahl, J. F. & Nadler, R. D. (1992). Genital swelling in females of the monogamous gibbon, Hylobates (*H. lar*). *American Journal of Physical Anthropology*, **89**, 101–8.

Dallmann, R. & Geissmann, T. (2001). Different levels of variability in the female song of wild silvery gibbons (*Hylobates moloch*). *Behaviour*, **138**, 629–48.

Daltry, J. C. & Momberg, F. (2000). First biological assessment of Cardamom mountains, south-western Cambodia, reveals a wealth of wildlife. *Oryx*, **34**, 227–8.

Dao Van Tien (1983). On the north Indochinese gibbons (*Hylobates concolor*) (Primates: *Hylobatidae*). *Journal of Human Evolution*, **12**, 367–72.

Davies, N. B. (1991). Mating systems. In *Behavioural Ecology: An Evolutionary Approach*, ed. J. R. Krebs & N. B. Davies, pp. 263–94. Oxford: Blackwell Science.

Dehua, Y. & Peikun, X. (1990). A preliminary study on the food of the *Hylobates concolor concolor*. *Primate Report*, **26**, 81–7.

Dempsey, E. W. (1940). The structure of the reproductive tract of the female gibbon. *American Journal of Anatomy*, **67**, 229–53.

Dunbar, R. I. M. (1988). *Primate Social Systems*. Ithaca, New York: Cornell University Press.

Dunbar, R. I. M. & Dunbar, E. P. (1980). The pairbond in klipspringer. *Animal Behaviour*, **28**, 219–29.

Ellefson, J. O. (1968). Territorial behavior in the common white-handed Gibbon, *Hylobates lar* Linn. In *Primates: Studies in Adaptation and Variability*, ed. P. C. Jay, pp. 180–99. New York: Holt Rinehart & Winston.

(1974). A natural history of white-handed gibbons in the Malayan Peninsular. In *Gibbon and Siamang, Natural History, Social Behavior, Reproduction, Vocalizations, Prehension*, Volume 3, ed. D. M. Rumbaugh, pp. 1–136. Basel: Karger.

Emlen, S. T. & Oring, L. W. (1977). Ecology, sexual selection and the evolution of mating systems. *Science*, **197**, 215–23.

Evans, T. D., Duckworth, J. W. & Timmins, R. J. (2000). Field observations of large mammals in Laos, 1994–1995. *Mammalia*, **64**, 55–99.

Fischer, J. O. & Geissmann, T. (1990). Group harmony in gibbons: comparison between white-handed gibbon

(*Hylobates lar*) and siamang (*H. syndactylus*). *Primates*, **31**, 481–94.

Fisher, D. O. & Owens, I. P. F. (2000). Female home range size and the evolution of social organization in macropod marsupials. *Journal of Animal Ecology*, **69**, 1083–98.

Foltz, D. W. (1981). Genetic evidence for long-term monogamy in a small rodent, *Peromyscus polionotus*. *American Naturalist*, **117**, 665–75.

Fuentes, A. (1999). Re-evaluating primate monogamy. *American Anthropologist*, **100**, 890–907.

Geissmann, T. (1991). Reassessment of age of sexual maturity in gibbons (*Hylobates* spp.). *American Journal of Primatology*, **23**, 11–22.

(1994). Systematik der Gibbons. *Zeitschrift des Kölner Zoos*, **2**, 65–77.

Geissmann, T. & Orgeldinger, M. (2000). The relationship between duet songs and pair bond in siamangs, *Hylobates syndactylus*. *Animal Behaviour*, **60**, 805–9.

Gibbs, H. L., Weatherhead, P. J., Boag, P. T., White, B. N., Tabak, L. M. & Hoysak, D. J. (1990). Realized reproductive success of polygynous red-winged blackbirds revealed by DNA markers. *Science*, **250**, 1394–7.

Gittins, S. P. (1978). The species range on the gibbon *Hylobates agilis*. In *Recent Advances in Primatology*, Volume 3, ed. D. J. Chivers & K. A. Joysey, pp. 319–21. London: Academic Press.

(1980). Territorial behavior in the agile gibbon. *International Journal of Primatology*, **1**, 381–99.

(1984*a*). The vocal repertoire and song of the agile gibbon. In *The Lesser Apes: Evolutionary and Behavioural Biology*, ed. H. Preuschoft, D. J. Chivers, W. Y. Brockelman & N. Creel, pp. 354–75. Edinburgh: Edinburgh University. Press.

(1984*b*). Territorial advertisement and defence in gibbons. In *The Lesser Apes: Evolutionary and Behavioural Biology*, ed. H. Preuschoft, D. J. Chivers, W. Y. Brockelman & N. Creel, pp. 420–4. Edinburgh: Edinburgh University Press.

Gittins, S. P. & Akonda, A. W. (1982). What survives in Bangladesh? *Oryx*, **16**, 275–81.

Gittins, S. P. & Raemaekers, J. J. (1980). Siamang, lar, and agile gibbons. In *Malayan Forest Primates: Ten Years' Study in Tropical Rain Forest*, ed. D. J. Chivers, pp. 63–105. New York: Plenum Press.

Goossens, B., Graziani, L., Waits, E. F., Magnolon, S., Coulon, J., Bel, M.-C., Taberlet, P. & Allainé, D. (1998). Extra-pair paternity in the monogamous Alpine marmot revealed by nuclear DNA microsatellite analysis. *Behavioral Ecology and Sociobiology*, **43**, 281–8.

Gowaty, P. A. (1996*a*). Battles of the sexes and origins of monogamy. In *Partnerships in Birds: The Study of Monogamy*, ed. J. M. Black, pp. 21–52. Oxford: Oxford University Press.

(1996*b*). Multiple mating by females selects for males that stay: another hypothesis for social monogamy in passerine birds. *Animal Behaviour*, **51**, 482–4.

Green, K., Mitchell, A. T. & Tennant, P. (1998). Home range and microhabitat use by the long-footed potoroo, *Potorous longipes*. *Wildlife Research*, **25**, 357–72.

Groves, C. P. (1967). Geographic variation in the hoolock or white-browed gibbon (*Hylobates hoolock*), Harlan 1834. *Folia Primatologica*, **7**, 276–83.

Gursky, S. L. (2000). Sociality in the spectral tarsier, *Tarsius spectrum*. *American Journal of Primatology*, **51**, 89–101.

Haimoff, E. H. (1984). Acoustic and organizational features of gibbon songs. In *The Lesser Apes: Evolutionary and Behavioural Biology*, ed. H. Preuschoft, D. J. Chivers, W. Y. Brockelman & N. Creel, pp. 333–53. Edinburgh: Edinburgh University Press.

Haimoff, E. H., Yang, X.-J., He, S.-J. & Chen, N. (1986). Census and survey of black-crested gibbons (*Hylobates concolor concolor*) in Yunnan Province, People's Republic of China. *Folia Primatologica*, **46**, 205–14.

(1987). Conservation of gibbons in Yunnan Province, China. *Oryx*, **21**, 168–73.

Harvey, P. H. & Clutton-Brock, T. H. (1985). Life history variation in primates. *Evolution*, **39**, 559–81.

Hasselquist, D. S. & Sherman, P. W. (2001). Social mating systems and extrapair fertilizations in passerine birds. *Behavioral Ecology*, **12**, 457–66.

Hayssen, V. T., van Tienhoven, A. & van Tienhoven, A. (1993). *Asdell's Patterns of Mammalian Reproduction; a Compendium of Species-Specific Data*. Ithaca, New York: Comstock Publishing Associates.

Hubrecht, R. C. (1985). Home range size and use and territorial behavior in the common marmoset, *Callithrix jacchus jacchus*, at the Tapacura Field Station, Recife, Brazil. *International Journal of Primatology*, **6**, 533–50.

Hughes, C. (1998). Integrating molecular techniques with field methods in studies of social behavior: a revolution results. *Ecology*, **79**, 383–99.

Islam, M. A. & Feeroz, M. M. (1992). Ecology of hoolock gibbon of Bangladesh. *Primates*, **33**, 451–64.

Kappeler, M. (1984). Diet and feeding behaviour of the moloch gibbon. In *The Lesser Apes: Evolutionary and Behavioural Biology*, ed. H. Preuschoft, D. J. Chivers, W. Y. Brockelman & N. Creel, pp. 228–41. Edinburgh: Edinburgh University Press.

Kleiman, D. G. (1977). Monogamy in mammals. *Quarterly Review of Biology*, **52**, 39–69.

(1981). Correlations among life history characteristics of mammalian species exhibiting two extreme forms of monogamy. In *Natural Selection and Social Behavior:*

Recent Research and Theory, ed. R. D. Alexander & D. W. Tinkle, pp. 332–44. New York: Chiron Press.

Knowlton, N. (1979). Reproductive synchrony, parental investment and the evolutionary dynamics of sexual selection. *Animal Behaviour*, **27**, 1022–3.

Kollias, G. V. & Kawakami, T. G. (1981). Factors contributing to the successful captive reproduction of white-handed gibbons (*Hylobates lar*). *Annual Proceedings of the American Association of Zoo Veterinarians*, 45–7.

Komers, P. E. & Brotherton, P. N. M. (1997). Female space use is the best predictor of monogamy in mammals. *Proceedings of the Royal Society of London, Series B*, **264**, 1261–70.

Lan, D.-Y. (1993). Feeding and vocal behaviours of black gibbons (*Hylobates concolor*) in Yunnan: a preliminary study. *Folia Primatologica*, **60**, 94–105.

Lan, D.-Y., He, S. & Shu, L. D. (1990). Preliminary observations on the group composition of the wild concolor gibbons (*Hylobates concolor*) in Yunnan, China. *Primate Report*, **26**, 89–96.

Leighton, D. R. (1987). Gibbons: territoriality and monogamy. In *Primate Societies*, ed. B. B. Smuts, D. L. Cheney, R. M. Seyfarth, R. W. Wrangham & T. T. Struhsaker, pp. 135–45. Chicago: University of Chicago Press.

Liu, Z., Jiang H., Zhang, Y., Liu, Y., Chou, T., Manry, D. & Southwick, C. (1987). Field report on the Hainan gibbon. *Primate Conservation*, **8**, 49–50.

Liu, Z., Zhang, Y., Jiang, H. & Southwick, C. (1989). Population structure of *Hylobates concolor* in Bawanglin Nature Reserve, Hainan, China. *American Journal of Primatology*, **19**, 247–54.

Ma, S. & Wang, Y. (1986). The taxonomy and distribution of gibbons in south China and its adjacent region, with description of three new subspecies. *Zoological Research*, **7**, 393–410.

Ma, S., Wang, Y. & Poirier, F. E. (1988). Taxonomy, distribution and status of gibbons (*Hylobates*) in Southern China and adjacent areas. *Primates*, **29**, 277–86.

Marshall, J. T. (1990). Salween river gibbon study area: Thailand and Burma. *Natural History Bulletin of the Siam Society*, **28**, 93–4.

Marshall, J. T. & Marshall, E. R. (1976). Gibbons and their territorial songs. *Science*, **193**, 235–7.

Mason, W. A. (1966). Social organization of the South American monkey, *Callicebus moloch*: a preliminary report. *Tulane Studies in Zoology*, **13**, 23–8.

Mather, R. (1992). A Field Study of Hybrid Gibbons in Central Kalimantan, Indonesia. Ph.D. thesis, University of Cambridge.

McCann, C. (1933). Notes on the colouration and habits of the white-browed gibbon or hoolock (*Hylobates hoolock* Harl.). *Journal of the Bombay Natural History Society*, **36**, 395–405.

Mitani, J. C. (1984). The behavioural regulation of monogamy in gibbons (*Hylobates muelleri*). *Behavioral Ecology and Sociobiology*, **15**, 225–9.

(1990). Demography of agile gibbons (*Hylobates agilis*). *International Journal of Primatology*, **11**, 411–24.

Mitani, J. C. & Rodman, P. S. (1979). Territoriality: the relation of ranging patterns and home range size to defendability, with an analysis of territoriality among primate species. *Behavioral Ecology and Sociobiology*, **5**, 241–51.

Møller, A. P. (1998). Sperm competition and sexual selection. In *Sperm Competition and Sexual Selection*, ed. T. R. Birkhead & A. P. Møller, pp. 55–90. San Diego: Academic Press.

Mootnick, A. R., Haimoff, E. H. & Nyunt-Lewin, K. (1987). Conservation and captive management of hoolock gibbons in the Socialist Republic of the Union of Burma. *American Zoo and Aquarium Association Annual Conference Proceedings*, 398–424.

Mukherjee, R. P. (1982). Survey of non-human primates of Tripura, India. *Journal of the Zoological Society of India*, **34**, 70–81.

(1986). The ecology of the hoolok gibbon, *Hylobates hoolock*, in Tripura, India. In *Primate Ecology and Conservation*, ed. J. G. Else & P. C. Lee, pp. 115–23. Cambridge: Cambridge University Press.

Mukherjee, R. P., Chaudhouri, S. & Murmu, A. (1992). Status and conservation problems of Hoolock gibbon (*Hylobates hoolock*) in some of its range of distribution in North Eastern India. *Primate Report*, **34**, 133–8.

Nadler, R. D., Dahl, J. F. & Collins, D. C. (1993). Serum and urinary concentrations of sex hormones and genital swelling during the menstrual cycle of the gibbon. *Journal of Endocrinology*, **136**, 447–55.

Napier, J. R. & Napier, P. H. (1967). *A Handbook of Living Primates*. New York: Academic Press.

Nettelbeck, A. R. (1993). Zur Öko-Ethologie freilebender Weißhandgibbons (*Hylobates lar*) in Thailand. M.Sc. thesis, University of Hamburg.

Neudenberger, J. (1993). Monogamie als Paarungssystem: Eine Fallstudie am Weißhandgibbon (*Hylobates lar*) im Khao Yai Nationalpark, Thailand. M.Sc. thesis, University of Göttingen.

Nunn, C. L. (1999*a*). The number of males in primate social groups: a comparative test of the socioecolgical model. *Behavioral Ecology and Sociobiology*, **46**, 1–13.

(1999*b*). The evolution of exaggerated swellings in primates and the graded-signal hypothesis. *Animal Behaviour*, **58**, 226–46.

Palombit, R. A. (1992). Pair Bonds and Monogamy in Wild Siamang (*Hylobates syndactylus*) and White-handed Gibbon (*Hylobates lar*) in Northern Sumatra. Ph.D. thesis, University of California, Davis.

(1994). Extra-pair copulations in a monogamous ape. *Animal Behaviour*, **47**, 721–3.

Raemaekers, J. J. (1979). Ecology of sympatric gibbons. *Folia Primatologica*, **31**, 227–45.

Raemaekers, J. J. & Raemaekers, P. M. (1985). Field playback of loud calls to gibbons (*Hylobates lar*): territorial, sex-specific and species-specific responses. *Animal Behaviour*, **33**, 481–93.

Raemaekers, J. J., Raemaekers, P. M. & Haimoff, E. H. (1984). Loud calls of the gibbon (*Hylobates lar*): repertoire, organisation and context. *Behaviour*, **91**, 146–89.

Reichard, U. (1995*a*). *Sozial- und Fortpflanzungsverhalten von Weißhandgibbons* (Hylobates lar). Göttingen: Cuvillier Verlag.

(1995*b*). Extra-pair copulations in a monogamous gibbon (*Hylobates lar*). *Ethology*, **100**, 99–112.

(1998). Sleeping sites, sleeping places, and pre-sleep behavior of gibbons (*Hylobates lar*). *American Journal of Primatology*, **46**, 35–62.

Reichard, U. H. (manuscript). Group structures and the mating system of wild gibbons (*Hylobates lar*).

Reichard, U. H. & Bracebridge, C. (in preparation). Infant death after male immigration: support for a male infanticide risk in gibbons?

Reichard, U. & Sommer, V. (1997). Group encounters in wild gibbons (*Hylobates lar*): agonism, affiliation, and the concept of infanticide. *Behaviour*, **134**, 1135–74.

Reichert, K. E., Heistermann, M., Hodges, J. K., Boesch, C. & Hohmann, G. (2002). What females tell males about their reproductive status: are morphological and behavioural cues reliable signals of ovulation in bonobos (*Pan paniscus*). *Ethology*, **108**, 583–600.

Richardson, P. R. K. (1987). Aardwolf mating system: overt cuckoldry in an apparently monogamous mammal. *South African Journal of Science*, **83**, 405–12.

Rutberg, A. T. (1983). The evolution of monogamy in primates. *Journal of Theoretical Biology*, **104**, 93–112.

Siddiqi, N. A. (1986). Gibbons (*Hylobates hoolock*) in the west Banugach reserved forest of Sylhet district, Bangladesh. *Tigerpaper*, **13**, 29–31.

Sommer, V. & Reichard, U. (2000). Rethinking monogamy: the gibbon case. In *Primate Males: Causes and Consequences of Variation in Group Composition*, ed. P. M. Kappeler, pp. 159–68. Cambridge: Cambridge University Press.

Srikosamatara, S. (1984). Ecology of pileated gibbons in South-east Thailand. In *The Lesser Apes: Evolutionary and Behavioural Biology*, ed. H. Preuschoft, D. J. Chivers, W. Y. Brockelman & N. Creel, pp. 242–57. Edinburgh: Edinburgh University Press.

Srikosamatara, S. & Brockelman, W. Y. (1987). Polygyny in a group of pileated gibbons via a familial route. *International Journal of Primatology*, **8**, 389–93.

Sterck, E. H. M., Watts, D. P. & van Schaik, C. P. (1997). The evolution of female social relationships in nonhuman primates. *Behavioral Ecology and Sociobiology*, **41**, 291–309.

Tenaza, R. R. (1975). Territory and monogamy among Kloss' gibbons (*Hylobates klossii*) in Siberut Island, Indonesia. *Folia Primatologica*, **24**, 60–80.

Treesucon, U. & Raemaekers, J. J. (1984). Group formation in gibbon through displacement of an adult. *International Journal of Primatology*, **5**, 387.

Trivers, R. L. (1972). Parental investment and sexual selection. In *Sexual Selection and the Descent of Man 1871–1971*, ed. B. Campbell, pp. 136–79. Chicago: Aldine Press.

Uhde, N. L. (1997). Das Raubfeindrisiko bei Gibbons (*Hylobates lar*). Eine Sozioökologische Studie im Regenwald des Khao Yai National Park. M.Sc. thesis, University of Göttingen.

Uhde, N. L. & Sommer, V. (2002). Antipredatory behavior in gibbons (*Hylobates lar*, Khao Yai/Thailand). In *Eat or Be Eaten: Predator Sensitive Foraging among Nonhuman Primates*, ed. L. M. Miller, pp. 268–91. Cambridge: Cambridge University Press.

van Noordwijk, M. A. & van Schaik, C. P. (2000). Reproductive patterns in eutherian mammals: adaptations against infanticide? In *Infanticide by Males and its Implications*, ed. C. P. van Schaik & C. H. Janson, pp. 322–60. Cambridge: Cambridge University Press.

van Schaik, C. P. (2000). Infanticide by male primates: the sexual selection hypothesis revisited. In *Infanticide by Males and its Implications*, ed. C. P. van Schaik & C. H. Janson, pp. 27–60. Cambridge: Cambridge University Press.

van Schaik, C. P. & Dunbar, R. I. M. (1990). The evolution of monogamy in large primates: a new hypothesis and some crucial tests. *Behaviour*, **115**, 30–62.

van Schaik, C. P. & Hörstermann, M. (1994). Predation risk and the number of adult males in a primate group: a comparative test. *Behavioral Ecology and Sociobiology*, **35**, 261–72.

van Schaik, C. P. & Kappeler, P. M. (1993). Life history, activity period and lemur social systems. In *Lemur Social Systems and their Ecological Basis*, ed. P. M. Kappeler & J. Ganzhorn, pp. 241–60. New York: Plenum Press.

(1997). Infanticide risk and the evolution of male–female association in primates. *Proceedings of the Royal Society of London, Series B*, **264**, 1687–94.

van Schaik, C. P. & van Hooff, J. A. R. A. M. (1983). On the ultimate causes of primate social systems. *Behaviour*, **95**, 91–117.

van Schaik, C. P. & van Noordwijk, M. A. (1989). The special role of male Cebus monkeys in predation avoidance and its effect on group composition. *Behavioral Ecology and Sociobiology*, **24**, 265–76.

van Schaik, C. P., Hodges, J. K. & Nunn, C. L. (2000). Paternity confusion and the ovarian cycles of female primates. In *Infanticide by Males and its Implications*, ed. C. P. van Schaik & C. H. Janson, pp. 361–87. Cambridge: Cambridge University Press.

van Schaik, C. P., van Noordwijk, M. A. & Nunn, C. L. (1999). Sex and the social evolution in primates. In *Comparative Primate Socioecology*, ed. P. I. Lee, pp. 204–40. Cambridge: Cambridge University Press.

von Leyhausen, P. (1988). Katzen. In *Grzimek's Enzyklopädie: Säugetiere*, Band 3, pp. 580–636. München: Kindler.

von Leyhausen, P., Grzimek, B. & Zhiwotschenko, V. (1988). Pantherkatzen und Verwandte. In *Grzimek's Enzyklopädie: Säugetiere*, Band 4, pp. 1–48. München: Kindler.

Wade, M. & Arnold, S. J. (1980). The intensity of sexual selection in relation to male sexual behavior, female choice and sperm precedence. *Animal Behaviour*, **28**, 446–61.

Whitten, A. J. (1980). The Kloss Gibbon in Siberut Rain Forest. Ph.D. thesis, University of Cambridge.

Wittenberger, J. F. & Tilson, R. L. (1980). The evolution of monogamy: hypotheses and evidence. *Annual Review of Ecology and Systematics*, **11**, 197–232.

Wrangham, R. W. (1980). An ecological model of female-bonded primate groups. *Behaviour*, **75**, 262–300.

CHAPTER 14

Pair living and mating strategies in the fat-tailed dwarf lemur (*Cheirogaleus medius*)

Joanna Fietz

INTRODUCTION

Primates show a great diversity in their social organizations and mating systems (ranging from single pair bonds to multi-male/multi-female groups), including monogamy, polygyny, and promiscuity (Smuts *et al.*, 1987; Jolly, 1995; Kappeler, 1997; Müller & Thalmann, 2000). Social monogamy is assumed to occur in 15% of all primate species (Kleiman, 1977; Rutberg, 1983; Wright, 1990; Müller & Thalmann, 2000). And even though male care in primates is often associated with pair living, the participation of the male is assumed to be the consequence and not the cause of pair living (Dunbar, 1995; Komers & Brotherton, 1997; Brotherton & Komers, chapter 3; van Schaik & Kappeler, chapter 4). To understand the evolution of monogamy in a certain species, information about cost:benefit ratios of reproductive strategies is necessary. In terms of individual reproductive output, molecular genetic methods are powerful tools for estimating potential costs and fitness benefits associated with different mating strategies (Hughes, 1998). Sociobiological models suggest that sociogenetic monogamy occurs only if fitness benefits (i.e., future reproduction of offspring) can compensate for the costs of missed additional matings (Hamilton, 1967; Trivers, 1985; Mock & Fujioka, 1990; Maynard Smith, 1991). However, paternity analyses, especially in bird species, have revealed that living with a permanent partner does not necessarily imply renunciation of additional matings with extra-pair mates, and that social monogamy masks a diverse range of genetic mating systems (Mock & Fujioka, 1990; Birkhead & Møller, 1992; Stutchbury & Morton, 1995; Reynolds, 1996; Hughes, 1998). The occurrence and frequency of extra-pair paternity (EPP), as well as the estimation of potential costs and fitness benefits from extra-pair copulations (EPCs), are therefore central to our understanding of the evolution and maintenance of monogamous mating systems (Brooker *et al.*, 1990; Petrie & Kempenaers, 1998).

The small, nocturnal, fat-tailed dwarf lemur (*Cheirogaleus medius*) lives in permanent pair bonds, together with the offspring from one or more breeding seasons (Müller, 1998, 1999*a*, *b*, *c*; Fietz, 1999*a*, *b*; Fietz *et al.*, 2000). To characterize and elucidate the evolution and maintenance of pair living in the fat-tailed dwarf lemur, behavioural field observations were combined with genetic parentage analysis.

SITE, TIME, AND SPECIES

Field studies were carried out in Kirindy Forest, a dry, deciduous forest of western Madagascar (60 km north-east of Morondava; for detailed description of the forest see Ganzhorn & Sorg, 1996). The study was carried out in a study area of about 25 ha (500 × 500 m) from 1995 to 1999. Two hundred Sherman traps (7.7 × 7.7 × 30.5 cm) were set each month for four consecutive nights at 50 m intervals on the intersections of a grid system of trails. Traps were baited with bananas and fixed in the vegetation at heights ranging from 40 to 200 cm. Traps were opened and baited in the late afternoon and checked early in the morning. Captured individuals were sexed, measured, and individually marked with subdermally injected transponders, then released at their capture site in the late afternoon of the capture day (for detailed information see Fietz, 1999*a*, *b*). Skin samples (1 × 2 mm) were collected from 134 individuals. Samples were taken from ear lobes, conserved in 70% ethanol, and stored at 4 °C. A total of 36 individuals (22 males and 14 females) were equipped with radio collars and observed for a total of 500 hours. In order to estimate defendability of the home ranges, the index of defendability was calculated according to Mitani and Rodman (1979), dividing nightly travel distance by the area of the observed nightly home range. Sleeping hole sites were checked daily (for detailed description see Fietz, 1999*a*, *b*). Pairs and social family groups were identified by observational data, and 53 individuals could be attributed to either pairs or family groups (Fietz, 1999*b*). Genetic parentage analyses were carried out with seven species-specific microsatellite markers with an exclusion probability of 96% (for a detailed description see Fietz *et al.*, 2000).

Fat-tailed dwarf lemurs (*Cheirogaleus medius*; Primates: Cheirogaleidae) are small (130 g), and are exceptional among primates because they spend seven months of the year hibernating in tree holes. Hibernation coincides with the dry season (May–November), when food and water are scarce (Petter, 1978; Hladik *et al.*, 1980; Fietz, 1999*a*, *b*; Fietz & Ganzhorn, 1999). Their diet consists of flowers, nectar, and fruits, along with a seasonally varying proportion of insect prey (Petter *et al.*, 1977; Petter, 1978; Hladik *et al.*, 1980; Fietz & Ganzhorn, 1999). Animals emerge in November and mate at the beginning of the rainy season during December (Hladik *et al.*, 1980; Fietz, 1999*a*). After a gestation period of 61–64 days, females usually give birth to two young (Foerg, 1982). Adults may start hibernating as early as April, after accumulating a great quantity of fat (Petter *et al.*, 1977; Petter, 1978; Petter-Rousseaux, 1980), while infants usually remain active until May (Hladik *et al.*, 1980; Müller, 1999*c*). Part of the fat is stored within the tail, which swells from 10 to *c*. 50 ml in volume, giving the fat-tailed dwarf lemur its name (Petter *et al.*, 1977; Petter, 1978; Hladik *et al.*, 1980; Petter-Rousseaux, 1980; Fietz & Ganzhorn, 1999). *C. medius* hibernates either alone or in small groups within tree holes. Physiological measurements in the field show that the body temperature of hibernating *C. medius* closely tracks variations in ambient temperature, resulting in high daily fluctuations of body temperature. Such fluctuations are very exceptional and were previously unknown in any hibernating mammal (Dausmann *et al.*, 2000).

RESULTS

Demography and social mating system

Cheirogaleus medius lives in small family groups consisting of the reproducing couple and the offspring from one or more breeding seasons (Figure 14.1; Table 14.1; Müller, 1998, 1999*a*, *b*, *c*; Fietz, 1999*a*, *b*; Fietz *et al.*, 2000). These families inhabit common territories of about 1–2 ha (1.6 ± 0.5 ha; $N = 17$) and use the same tree holes as sleeping sites. Radio-collared pairs ($N = 9$) were sharing a sleeping hole 69.9 ± 5.1% of the times they were checked (Fietz, 1999*a*). Couples establish lifelong pair bonds and separate only if one partner dies. Territories are defended and olfactorily marked by the adult couple, especially along the territory borders (Müller, 1998; Fietz, 1999*a*, *b*; Wiedemann, unpublished data). The index of defendability of the home ranges in *C. medius* ranged between 2.5 and 21.5, with a mean of 10.38 ± 3.8 ($N = 79$). An index of 1 or greater indicates the possibility of crossing the home range at least once each night, which is regarded as the lower limit for monitoring the territory's boundaries and maintaining its exclusivity (Mitani & Rodman, 1979). Thus *C. medius* should easily be able to defend exclusive territories.

Intraspecific fights, in which the sex of each participant could be determined, revealed that they occur between adult individuals of the same sex and rarely between males and females. Offspring remain within their family groups until they are sexually mature (within the second year of life), but may stay with their families even as adults. They are tolerated when new offspring are born, but were never observed helping to raise their kin. This delayed emigration in *C. medius* offspring might be attributed to high population densities and an associated lack of vacant territories within the study area. In Kirindy Forest, from 50 to 70 adult individuals were captured on the 25-ha study site between 1995 and 1999. Mean population densities of all adult individuals therefore equals 2.18 ± 0.34 individuals per ha. Thus, the study area was highly saturated with densely packed territories that abutted one another (Figure 14.1). If one or both partners of the reproducing couple died or disappeared, their place was taken over within the same or the following reproductive period (Fietz, 1999*a*, *b*; Fietz *et al.*, 2000). Wild *C. medius* reach sexual maturity within their second year of life, usually after emigrating from their family group, while captive *C. medius* mature within the first year (Foerg, 1982). This suggests that yearlings in the field are sexually suppressed while remaining in their family groups, although endocrinological evidence is lacking.

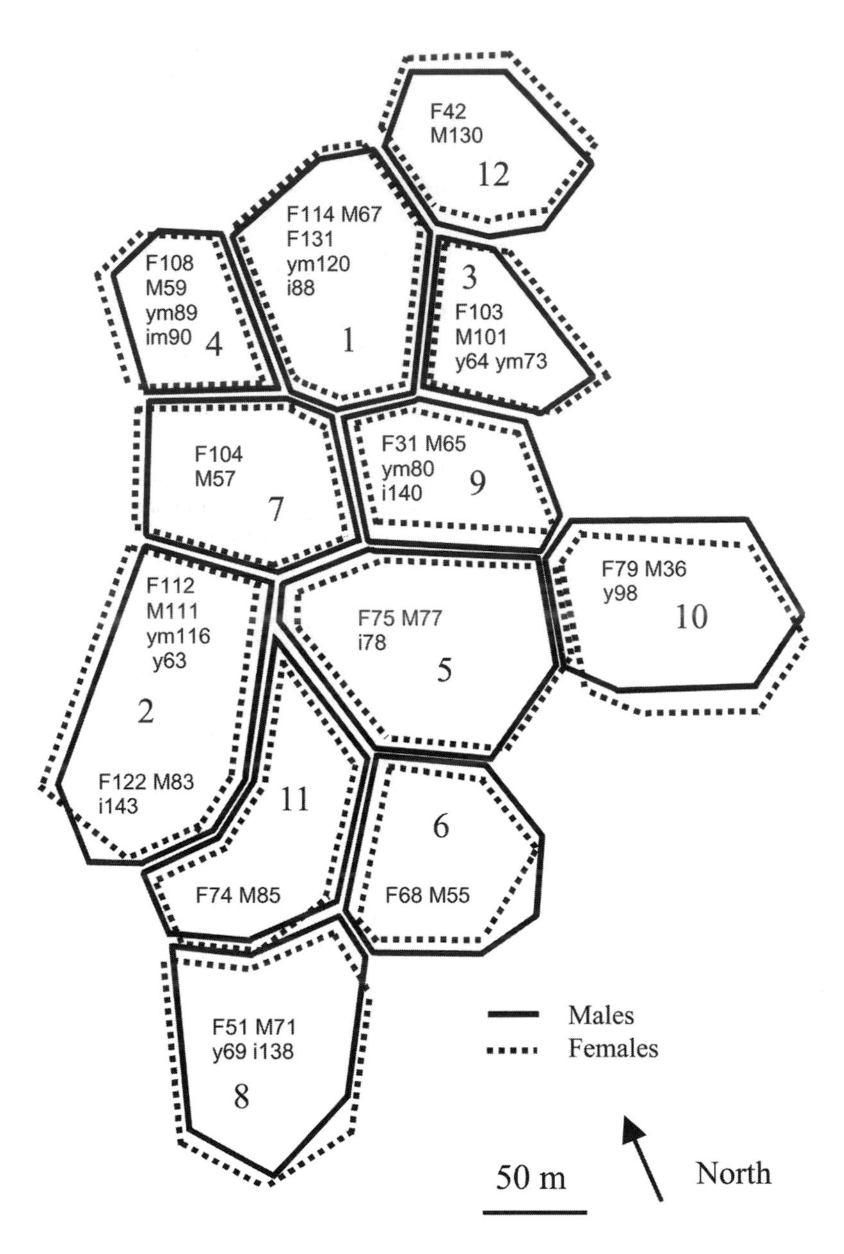

Figure 14.1. Territories of *Cheirogaleus medius* in Kirindy Forest/CFPF over the entire study period. Polygons denote schematic borderlines of territories (modified after Fietz *et al.*, 2000).

Sexual dimorphism and testis size

C. medius is sexually monomorphic (Table 14.2), suggesting weak intrasexual competition. During the mating season in December, wild male dwarf lemurs had a mean testis volume of 222.3 ± 14 mm^3 ($N = 45$) and a mean body mass of 119.4 ± 14.6 g ($N = 56$: Fietz, 1999*a*). The predicted testis volume for a strepsirhine primate with this body mass is about 680 mm^3. Therefore, testis volume of fat-tailed dwarf lemurs is 67% below the value predicted for strepsirhine primates (Fietz, 1999*a*).

Parental care

Offspring are born and cached within the nest holes and are usually not carried by their parents. The only occasion when a female was observed carrying her young was when she changed her nest hole during the night. On this occasion, she carried both infants in

Table 14.1. *Social family groups of 53 genotyped* Cheirogaleus medius *during the respective study years*

	Individual			
Territory	1995/96	1996/97	1997/98	1998/99
1	F114	F114	F114	F114
	M67	M67	M67[d]	M125
	F115[c]		**ym120**	**M120**
	i 88[NR]			**F131**
	i			
2	F112	F112	F112[NR]	F122
	M111	M111[d]	M83	M83
	y97[NR]	**y63[NR]**		**i143**
	ym116[c]			
3	F103	F103[d]		F107
	M101	M101[NR]		M124
	ym73	**M73[NR]**		
	y64[NR]			
4	F108[d]	F115[d]		F46
	M59[NR]	M116[NR]		M123
	ym89			
	im90			
5	F75	F75	F75[NR]	F16
	M77	M77[NR]	M117	M133
		i78[NR]		
6			F68	
			M55[NR]	
7			F104	F104
			M57	M57
8			F51	F51
			M71	M71
			y69[NR]	M126
				i138
9			F31	F31
			M65	M65
			ym80	**M80**
				i140
10			F79	F79
			M36	M36
			y98[NR]	
11			F74[NR]	
			M85[NR]	
12				F42
				M130

M, adult male; F, adult female; i, infant; y, yearling; m, male; f, female; c: individual changed group; d: individual died; NR: individual no longer recorded. Bold letters indicate offspring that were included in paternity analyses.
Modified after Fietz *et al.*, 2002.

her mouth (they were only a few days old), one after the other, to the new nest hole. Newborn infants have open eyes and are completely covered with fur (Foerg, 1982). Both males and females take part in raising their offspring (Figure 14.2). During the first two weeks after birth, infants remain in the nest holes, and both adults take turns in baby-sitting their young. In two different pairs, the males died, and neither female was able to rear her offspring successfully; the infants died shortly after birth, despite measurable compensatory effort (Fietz, 1999*a*). One of the infants, whose father had disappeared shortly after birth, was found dead in its nest hole with severe injuries (Fietz, 1999*b*). On the other hand, K. Dausmann (personal communication) observed a female successfully defeating a snake close to a nest hole containing her offspring.

As soon as infants start to leave their nest holes, either the male or female accompanies them on their excursions, guiding them back at the end of the night (Fietz, 1999*a*). Helpers other than the social parents were never observed taking care of offspring. In one instance, two adult females and one adult male were observed living together. Genetic analysis could not exclude the possibility that the females were closely related (e.g., mother and daughter or sisters), but since there is no information about the genetic father of either female, no final conclusion could be drawn (Fietz, unpublished data). These two females reproduced and raised their offspring together. Offspring were born in the same tree hole, and both females returned regularly to their joint nest to look for and suckle their babies. During the day, they took turns sleeping with their young, while the other female slept in a tree hole nearby. When the juveniles left their sleeping holes, they followed one or the other of the females, or were guided by both of them (Fietz, 1999*a*). As soon as a neighbouring territory became vacant, one of the two females moved to the adjoining territory and mated there with another male. But even when living in separate territories, these females met at the borders of their territories and had social contact (Fietz, 1999*a*).

Floaters

A high population density, in combination with a male-biased sex ratio in some of the study years (Figure 14.3), led to the number of males exceeding the number of possible territories. These surplus males (floaters), even

Table 14.2. *Comparison of body measurements of males and females of* Cheirogaleus medius

Variables (mm; g)	Males median	Males N	Quartiles 25/75%	Females median	Females N	Quartiles 25/75%	Mann–Whitney U-test z	Mann–Whitney U-test P
Head length	43.5	59	42.5/44.2	43.8	29	43.2/44.3	1.33	n.s.
Head width	26.4	59	25.7/27.2	26.8	29	26.2/27.4	1.49	n.s.
Ear length	18.5	30	17.5/19.0	17.8	14	17/18.5	1.46	n.s.
Tail length	185	57	176/192	182	28	178/191	0.02	n.s.
Tibia length	47.6	30	46.2/48.3	47.1	14	46/48.7	0.16	n.s.
Hind foot length	38	58	36.8/39.0	37.8	29	36.8/38.6	0.55	n.s.
Body mass	124	55	112/134	136	29	129/142	3.54	0.0004

n.s., not significant.
Modified after Fietz, 1999*a*.

though they were sexually mature, were not associated with a given female. According to behavioural traits, males could be assigned to two different categories: floating males and territory holders (Table 14.3). In this study six floaters could be identified (Fietz, 1999*b*). While territory holders live in stable pair bonds within relatively small, defended territories, floaters were observed to move within significantly larger home ranges of up to 11 ha; these ranges overlapped with each other as well as with the territories of pairs (Figure 14.4). Despite systematic searching, floaters were not always present in the study area: thus, the sizes of their home ranges were underestimated. Home ranges were never observed being marked or defended against conspecifics. Furthermore, floating males differed significantly from territory holders with regard to their sleeping habits. They used different sleeping holes more often than territorial males, and were never found sleeping with conspecifics, which is quite common in territory holders (Table 14.3). In fights with territory holders, floaters were generally defeated. While body mass and body measurements did not differ between territorial males and floaters, testis volume was significantly smaller in floating males during the mating season, which could be induced by increased stress or hormonal suppression. Age of floaters ranged from two years, which is the first year of reproductive activity in the field, to at least four years. Territorial males included individuals between two and at least five years old (Table 14.4). The occurrence of floating males does not seem to be restricted to the Kirindy population. In a population of *C. medius* in Ampijoroa, one young male showed the typical traits of a floater (Müller, 1998).

Genetic mating system

For 16 offspring, information about their social parents was available from field observations (Figure 14.5). For all offspring, social mothers were identified to be the true dams. Nine of these young (i.e., 56%) were sired by their social fathers, whereas for seven sibs (44%) parentage analyses revealed that they resulted from copulations with extra-pair males (Figure 14.5). In two cases (i88 and ym120) the genetic father (M59 and M36, respectively) was a territorial male living in the nearest and the second-nearest neighbouring territory to the extra-pair female (F114: see Figure 14.6). Males M59 and M36 sired extra-pair young (ym120 and i88), while at the same time having infants (im90 and y98) with their social partners (F108 and F79; Figures 14.1 and 14.6). Thus males that sired extra-pair young (EPY) did not lose paternity with their own mate, suggesting that male quality and reproductive success varies within the population. Infants y64 and ym73 are full sibs, the parents being F103 and M101. In the case where two females (F114 and F115) lived with one male (M67), an infant born and raised within this group (i88) was eliminated as being the genetic infant of the social father. This suggests that by living with more than one female a male might incur costs in terms of lost paternity. On the other hand, males living with only one female could just as well be excluded from paternity (Fietz *et al.*, 2000). Males were observed investing in litters containing EPY. All

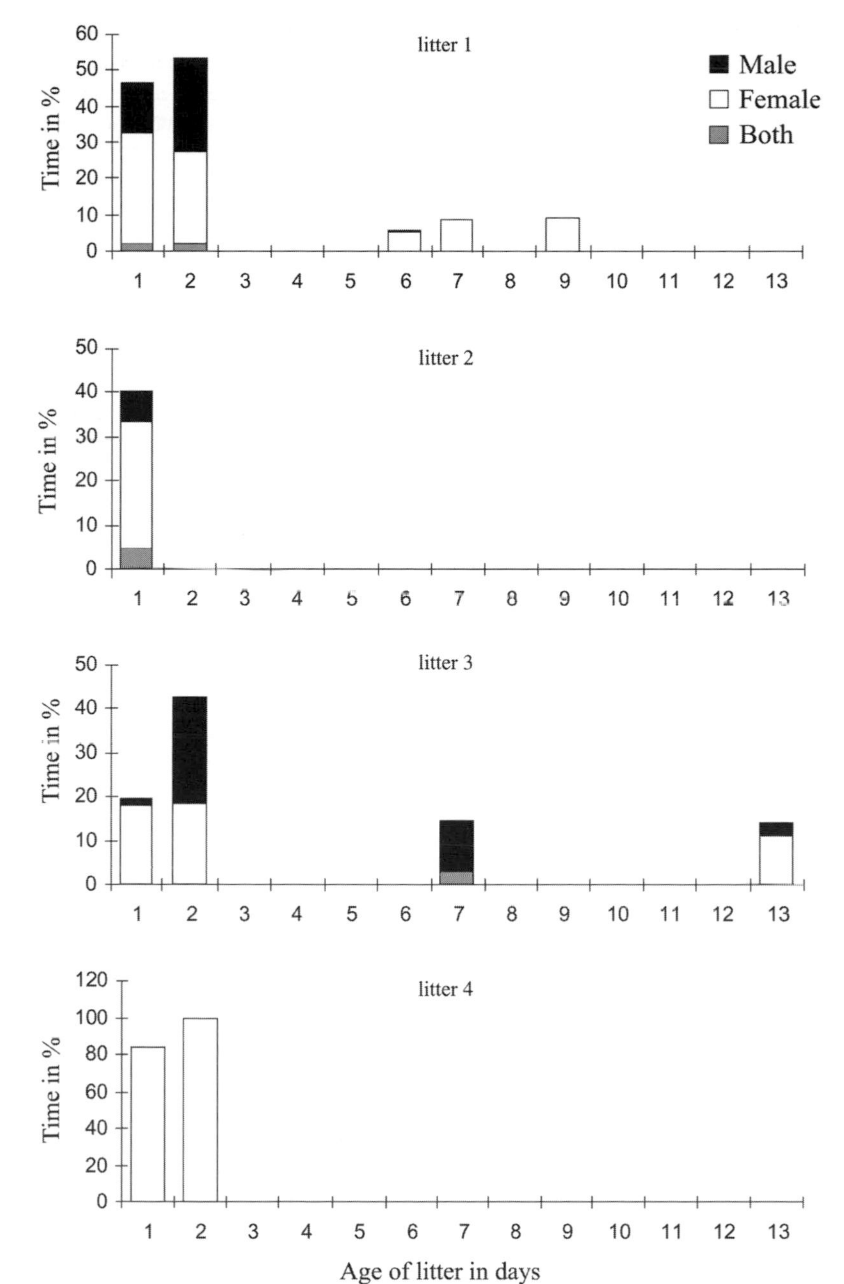

Figure 14.2. Time (per cent) that males, females, or both were baby-sitting (includes days without data collection) (modified after Fietz, 1999*a*).

six floating males could be excluded from having sired offspring (Fietz *et al.*, 2000).

DISCUSSION

Behavioural observations in combination with genetic parentage analysis revealed that the fat-tailed dwarf lemur lives in stable pairs (Müller, 1998, 1999*a*, *b*, *c*; Fietz, 1999*a*, *b*), with an extraordinary high incidence of EPY (Fietz *et al.*, 2000). This high proportion of offspring not sired by the social father is unusual in comparison with other mammal (Sillero-Zubiri *et al.*, 1996; Girman *et al.*, 1997; Goossens *et al.*, 1998; Spencer *et al.*, 1998) and even bird species (Birkhead & Møller, 1992;

Table 14.3. *Comparison of home range size and sleeping habits between floaters and territorial males of* Cheirogaleus medius. *Values are medians and 25/75% quartiles*

	Floater		Territorial male		Mann–Whitney *U*-test	
Variable	mean	*N*	mean	*N*	*z*	*P*
Home range size (m^2)	39302 25425/115889	4	14007 9421/19519	8	2.7	0.007
Relative number of sleeping holes	0.2 0.1/0.3	4	0.09 0.03/0.1	6	2.1	0.03
Rate of communal sleeping (%)	0 0/0	4	56.3 23.5/89.7	5	2.1	0.03
Exclusivity of sleeping hole use (%)	14.8 13.6/26.3	4	0 0/0	7	2.3	0.02
Absence from study area (%)	9.2 7.6/24.7	4	0 0/0.2	6	2.8	0.006

Modified after Fietz, 1999*b*.

Stutchbury & Morton, 1995), especially in regard to the occurrence of male care in this species (Mock & Fujioka, 1990; Reynolds, 1996).

Mating strategies: some background

To understand the evolution and occurrence of mating strategies in mammals, historical pathways as well as knowledge about the origin and adaptiveness of these traits are needed (van Schaik & Kappeler, chapter 4). Reconstructions of historical pathways are generally very difficult to assess and evaluate because assumptions about ancient social and mating systems are rather speculative. Extant social and ecological conditions might be substantially different from those experienced by ancestors during the period when monogamy evolved.

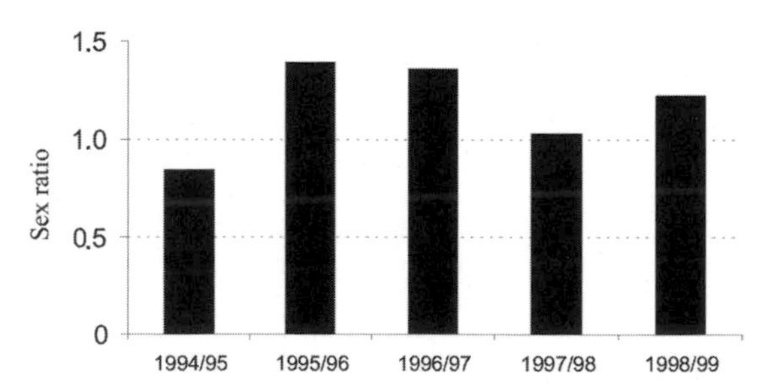

Figure 14.3. Sex ratio of captured adult *C. medius* (number of males/number of females).

And the fact that in only a few species information about mating systems is complete must also be taken into account. Especially within the nocturnal taxa (that represent the major segment of the mammalian radiation), knowledge about mating strategies is more or less unknown. Consequently, the information used for these phylogenetic comparisons is incomplete and biased towards the better-studied diurnal species. This bias and lack of information on social and breeding systems, as well as the limited number of paternity studies for mammals, leaves us with a substantial degree of uncertainty as to the accuracy of current ideas on the occurrence and evolution of monogamy in mammals. Variations in mating systems, even within populations of the same species in response to immediate socioecological conditions (Clutton-Brock, 1989; Davies, 1992; Sun, chapter 9), make it even harder to characterize and define clearly the mating systems in both ancestors and extant species.

In addition to all these limitations in assessing and understanding the evolution of mating systems, pair living could have evolved only if alternative mating strategies had failed and pair living represented the Evolutionary Stable Strategy (ESS) (Hamilton, 1967; Wittenberger & Tilson, 1980; Trivers, 1985; Mock & Fujioka, 1990; Maynard Smith, 1991). This volume includes current discussions about different agents considered to be selective factors favouring and maintaining

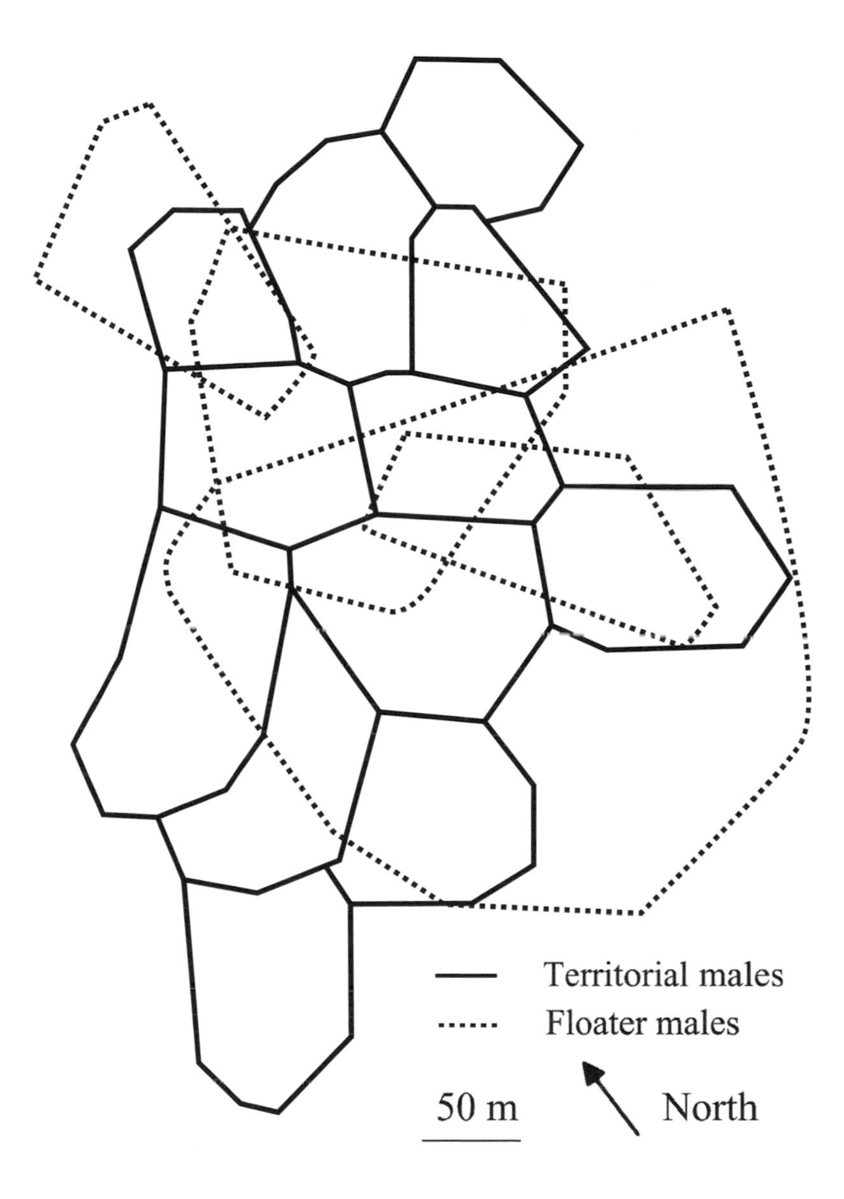

Figure 14.4. Territories of territorial males and home ranges of male floaters. Polygons denote schematic borderlines of territories and home ranges (modified after Fietz, 1999*b*).

pair living in mammals (Brotherton & Komers, chapter 3; van Schaik & Kappeler, chapter 4; Ribble, chapter 5; Sommer, chapter 7).

Ecological concerns: food and other resources

One reason for a male to stay with a single female could be that the male defends access to food or other valuable resources, indirectly improving his own fitness (van Schaik & Kappeler, chapter 4). Consequently the number of the females within a territory would be limited by its quality, and rich habitats should contain more than a single female. *C. medius* spends seven months hibernating within tree holes, and good quality hibernacula (e.g., insulation capacities; shelter against predation) probably constitute a critical resource for this obligate hibernator. Variance in tree-hole quality is reflected by the frequency with which a tree hole is used over several years (Dausmann & Fietz, unpublished data). Another crucial resource is adequate feeding trees. During December, directly after emergence from hibernation, *C. medius* feeds on fruits of only three different tree species, with fruits of one of them eaten in over 80% of observed feeding bouts within a period of two weeks (Fietz & Ganzhorn, 1999). Because fruiting seasons of

Table 14.4. *Comparison of body measurements between floaters and territorial males of* Cheirogaleus medius. *Values are medians and 25/75% quartiles*

	Floater		Territorial male		Mann–Whitney *U*-test	
Variable	mean	*N*	mean	*N*	*z*	*P*
Testis volume (mm^3)	110.6 83.2/154.8	4	201.0 139.7/363.5	11	2.1	0.04
Body mass (g)	116 106.5/123	5	128.0 112/160	15	0.7	n.s.
Head width (mm)	26.9 25.6/27.5	5	27.0 26.5/27.6	15	0.7	n.s.
Head length (mm)	42.7 40.7/43.8	5	43.9 42.6/45.2	15	1.3	n.s.
Hind foot length (mm)	35.1 32.7/36.6	4	36.4 34.4/39.5	15	1.3	n.s.

n.s., not significant.
Modified after Fietz, 1999*b*.

the feeding plants are short, *C. medius* depends on a combination of certain tree species within its territory to provide fruits throughout the whole activity period, with some species representing the staple resource.

However, in *C. medius*, food resources and nest holes do not seem to be limited, as offspring remain with their parents sometimes even after adulthood and family groups may contain up to five individuals (Table 14.1). Thus, the carrying capacity of a territory seems to be sufficient for a number of additional individuals beyond those of a single pair. Territorial boundaries generally remained unchanged over the course of the study period, irrespective of the number of individuals living within them. Polygyny was observed in *C. medius* in one instance, but only temporarily since one of the females switched to an adjoining territory as soon as it became vacant. Thus, neither food resources nor nest holes seem to constrain *C. medius* to pair living.

Pair living and territoriality

Large and dispersed female territories, of which males cannot defend more than one, have, in addition to paternal care, typically been suggested as being responsible for pair living in mammals (Kleiman, 1977; Clutton-Brock, 1989; Komers & Brotherton, 1997). In *C. medius*, territories are relatively small (1–2 ha) and abut one another. This spatial clumping of small exclusive territories and the high index of defendability suggests that males should be able to monopolize more than one female. However, no male was ever known to increase his territory by adopting a neighbouring one, either vacant or occupied by a female. So it is very unlikely that female dispersion in *C. medius* constrains males into monogamy.

Pair living and infanticide

Infanticide by males should only be adaptive when the male kills offspring he did not sire, the female returns to receptivity earlier than she would have otherwise, and the male has a high probability of siring the female's next infant (van Schaik & Dunbar, 1990; van Schaik & Kappeler, 1993, 1996, 1997, chapter 4). No infanticide attempts were ever observed in *C. medius*, either in males or in females, though this is not nessessarily unexpected if counterstrategies work effectively (van Schaik & Kappeler, chapter 4). Infanticide would probably not pay for *C. medius* males in any case, since females would not be able to raise a new set of offspring within the same reproductive period if their infants were killed. Depending on climatic conditions, *C. medius* in Kirindy forest generally emerges from hibernation in November, mates in December and January, and offspring are born in February and March. Adults begin hibernating in the middle of April (Petter *et al.*, 1977; Petter, 1978; Petter-Rousseaux, 1980; Fietz & Ganzhorn, 1999). Time is too limited to raise a second litter of offspring before the onset of hibernation. Even in the following year, successful

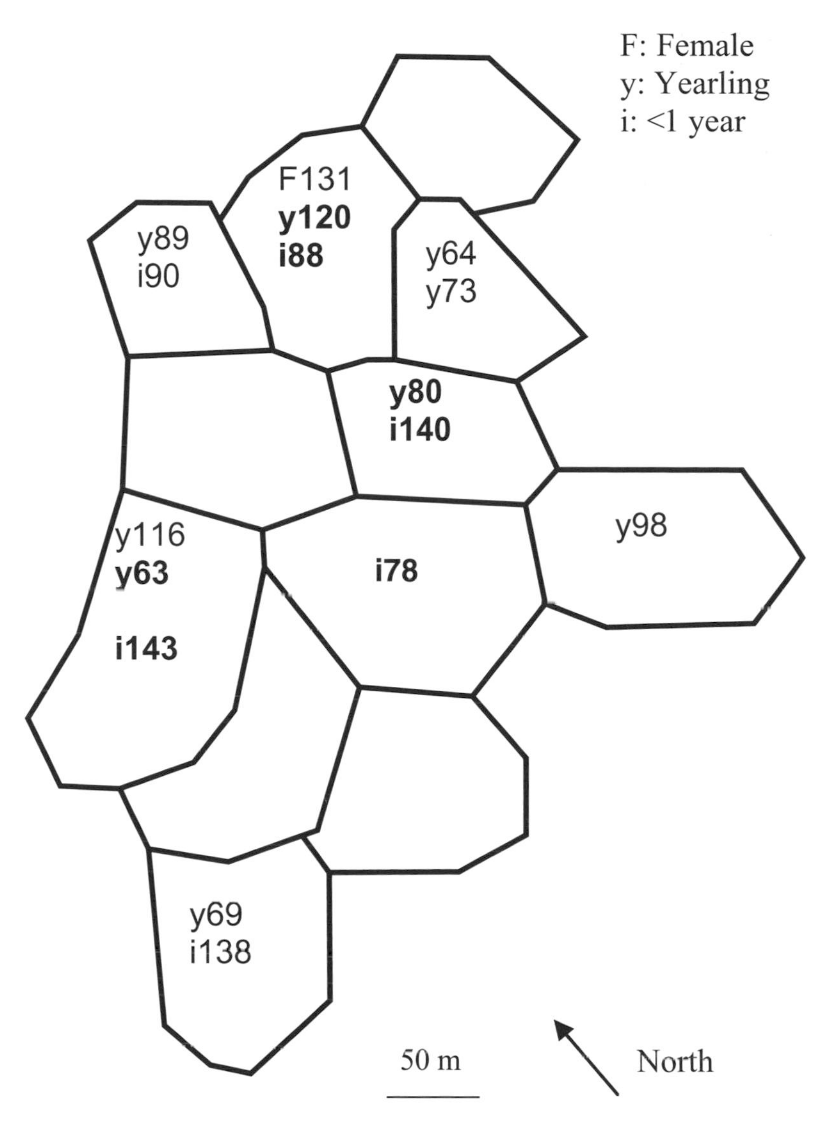

Figure 14.5. Offspring for which paternity was analysed, and the respective territories of their social parents. Polygons denote schematic borderlines of territories. Bold letters indicate extra-pair young identified by parentage analysis (modified after Fietz *et al.*, 2000).

reproduction of the male with the respective female is very unlikely because *C. medius* shows strong variations in reproductive output, with the birth of almost all offspring occurring every other year (Fietz, 1999*b*). Thus, in this species infanticide seems unlikely to function as a selective pressure forcing a male to stay with a single female. This is in accordance with the expectation that solitary foraging, infant-caching lemur species are generally thought to be invulnerable to male infanticide (van Schaik & Kappeler, chapter 4). *Avahi*s are also nocturnal pair-living lemurs, but, in contrast to *C. medius*, females do not cache their young but carry them. In *Avahi*s, males and females are usually observed in pairs during their activity period (Ganzhorn *et al.*, 1985; Harcourt, 1991; Warren & Crompton, 1997; Thalmann, 2001), which is interpreted as an infanticide avoidance strategy by the males (van Schaik & Kappeler, 1993).

Pair living and male care

Theoretical considerations and phylogenetic comparisons suggest that male care followed the evolution of pair living and that male benefits became significant only after major changes in the life histories of the females and infants (Dunbar, 1995; Komers & Brotherton,

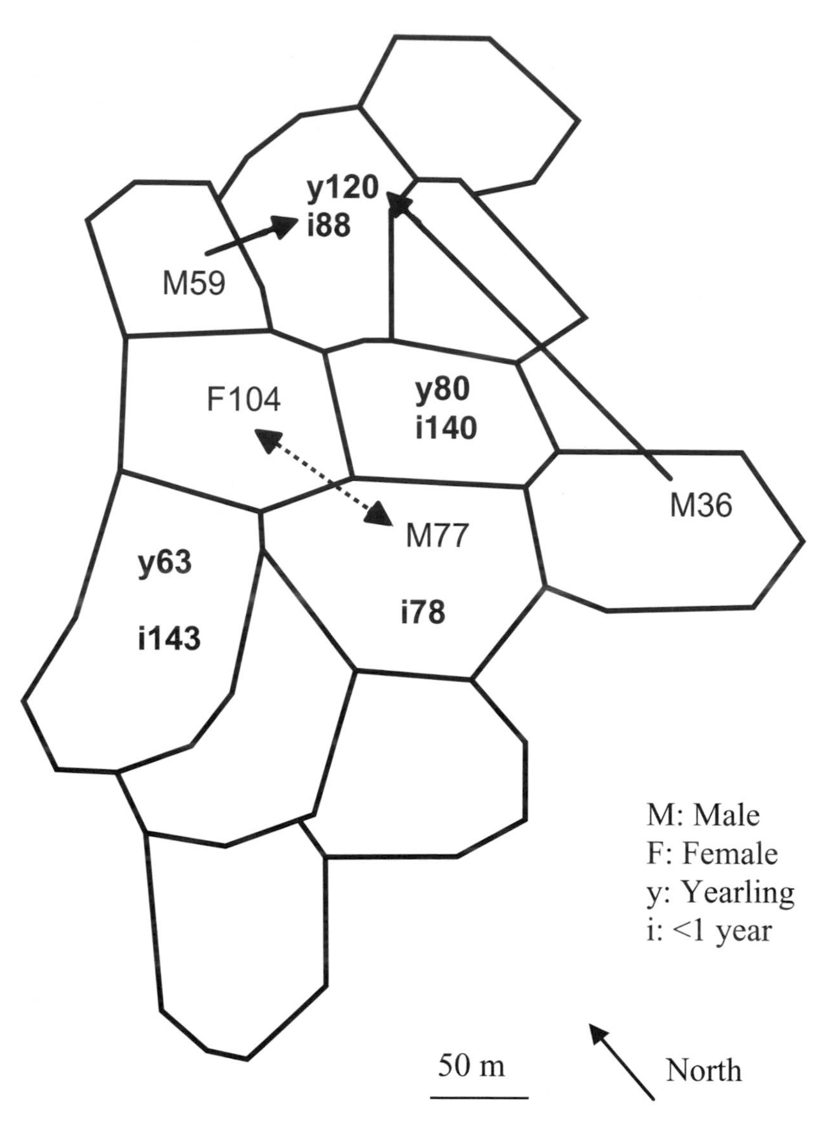

Figure 14.6. Extra-pair young within the territories of their social parents. Arrows indicate extra-pair paternity of respective young. Dotted arrow indicates EPC, revealed by vaginal plug. Note that offspring are born in different years (modified after Fietz *et al.*, 2000).

1997; Brotherton & Komers, chapter 3; van Schaik & Kappeler, chapter 4). Male care, as found in the extreme case of the callitrichids (Goldizen, 1987, Dunbar, 1995), is therefore thought to explain not the evolution of pair living but rather its maintenance.

C. medius males take care of their offspring in the form of baby-sitting, playing, travelling, sleeping with them in nest holes during the day, and responding to their calls. While baby-sitting, males stay in the nest holes with their offspring for considerable periods of the night, and during these times they cannot forage. Rearing of young and pre-hibernation fattening coincide in *C. medius*. Males not able to compensate for the shorter periods of foraging might not be able to accumulate sufficient fat stores before the onset of hibernation, resulting in higher winter mortality. Thus, by taking care of their young males may incur extremely high costs, and payment may be a decreased chance of surviving the following winter. Male care is generally divided into two categories: direct (infant carrying, provisioning) and indirect (territorial defence, warming, protection from predation; van Schaik & Kappeler, chapter 4; Ribble, chapter 5; Sommer, chapter 7; Reichard, chapter 13). In the special case of the callitrichids, it has been suggested that the carrying of infants by males enables females to maximize their reproductive output (Dunbar, 1995). Nevertheless, it is still questionable whether these two forms of male care should be separated since males incur energetic costs in both.

In *C. medius*, reasonable explanations for the baby-sitting behaviour of both parents could be keeping infants warm and protecting them from predators. Although the development of thermoregulation in *C. medius* has not yet been studied, it can be assumed that *C. medius* offspring are poikilothermic, as are offspring of small mammals in general. Thus, as shown for *Peromyscus californicus* (Ribble, chapter 5), and *Mus musculus* (König *et al.*, 1988), warmth provided by the parents should be beneficial for the development, survival, and growth of *C. medius* offspring. This is particularly relevant in terms of *C. medius* litter size, which is small (1–2 young): smaller litters are in general energetically less efficient because of greater heat loss (König *et al.*, 1988). In *C. medius,* parents spend more time with their offspring during the first days after birth than later on, suggesting that thermoregulatory demands in newborns might favour baby-sitting shortly after birth. On the other hand, baby-sitting occurs 13 days after birth as well, and males guide older infants on excursions. This suggests that thermoregulation is not the only reason for parents to guard their young.

Direct observations, and the fact that predation pressure is relatively high (20% of radio-collared individuals were predated: Fietz & Dausmann, unpublished data), suggest that protection against predators could also be an important reason for biparental care in this species. Male care seems to be obligate since without it infants died within the first days after birth. Unfortunately, small sample size prevents quantitative assessment of the causes and importance of biparental care in *C. medius*. An experimental removal of males in order to test whether male care influences infant survival, as done in *P. californicus* (Ribble, chapter 5), is not possible in a protected primate species such as *C. medius*. Male care seems to at least maintain pair living in *C. medius*. But the answer to the question of whether it evolved before or after the evolution of pair living remains elusive.

Territorial and floating males

If males achieve higher reproductive success by staying and mating with only one female for successive reproductive periods than by leaving the female after copulation and searching for other receptive partners, then this may indicate the evolution of monogamy as a mate strategy of male mammals (Wittenberger & Tilson, 1980; Brotherton & Komers, chapter 3). This higher male reproductive rate should occur if the costs of roving and searching for a new mating partner are higher than the costs of renouncing additional matings by staying with a single female. Costs of roving are influenced by predation, fights to gain access to females, and the information concerning length of female oestrous cycles (Brotherton & Komers, chapter 3).

In *C. medius*, roving males (i.e., floaters) indeed occur, but since they could be excluded from all paternities tested, they do not seem to have any reproductive success. Thus floating *C. medius* males are unlikely to be following an alternative reproductive strategy, as shown for floater male tree swallows (*Tachycineta bicolor*: Kempenaers *et al.*, 2001), but are rather the consequence of high population densities and a male-biased sex ratio. Because predation risk in *C. medius* is relatively high, information about safe hiding and sleeping places might be significant for survival. Floaters use large home ranges and do not have exclusive nest holes; thus, it can be assumed that information about and access to safe sleeping and hiding places is inferior to that of territorial individuals. Consequently, there seems to be no good reason for a male fat-tailed dwarf lemur to become a floater, and males should therefore clearly prefer to be territorial.

Reproductive success with females depends, in addition to other factors, on information about their oestrous states, oestrus lengths, and ovarian synchrony (Birkhead & Møller, 1992; Stutchbury & Morton, 1995; Saino et al., 1999), especially if the female concerned is not the social partner. Even though *C. medius* is a very seasonal breeder and oestrus lasts only one day (Foerg, 1982), oestrus was not strictly synchronized within the population (Fietz, 1999a).

Mixed mating strategies

High population densities, in combination with small territories and oestrus asynchrony within an extremely limited time period, should provide males with opportunities to engage in EPCs and reproduce with additional females, if they have information about the oestrous states of those females. The high rates of EPY revealed by paternal analysis show that *C. medius* frequently mates outside the pair bond and that males must obviously have the information they need about the oestrous states of other females. Based on this data, it seems that *C. medius* males adopt a mixed strategy. They live in lifelong pair bonds and stay with one female, which seems to be beneficial, and they attempt to mate with neighbours to maximize reproductive output. Utilizing this approach, they do not incur the increased costs of roving. However, these males are subject to information asymmetry concerning the oestrous state of the females. Social pairs spend most of their days together within the same nest holes. As a result, information about a female's oestrous state should be easily accessible to her social partner, but not necessarily to surrounding males. Consequently, males are only likely to achieve additional matings if females are willing to mate with them.

EXTRA-PAIR COPULATIONS

Female benefits and costs

Brotherton and Komers (chapter 3) argue that females accept monogamy because of high costs of EPC, for example, harassment by multiple males. In *C. medius*, harassment of a female was never observed. Two further assumptions are that females play an active role in seeking EPCs since males in captivity were not able to coerce copulations (Foerg, 1982), and that females benefit from EPCs. One of the potential benefits for females would be decreased risk of infanticide (Hrdy, 1977; Borries, 1997; van Schaik & Kappeler, 1997). But, as already discussed above, male infanticide is very unlikely to be adaptive in *C. medius*. The most widely accepted explanation for females seeking EPCs is that they do so to obtain genetic benefits (Trivers, 1985; Birkhead & Møller, 1992; Kempenaers *et al.*, 1992; Petrie & Kempenaers, 1998), their main incentive being to secure the superior genetic quality of extra-pair males compared with their male partners (Genetic Quality Hypothesis: Birkhead & Møller, 1992; Kempenaers *et al.*, 1992; Petrie & Kempenaers, 1998; Spencer *et al.*, 1998). Alternatively, females might enhance their fitness by diversifying the genotypes of their offspring (Genetic Diversity Hypothesis: Brooker *et al.*, 1990; Birkhead & Møller, 1992; Kempenaers & Dhondt, 1993; Dunn *et al.*, 1994; Petrie & Kempenaers, 1998).

None of the analyses performed could assign any of the floaters as fathers to any of the infants who were genotyped. This could, however, be due to the relatively small sample size. But observations revealed that floaters were generally defeated in fights and had smaller testes, suggesting that floaters are of inferior quality in comparison to territorial males and that the occupation of a territory and reproduction might depend on male quality. Paternity analyses suggest that females prefer territory holders for reproduction, which again supports the idea that for females, one of the main benefits of soliciting EPCs could be the superior genetic quality of the extra-pair males over male partners (Fietz *et al.*, 2000). Nevertheless, the traits enabling one male to occupy a territory and to be chosen as a mating partner by any female remain unclear. Future analyses, using larger sample sizes and focusing on female mate choice and individual reproductive success, will hopefully elucidate female choice in this species as far as male quality and territoriality in *C. medius* are concerned.

The most likely cost of EPCs for females is reduced male care, if males can assess the probability of their paternity or relatedness of young. In birds, the degree of paternal care was indeed shown to be positively correlated with the certainty of paternity (Møller & Birkhead, 1993; Dixon *et al.*, 1994). However, in *C. medius*, paired males took care of unrelated young, which suggests that a female that conceives with males other than her social partner does not run the risk of reduced male care from that social partner.

Male benefits and costs

For males, the main potential benefit of EPCs is an increase in reproductive success, while the risk of cuckoldry seems to be the only cost sufficiently important to modify male reproductive behaviour. In *C. medius*, at least some males obviously adopt a strategy of mating with one female, increasing their reproductive output by engaging in EPCs without the increased costs of roving. As these males do not necessarily run the risk of reduced reproductive success within their pair bonds, male reproductive success should vary within the population (Fietz *et al.*, 2000). Up to now sample size has unfortunately been too small to quantify male reproductive output to a degree sufficient to be correlated with morphological or behavioural traits.

In order to ensure paternity within the pair bond, males should try to conceive first with their social partner and then to increase their reproductive success by copulating with extra-pair females. Regrettably, information about the timing of the fertilization of intra-pair young (IPY) and EPY is not available, as it could elucidate the mating tactics of males that reproduce outside their pair bond. Nevertheless, not all males are able to gain access to extra-pair females, and since males run the risk of being cheated on by their females, they should have counterstrategies to prevent these females from being fertilized by other males.

Sexual monomorphism, in combination with relatively small testes in floating males, suggests a weak intrasexual competition and a low intensity of sperm competition, as is usually expected in a pair-living species (Clutton-Brock *et al.*, 1977; Leutenegger, 1978; Harcourt *et al.*, 1981; Cheverud *et al.*, 1985; Kenagy & Trombulak, 1986; Heske & Ostfeld, 1990; Ostfeld, 1990). However, the high incidence of EPPs in *C. medius* implies sperm competition and, consequently, relatively large testes. This discrepancy might be explained by a generally low copulation frequency even between social partners. Indeed, during 600 hours of direct observation, only two copulations were observed, both of which occurred during the mating season. Thus, copulations in *C. medius* seem to be infrequent, which would explain the relatively small testes. Furthermore, *C. medius* probably does not copulate outside the breeding season, suggesting that copulations do not serve a pair-bonding function like that found in monogamous birds (Wagner, chapter 6).

To ensure paternity, males can prevent females from copulating with extra-pair males by strict mate guarding or by applying vaginal plugs (Trivers, 1985; Kempenaers & Dhondt, 1993). Outside the mating season, pairs mainly range as solitaries and meet three to five times during a night (Wiedemann, unpublished data), while during oestrus *C. medius* males spend about 60% of the night with their females (Fietz, 1999*b*). Thus, even though males spent much more time in the proximity of the female when she was in oestrus than when she was not, males did not strictly mate-guard. Van Schaik and Kappeler (chapter 4) suggest that a high incidence of EPCs might be a common trait of dispersed pairs since mate guarding is much harder to maintain under those circumstances.

Energetically, strict mate guarding might be very costly for the male, and the application of vaginal plugs after copulation seems to be a much easier and cheaper way for the male to ensure mating success. Male *C. medius* are known to apply vaginal plugs after copulation (Foerg, 1982; Fietz *et al.*, 2000). This may at first seem surprising in a pair-living species. But since EPPs are so frequent in *C. medius*, post-copulatory mechanisms might have a potentially strong impact on male reproductive success (especially because mate guarding is constrained). Nevertheless, females of a closely related lemur species (*Microcebus murinus*) were observed easily removing vaginal plugs themselves (M. Eberle, personal communication, 1999), and it may be assumed that females of *C. medius* have the same ability. As long as females play an active role in seeking EPCs, which seems to be true for *C. medius*, vaginal plugs seem rather unreliable, at least for males. Females, on the other hand, could ensure insemination by a certain male by not removing his plug after copulation. In conclusion, *C. medius* males do not seem to have any effective tactics for preventing their females from engaging in EPCs.

Male care of current young can only evolve if it ensures the highest fitness return, and if they are able to detect them, males should not invest in unrelated young, (Hamilton, 1964*a*, *b*; Maynard Smith, 1977; Trivers, 1985; but see Freeman-Gallant, 1997; Hughes, 1998).

Even so, in *C. medius*, males were known to invest in litters containing EPY (Fietz *et al.*, 2000). There are several explanations why males take care of unrelated young. Male care can be an indicator of male quality and so males may benefit indirectly by investing in EPY (Freeman-Gallant, 1997). However, in cases of mixed paternity (Birkhead & Møller, 1993), males may jeopardize the survival of their own young by providing less care to these kinds of heterogeneously-sired broods (Hughes, 1998). This seems to present a problem only if a male is not able to identify an individual EPY as opposed to his own, but can assess the likelihood of his paternity. If *C. medius* males were able to detect EPY, they could exclude them from baby-sitting on an individual basis either by expelling them from the nest hole or by killing them as soon as possible. But *C. medius* social fathers even tolerated unrelated yearlings within their territories and nest holes, which suggests that they cannot discriminate between related and unrelated young. The question of whether or not males can assess their own paternity remains unresolved. In contrast to *C. medius*, beavers (*Castor canadiensis*) are able to recognize different degrees of relatedness by a genetically controlled kinship pheromone contained in the anal gland secretion (Sun, chapter 9). Thus, EPY in beavers would probably be detected and killed by the respective male; as a result, beavers are expected to have a genetically monogamous mating system (Sun, chapter 9). However, up to now no mixed paternity has been detected in *C. medius* (Fietz *et al.*, 2000). Finally, if males were the philopatric sex, and neighbouring males were therefore closely related, males could even invest in their inclusive fitness by raising EPY sired by their neighbour. And while female *C. medius* indeed seem preferentially to solicit EPCs with neighbouring territory holders (Figure 14.6; Fietz *et al.*, 2000), microsatellite analysis did not reveal close relatedness of neighbouring males.

CONCLUSION

The discrepancy between high rates of EPPs and the existence of male care in *C. medius* remains unexplained. One possibility is that sample size in this study is too small and does not represent the real incidence of EPY. Another possible explanation is that males do not expel EPY because they cannot detect individual relatedness. Female choice, on the other hand, might depend on male quality, which is assumed to vary within the population. The 'best' males gain territories and maximize their reproductive output through extra-pair fertilizations (EPFs), without the increased costs of roving or caring for EPY. However, successful mating outside the pair bond depends on the oestrous states of the surrounding females, as well as on the relative quality of the surrounding males. Males of inferior quality do not acquire a territory but become floaters. Some males do secure a territory but are frequently cheated on by their females. And even though they reproduce only once in a while, this state is still preferable to that of being a floater.

More information about dispersal patterns and philopatry, and future genetic analysis with a larger sample size that focuses on mate choice, individual reproductive success, and the paternity of sibs will hopefully increase our knowledge of female choice in respect to male quality and territoriality in *C. medius*.

Acknowledgements

I am grateful to the 'Commission Tripartite' of the Malagasy Government, the Laboratoire de Primatologie et des Vértebrés de l'Université d'Antananarivo, the Ministère pour la Production Animale, and the Départment des Eaux et Forêts for permits to work in Madagascar. I also thank the Centre de Formation Professionelle Forestière de Morondava for their hospitality and permission to work on their concession. J. U. Ganzhorn, P. M. Kappeler, K. H. Dausmann, H. Zischler, C. Schwiegk, L. Razafimanantsoa, R. Rasoloarison, J. Burkhardt, E. Bienert, D. Schwab, J. Schmid, H. Klensang, B. Rakotosamimanana, C. Ravoarinoromanga, and K. Schmidt-Koenig supported the project in numerous ways. I thank the DPZ and Peter M. Kappeler for the opportunity to work at the field station. Financial aid from the Deutsches Primatenzentrum, the HSPII, and the Deutsche Forschungsgemeinschaft to the Graduiertenkolleg GRK 289/2 'Primatologie' made this study possible. Last but not least, I want to thank Ulrich Reichard and Christoph Boesch for the organization and invitation to the 'monogamy' workshop, and all the participants for an interesting workshop and nice atmosphere.

References

Birkhead, T. R. & Møller, A. P. (1992). *Sperm Competition in Birds: Evolutionary Causes and Consequences*. London: Academic Press.

(1993). Why do male birds stop copulating while their partners are still fertile? *Animal Behaviour*, **45**, 105–18.

Borries, C. (1997). Infanticide in seasonally breeding multimale groups of Hanuman langurs (*Presbytis entellus*) in Ramnagar (South Nepal). *Behavioral Ecology and Sociobiology*, **41**, 139–50.

Brooker, M. G., Rowley, I., Adams, M. & Braverstock, P. R. (1990). Promiscuity: an inbreeding avoidance mechanism in a socially monogamous species? *Behavioral Ecology and Sociobiology*, **26**, 191–9.

Cheverud, J. M., Dow, M. M. & Leutenegger, W. (1985). The quantitative assessment of phylogenetic constraints in comparative analyses: sexual dimorphism in body weight among primates. *Evolution*, **39**, 1335–51.

Clutton-Brock, T. H. (1989). Mammalian mating systems. *Proceedings of the Royal Society of London, Series B*, **236**, 339–72.

Clutton-Brock, T. H., Harvey, P. H. & Rudder, B. (1977). Sexual dimorphism, socionomic sex ratio and body weight in primates. *Nature*, **269**, 797–800.

Dausmann, K. H., Ganzhorn, J. U. & Heldmaier, G. (2000). Body temperature and metabolic rate of a hibernating primate in Madagascar: preliminary results from the field study. In *Life in the Cold: Eleventh International Hibernation Symposium*, ed. G. Heldmaier & M. Klingenspor, pp. 41–7. Berlin: Springer.

Davies, N. B. (1992). *Dunnock Behaviour and Social Evolution*. Oxford: Oxford University Press.

Dixon, A., Ross, D., O'Malley, S. L. C. & Burke, T. (1994). Paternal investment inversely related to degree of extra-pair paternity in the reed bunting. *Nature*, **371**, 698–700.

Dunbar, R. I. M. (1995). The mating system of callitrichid primates: I. Conditions for the coevolution of pair bonding and twinning. *Animal Behavior*, **50**, 1057–70.

Dunn, P. O., Robertson, R. J., Michaud-Freeman, D. & Boag, P. T. (1994). Extra-pair paternity in tree swallows: why do females mate with more than one male? *Behavioral Ecology and Sociobiology*, **25**, 273–81.

Fietz, J. (1999*a*). Monogamy as a rule rather than exception in nocturnal lemurs: the case of the fat-tailed dwarf lemur, *Cheirogaleus medius*. *Ethology*, **105**, 259–72.

(1999*b*). Demography and floating males in a population of *Cheirogaleus medius*. In *New Directions in Lemur Studies*, ed. B. Rakotosamimanana, H. Rasaminanana & J. U. Ganzhorn, pp. 159–72. New York: Kluwer Academic/ Plenum.

Fietz, J. & Ganzhorn, J. U. (1999). Feeding ecology of the hibernating primate *Cheirogaleus medius*: how does it get so fat? *Oecologia*, **121**, 157–64.

Fietz, J., Zischler, H., Schwiegk, C., Tomiuk, J., Dausmann, K. H. & Ganzhorn, J. U. (2000). High rates of extra-pair young in the pair-living fat-tailed dwarf lemur, *Cheirogaleus medius*. *Behavioral Ecology and Sociobiology*, **49**, 8–17.

Foerg, R. (1982). Reproduction in *Cheirogaleus medius*. *Folia Primatologica*, **39**, 49–62.

Freeman-Gallant, C. R. (1997). Parentage and paternal care: consequences of intersexual selection in savannah sparrows? *Behavioral Ecology and Sociobiology*, **40**, 395–400.

Ganzhorn, J. U. & Sorg, J. P. (1996). Ecology and economy of a tropical dry forest in Madagascar. *Primate Report*, **46–1**, 1–382.

Ganzhorn, J. U., Abraham, J. P. & Razanahoera-Rakotomalala, M. (1985). Some aspects of the natural history and food selection of *Avahi laniger*. *Primates*, **26**, 452–63.

Girman, D. J., Mills, M. G. L., Geffen, E. & Wayne, R. K. (1997). A molecular genetic analysis of social structure, dispersal, and interpack relationships of the African wild dog (*Lycaon pictus*). *Behavioral Ecology and Sociobiology*, **40**, 187–98.

Goldizen, A. W. (1987). Tamarins and marmosets: communal care of offspring. In *Primate Societies*, ed. B. B. Smuts, D. L. Cheney, R. M. Seyfarth, R. W. Wrangham & T. T. Struhsaker, pp. 34–43. Chicago: University of Chicago Press.

Goossens, B., Graziani, L., Waits, L. P., Farand, E., Magnolon, S., Coulon, J., Bel, M.-C., Taberlet, P. & Allainé, D. (1998). Extra-pair paternity in the monogamous alpine marmot revealed by nuclear DNA microsatellite analysis. *Behavioral Ecology and Sociobiology*, **43**, 281–8.

Hamilton, W. D. (1964*a*). The genetical evolution of social behaviour. I. *Journal of Theoretical Biology*, **7**, 1–16.

(1964*b*). The genetical evolution of social behaviour. II. *Journal of Theoretical Biology*, **7**, 17–52.

(1967). Extraordinary sex ratios. A sex-ratio theory for sex linkage and inbreeding has new implications in cytogenetics and entomology. *Science*, **156**, 477–88.

Harcourt, C. (1991). Diet and behavior of a nocturnal lemur *Avahi laniger* in the wild. *Journal of Zoology, London*, **223**, 667–74.

Harcourt, C., Harvey, P. M., Larson, S. G. & Short, R. V. (1981). Testis size, body weight and breeding system in primates. *Nature*, **293**, 55–7.

Heske, E. J. & Ostfeld, R. S. (1990). Sexual dimorphism in size, relative size of testes and mating systems in North American voles. *Journal of Mammalogy*, **71**, 510–19.

Hladik, C. M., Charles-Dominique, P. & Petter, J. J. (1980). Feeding strategies of five nocturnal prosimians in the dry forest of the west coast of Madagascar. In *Nocturnal Malagasy Primates: Ecology, Physiology and Behaviour*, ed. P. Charles-Dominique, H. M. Cooper, A. Hladik, C. M. Hladik, E. Pages, G. F. Pariente, A. Petter-Rousseaux, J. J. Petter & A. Schilling, pp. 41–73. New York: Academic Press.

Hrdy, S. B. (1977). *The Langurs of Abu*. Cambridge: Harvard University Press.

Hughes, C. (1998). Integrating molecular techniques with field methods in studies of social behavior: a revolution results. *Ecology*, **79**, 383–99.

Jolly, A. (1995). *The Evolution of Primate Behavior*. New York: Macmillan.

Kappeler, P. M. (1997). Determinants of primate social organization: comparative evidence and new insights from Malagasy lemurs. *Biological Reviews of the Cambridge Philosophical Society*, **72**, 111–51.

Kempenaers, B. & Dhondt, A. A. (1993). Why do females engage in extra-pair copulations? A review of hypotheses and their predictions. *Belgian Journal of Zoology*, **123**, 93–123.

Kempenaers, B., Everding, S., Bishop, C., Boag, P. & Robertson, R. J. (2001). Extra-pair paternity and the reproductive role of male floaters in the tree swallow (*Tachycineta bicolor*). *Behavioral Ecology and Sociobiology*, **49**, 251–9.

Kempenaers, B., Verheyen, G. R., van der Broeck, M., Burke, T., van Broeckhoven, C. & Dhondt, A. A. (1992). Extra-pair paternity results from female preference for high-quality males in the Blue Tit. *Nature*, **357**, 494–6.

Kenagy, G. J. & Trombulak, S. C. (1986). Size function of mammalian testes in relation to body size. *Journal of Mammalogy*, **67**, 1–22.

Kleiman, D. G. (1977). Monogamy in mammals. *Quarterly Review of Biology*, **52**, 39–69.

Komers, P. E. & Brotherton, P. N. M. (1997). Female space use is the best predictor of monogamy in mammals. *Proceedings of the Royal Society of London, Series B*, **264**, 1261–70.

König, B., Riester, J. & Markl, H. (1988). Maternal care in house mice (*Mus musculus*): II. The energy cost of lactation as a function of litter size. *Journal of Zoology, London*, **216**, 195–210.

Leutenegger, W. (1978). Scaling of sexual dimorphism in body size and breeding system in primates. *Nature*, **272**, 610–11.

Maynard Smith, J. (1977). Parental investment: a prospective analysis. *Animal Behavior*, **25**, 1–9.

(1991). *Evolution and the Theory of Games*. Cambridge: Cambridge University Press.

Mitani, J. C. & Rodman, P. S. (1979). Territoriality: the relation of ranging pattern and home range size to defendability, with an analysis of territoriality among primate species. *Behavioral Ecology and Sociobiology*, **5**, 241–51.

Mock, D. W. & Fujioka, M. (1990). Monogamy and long-term pair bonding in vertebrates. *Trends in Ecology and Evolution*, **5**, 39–43.

Møller, A. P. & Birkhead, T. R. (1993). Certainty of paternity covaries with paternal care in birds. *Behavioral Ecology and Sociobiology*, **33**, 261–8.

Müller, A. E. (1998). A preliminary report on the social organisation of *Cheirogaleus medius* (Cheirogaleidae; Primates) in north-west Madagascar. *Folia Primatologica*, **69**, 160–6.

(1999*a*). The social organisation of the fat-tailed dwarf lemur *Cheirogaleus medius* (Lemuriformes, Primates). Ph.D. thesis, University of Zürich.

(1999*b*). Social organization of the fat-tailed dwarf lemur (*Cheirogaleus medius*) in Northwestern Madagascar. In *New Directions in Lemur Studies*, ed. B. Rakotosamimanana, H. Rasaminanana, J. U. Ganzhorn & S. M. Goodman, pp. 139–57. New York: Kluwer Academic/Plenum.

(1999*c*). Aspects of social life in the fat-tailed dwarf lemur (*Cheirogaleus medius*): inferences from body weights and trapping data. *American Journal of Primatology*, **49**, 265–80.

Müller, A. E. & Thalmann, U. (2000). Origin and evolution of primate social organisation: a reconstruction. *Biological Reviews*, **75**, 405–35.

Ostfeld, R. S. (1990). The ecology of territoriality in small mammals. *Trends in Ecology and Evolution*, **5**, 411–15.

Petrie, M. & Kempenaers, B. (1998). Extra-pair paternity in birds: explaining variation between species and populations. *Trends in Ecology and Evolution*, **13**, 52–8.

Petter, J. J. (1978). Ecological and physiological adaptations of five sympatric nocturnal lemurs to seasonal variations in food production. In *Recent Advances in Primatology*, Volume 1, ed. D. J. Chivers & J. Herbert, pp. 211–23. New York: Academic Press.

Petter, J. J., Albignac, R. & Rumpler, R. (1977). *Faune de Madagascar: Mammifères Lémuriens*. Paris: ORSTOM/CNRS.

Petter-Rousseaux, A. (1980). Seasonal activity rhythms, reproduction, and body weight variations in five sympatric nocturnal prosimians, simulated light and climatic conditions. In *Nocturnal Malagasy Primates: Ecology, Physiology and Behaviour*, ed. P. Charles-Dominique, H. M. Cooper, A. Hladik, C. M. Hladik, E. Pages, G. F. Pariente, A. Petter-Rousseaux, J. J. Petter & A. Schilling, pp. 137–52. New York: Academic Press.

Reynolds, J. D. (1996). Animal breeding systems. *Trends in Ecology and Evolution*, **11**, 68–72.

Rutberg, A. T. (1983). The evolution of monogamy in primates. *Journal of Theoretical Biology*, **104**, 93–112.

Saino, N., Primmer, C. R., Ellergren, H. & Møller, A. P. (1999). Breeding synchrony and paternity in the barn swallow (*Hirundo rustica*). *Behavioral Ecology and Sociobiology*, **45**, 211–18.

Sillero-Zubiri, C., Gottelli, D. & Macdonald, D. W. (1996). Male philopatry, extra-pack copulations and inbreeding avoidance in ethiopian wolves (*Canis simensis*). *Behavioral Ecology and Sociobiology*, **38**, 331–40.

Smuts, B. B., Cheney, D. L., Seyfarth, R. M., Wrangham, R. W. & Struhsaker, T. T. (ed.) (1987). *Primate Societies*. Chicago: University of Chicago Press.

Spencer, P., Horsup, A. & Marsh, H. (1998). Enhancement of reproductive success through mate choice in a social rock-wallaby, *Pterogale assimilis* (Macropodidae) as revealed by microsatellite markers. *Behavioral Ecology and Sociobiology*, **43**, 1–9.

Stutchbury, B. J. & Morton, E. S. (1995). The effect of breeding synchrony on extra-pair mating systems in songbirds. *Behaviour*, **132**, 675–90.

Thalmann, U. (2001). Food resource characteristics in two nocturnal lemurs with different social behavior: *Avahi occidentalis* and *Lepilemur edwardsi*. *International Journal of Primatology*, **22**, 287–324.

Trivers, R. L. (1985). *Social Evolution*. Menlo Park, California: Benjamin Cummings.

van Schaik, C. P. & Dunbar, R. I. M. (1990). The evolution of monogamy in large primates: a new hypotheses and some crucial tests. *Behaviour*, **115**, 30–61.

van Schaik, C. P. & Kappeler, P. M. (1993). Life history, activity pattern and lemur social systems. In *Lemur Social Systems and their Ecological Basis*, ed. P. M. Kappeler & J. U. Ganzhorn, pp. 241–60. New York: Plenum Press.

(1996). The social systems of gregarious lemurs: lack of convergence with anthropoids due to evolutionary disequilibrium? *Ethology*, **102**, 915–41.

(1997). Infanticide risk and the evolution of male–female association in primates. *Proceedings of the Royal Society of London, Series B*, **264**, 1687–94.

Warren, R. & Crompton, R. H. (1997). A comparative study of the ranging behavior, activity rhythm and sociality of *Lepilemur edwardsi* (Primates, Lepilemuridae) and *Avahi occidentalis* (Primates, Indridae) at Ampijoroa, Madagascar. *Journal of Zoology, London*, **243**, 397–415.

Wittenberger, J. F. & Tilson, R. L. (1980). The evolution of monogamy: hypotheses and evidence. *Annual Review of Ecological Systems*, **11**, 197–232.

Wright, P. C. (1990). Patterns of paternal care in primates. *International Journal of Primatology*, **11**, 89–102.

CHAPTER 15

Social monogamy and its variations in callitrichids: do these relate to the costs of infant care?

Anne W. Goldizen

INTRODUCTION

When I studied saddle-back tamarins (*Saguinus fuscicollis*) by following wild groups of individually marked tamarins around the rainforest in Manu National Park in southeastern Peru, one group particularly intrigued me because of its unusual family structure. Tamarins usually have twin births and infants are carried by adults for their first three months. In this group, when an adult male had tired of carrying the twins and scraped them off his back onto a branch, it was usually not the twins' mother who finally came to retrieve the screaming young. Instead, it was generally a second adult male who picked up the infants and carried them for a period of time, just as solicitously as the first male had done. When I later compared the infant-carrying contributions of the two males and the female, I found that the males had done roughly equal amounts of carrying, and each had done much more than the female. I also found that both males had mated frequently with the female earlier in the year, before the female became pregnant with the twins, with no clear indication that one was dominant over the other and was more likely to have achieved parentage.

None of these observations fits with the traditional view of callitrichids (tamarins and marmosets) as socially and/or genetically monogamous primates. This assumption had been based on the lack of sexual dimorphism in callitrichid species, the fact that only single females bred in captive groups, and the paternal care shown by males (Kleiman, 1977). The assumption of sociogenetic monogamy was also consistent with the fact that in captivity callitrichids bred well when housed in pairs.

There are four genera of callitrichids, including the tamarins (*Saguinus*), lion tamarins (*Leontopithecus*), marmosets (*Callithrix*), and the genus containing the pygmy marmoset (*Cebuella pygmaea*). A fifth genus of primates that is closely related to callitrichids is *Callimico*, and it only includes a single species, Goeldi's monkey (*C. goeldii*), but the phylogenetic position of this genus has long been debated and is still not clear (reviewed by Rylands *et al.*, 1993; see also Kay, 1990; Canavez *et al.*, 1999). One feature that makes callitrichids unique among primates is that most litters consist of twins (Hampton & Hampton, 1965). These twins together weigh up to 20% of their mother's weight at birth, causing the costs of parental care to be high (Kleiman, 1977). Goeldi's monkeys have single infants, rather than twins, which makes them fundamentally different from the four genera of callitrichids. I therefore do not cover *Callimico* here, as the focus of my discussion is how the mating patterns of callitrichids relate to the costs of the rearing of their twins.

The four genera of callitrichids are small, arboreal, group-living, territorial primates with omnivorous diets. Tamarin species range in weight from about 320 to 560 g and primarily eat fruits, insects, and small vertebrates (reviewed by Garber, 1993). Lion tamarins are larger, with adults weighing over 500 g; like tamarins, lion tamarins primarily eat fruits and insects (reviewed by Rylands, 1993). The marmosets and the pygmy marmosets are smaller and their diets contain varying amounts of exudates, as well as fruits and insects (reviewed by Rylands & de Faria, 1993; Soini, 1993). These two genera have specialized dentition (large lower incisors with patterns of enamel that give them a chisel shape) that allows them to gouge holes in trees to stimulate the flow of exudates (Coimbra-Filho & Mittermeier, 1978).

The mating pattern exhibited by my saddle-back tamarin group in Peru can be called sociosexual cooperative polyandry. In such groups, only a single female breeds, but the female mates with more than one male, each of whom then provides care to the female's subsequent offspring. A female and her mates thus jointly care for a single set of offspring, each of which could have been fathered by any of the female's mates (Faaborg &

Patterson, 1981). This form of polyandry has been called 'cooperative polyandry' to distinguish it from 'classical polyandry,' in which a female has two or more mates, each of which defends his own territory and rears a separate set of young (Faaborg & Patterson, 1981). There are still no published genetic data on patterns of paternity in callitrichid groups with multiple mating males, so we do not yet know how equal the sociosexually polyandrous males' chances of fathering young may be.

Sociosexual cooperative polyandry has been found to occur frequently in only a small number of species, including some birds (e.g., Galapagos hawk (*Buteo galapagoensis*): Faaborg *et al.*, 1980, 1995; acorn woodpecker (*Melanerpes formicivorus*): Koenig & Mumme, 1987; Koenig & Stacey, 1990; dunnock (*Prunella modularis*): Davies, 1985, 1992; Davies & Lundberg, 1984; Tasmanian native hen (*Gallinula mortierii*): Maynard Smith & Ridpath, 1972; Goldizen *et al.*, 1998, 2000; trumpeter (*Psophia leucoptera*): Eason & Sherman, 1995; Sherman, 1995; pukeko (*Porphyrio porphyrio*): Jamieson *et al.*, 1994) and some mammals (e.g., kinkajou (*Potos flavus*): Kays, chapter 8; white-handed gibbons (*Hylobates lar*): Reichard, chapter 13; some species of tamarins and marmosets, reviewed below). In most of these species, cooperative polyandry is only one of several mating patterns found in populations. In some species and populations, both sociosexual monogamy and polyandry occur, while in others, sociosexually monogamous, polyandrous, polygynous, and even polygynandrous groups are found (Ridpath, 1972; Davies & Lundberg, 1984; Faaborg, 1986; Goldizen *et al.*, 2000). Groups in some of the bird species listed above have been shown to be reproductively as well as socially polyandrous (e.g., Burke *et al.*, 1989; Jamieson *et al.*, 1994; Faaborg *et al.*, 1995).

In socially monogamous pairs, males are paired with single females and provide parental care only to the young of those females. Because males exhibiting sociosexual cooperative polyandry are also paired with single females but share those mates with other males, cooperative polyandry can be viewed as an even more extreme version of monogamy, at least from the males' perspective. Why, then, would any species have evolved a tendency to exhibit sociosexual cooperative polyandry? How might males ever benefit from this form of mate sharing? Why do polygynous groups with two breeding females occasionally occur in populations that also exhibit sociosexual monogamy and polyandry?

In this chapter, I review data on the mating patterns and ecology of wild callitrichids and the current understanding of the proximate causes of sociosexual monogamy and its variations (polygyny and polyandry), in tamarins and marmosets. I begin by reviewing our knowledge of group compositions and mating patterns in the four callitrichid genera and the proposed link between the costs of infant care and callitrichid mating systems. I then review data on the costs of infant care as well as variations in ecological characteristics among callitrichid genera and species that might affect infant care costs and thus mating patterns, particularly patterns of polyandry. Finally, I review possible links between ecological patterns of female reproductive suppression and the frequency of reproductive polygyny in different callitrichid genera.

In this data review I use four terms to discuss mating patterns, social, and sexual behaviour.

1 *Social* qualifies mating patterns (e.g., *social monogamy*) when it is known that particular combinations of adult individuals associated together in groups, but information on mating patterns and genetic parentage is not available.
2 *Sociosexual* mating patterns (e.g., *sociosexual polyandry*) refers to groups for which there was knowledge of patterns of both social and sexual behaviour but not of actual genetic parentage.
3 *Reproductive polygyny* refers to groups in which multiple females were known to conceive.
4 *Reproductive polyandry* refers to groups in which two or more males were thought to have had the potential to father offspring.

GROUP COMPOSITIONS AND MATING PATTERNS OF CALLITRICHIDS

Tamarins

Data on group compositions and mating patterns in wild callitrichids are presented in Table 15.1. The first reports that wild tamarins might not all be sociosexually monogamous appeared in the mid-1980s (Garber *et al.*, 1984; Terborgh & Goldizen, 1985; Goldizen, 1987). Garber *et al.* (1984) reported that groups of moustached tamarins (*S. mystax*) frequently contained more than one adult of each sex. Among 18 groups on Padre Isla in northeastern Peru, only two comprised a single adult of each sex. There was no evidence from these 18 groups

Table 15.1. *Group compositions and mating patterns of wild callitrichids*

Species	Study site	Number of groups studied	Average group size	Presence/absence of >1 breeding female per group	Average number adult males per group	Polyandrous matings seen?	Paternal care by >1 male observed?	References
Moustached tamarin (*Saguinus mystax*)	Padre Isla, Peru	18	5.2 ± 1.5	$\leq$1 female had young in any group/year	1.8 ± 0.7	No	All adult males carried young	Garber *et al.*, 1984
	Padre Isla, Peru	13	7.0 ± 2.4	23% of groups had 2 reproductive females; no groups had two sets of young concurrently	2.2 ± 0.7	Polyandrous matings observed in the two intensively studied groups	No information provided	Garber *et al.*, 1993
	Quebrado Blanco, Peru	2	4.0 ± 1.4	No young born during study, thus no information	2 in each group	Few copulations seen; most by single male	No young present	Heymann, 1996
Saddle-back tamarin (*Saguinus fuscicollis*)	Manu National Park, Peru	47 group/years	5.1 ± 1.6 (excluding infants)	No information provided	1.9 ± 0.6	Yes	Yes	Goldizen, 1987; Goldizen *et al.*, 1996
Panamanian tamarin (*Saguinus oedipus geoffroyi*)	Panama Canal Zone, Panama	5	6.2 ± 0.3 (excluding infants)	Only one set of infants in any group at any one time	2.4 ± 0.4	No information provided	No information provided	Dawson, 1978
Cotton-top tamarin (*Saguinus oedipus*)	Tolu, Colombia	6	5.8 ± 1.7 (excluding infants)	No more than 2 juveniles present in any group at any one time	2.7 ± 0.8	No information provided	No information provided	Neyman, 1978
Black-mantled tamarin (*Saguinus nigricollis*)	Rio Caqueta, Colombia	10	5.2 ± 1.1 (excluding infants)	Only 1 breeding female per group	1.3 ± 0.5	No information provided	No information provided	Izawa, 1978
Common marmoset (*Callithrix jacchus*)	EFLEX-IBAMA, Brazil	3	8.7 ± 3.2	2 breeding females in each group	2.7 ± 1.1	Females mated with single males only within groups; extra-pair copulations observed	No information provided	Digby & Barreto, 1993; Digby & Ferrari, 1994; Digby, 1999
Pygmy marmoset (*Cebuella pygmaea*)	Maniti River, Peru	21	5.0 ± 1.8	1 breeding female in each group	1.1 ± 0.3	No, but subordinate male seen trying to mount female	Yes	Soini, 1988
Golden lion tamarin (*Leontopithecus rosalia*)	Poço das Antas Reserve, Brazil	211	5.4 ± 2.1 (including infants)	10.6% of groups had >1 parous female	No information provided	Yes, but dominant males appeared to monopolize oestrous females	No information provided	Baker *et al.*, 1993; Dietz & Baker, 1993

that more than one female ever bred at the same time, but multiple adult males were observed carrying the same sets of young, and in a larger sample of 47 groups Garber (1997) showed that infant survivorship and the numbers of adult males were related. These observations suggested that moustached tamarins might exhibit sociosexual polyandry, with individual females sometimes mating with more than one male, each of which then helped to rear their female's young (Garber *et al.*, 1984; Garber, 1997).

Groups with multiple adults of both sexes were also common among saddle-back tamarins at Cocha Cashu in Manu National Park (Terborgh & Goldizen, 1985; Goldizen, 1987; Goldizen *et al.*, 1996). The study of saddle-back tamarins at Cocha Cashu in Peru lasted from 1979 to 1992. During this period, groups contained an average of 1.87 adult males, 1.32 adult females, 0.85 subadults and 1.09 juveniles ($N = 47$ group/year compositions). Individuals were considered to be juveniles from weaning at about three months to 12 months of age, subadults until two years, and adults from two years of age. Some adults were still in their natal groups. Lone individuals were very rare; one male spent at least a few days alone while transferring between groups and another male was seen alone occasionally over a period of weeks.

Adult males that belonged to the same group sometimes appeared to be unrelated, though we did not know the histories of the adults in most groups, and in groups with multiple adult males, all adult males carried their group's infants. In the seven two-male groups in which we observed any copulations, both of the males were observed copulating in six of the groups (Goldizen *et al.*, 1996). These observations suggested that these groups exhibited sociosexual cooperative polyandry.

Most groups had only single breeding females, though in six out of the 47 group/years there was evidence that two females bred or at least copulated (Goldizen *et al.*, 1996). In each of four of these groups, two females produced young that survived to several months. The two females always had their young at least three months apart, so that the sets of young would not have required carrying at the same time. Despite these cases of successful reproductive polygyny or polygynandry, the evidence suggested that such cases of multiple female breeding were not stable. In five of the six groups, one of the females dispersed or disappeared within a year or two after the time when both females bred; the fates of the females in the sixth group were not known.

John Terborgh and I proposed that the occurrences in tamarins of both sociosexual polyandry and helping by non-reproductive older offspring are related to the fact that tamarins almost always produce twin offspring (Terborgh & Goldizen, 1985). We suggested that the costs of lactation are so high that a female would be unable to do much infant carrying, and the combined weight of twins would make it impossible for a single male to do most of the carrying. Thus, monogamous pairs of tamarins with older offspring who were able to contribute to infant carrying would be able to rear young, whereas lone pairs would be unlikely to succeed. In fact, lone pairs of saddle-back tamarins had never been observed at Cocha Cashu. We proposed that pairs without older offspring might have higher reproductive success by allowing a second adult male to join them and share in the chances of paternity, in exchange for helping to care for the young (Terborgh & Goldizen, 1985).

I further explored this hypothesis by collecting detailed data on the behaviour of two saddle-back tamarin groups (Goldizen, 1987). One of these groups contained three adults and twin infants and was sociosexually polyandrous, with copulations shared between the two males (25 vs. 31 copulations over nine months). The two adult males did 43.4% and 36.7% of the carrying of the group's infants, respectively, while the adult female did 19.9%. The other group contained an adult pair, one three-year-old adult daughter, one two-year-old adult daughter, and two one-year-old subadults, and was reproductively polygynous. The adult male had not immigrated into the group until after the births of the two adult daughters. While the oldest daughter reached a late stage of pregnancy, her infants must have died within days of birth. Thus, the parental care patterns of this group were assumed to be representative of those of reproductively monogamous groups. I found that this group's offspring, especially the two adult daughters, provided significant amounts of assistance with infant care. The adult male did 37.1% of the carrying, while the mother did 22.8%, the three-year-old daughter 24.4%, the two-year-old daughter 12.2%, and the two subadults 3.0% and 0.5%, respectively. Furthermore, most of the time that mothers 'carried' young in these two groups the young were nursing, and individuals carrying infants spent very little time feeding, presumably because feeding with infants would have increased the risk of predation. A model incorporating the costs of lactation and infant carrying (in terms of feeding time), as well as information on the division of infant carrying among group members, reinforced the suggestion that a lone pair would not be able to carry infants and still have enough time to meet their own nutritional demands, as well as those of lactation (Goldizen, 1987).

Dispersal patterns differed between the sexes in this population, reflecting different strategies for the acquisition of breeding positions in groups (Goldizen *et al.*, 1996). The evidence suggested that females rarely, if ever, dispersed between groups more than once in their lifetime, whereas such secondary dispersal was common among males. In addition, only two of 10 female immigrants moved into groups already containing adult females, while 20 of 24 male immigrants migrated into groups that already contained probable male breeders. These data suggest that females in this population took the first opportunity available to them to become primary breeders, usually in their natal group or a neighbouring group, and then kept that position for the rest of their lives (*c.* 3 years on average). Females that left their groups to search for breeding positions would have been unlikely to find them. Thus, female reproductive strategies in this population were based on the attainment of the primary breeding female position in a group, with age-related reproductive dominance hierarchies usually evident among females.

Male saddle-back tamarins appeared to have highly flexible reproductive strategies (Goldizen *et al.*, 1996). We do not know how frequently more than one male in a group fathered offspring, or how co-breeding males' chances of paternity compared. However, the observation that most male immigrants joined groups that already contained adult males, combined with our observations of polyandrous matings, suggested that sociosexual polyandry was frequent. These data also suggested that, compared with females, males that wandered away from their natal groups in search of breeding positions were more likely to succeed.

Published data on group compositions exist for two other species of tamarins, the cotton-top tamarin (*S. oedipus*) and the black-mantled tamarin (*S. nigricollis*). Cotton-top tamarin groups often contain more than one adult male, and sometimes two pregnant females, although only one female per group has been observed to have young at a time (Neyman, 1978; Savage *et al.*, 1996*a*), suggesting a similarity to saddle-back and moustached tamarins. However, copulations have rarely been observed in cotton-top tamarins, so it is not known whether sociosexual polyandry occurs in that species. Black-mantled tamarins appear to be somewhat different from other *Saguinus* species, with only 1.3 adult males per group, on average, in a population studied in Colombia in the 1970s, and virtually all infant carrying reported to be performed by infants' mothers rather than by adult males (Izawa, 1978). Further study of black-mantled tamarins is needed.

Golden lion tamarins

Golden lion tamarins (*Leontopithecus rosalia*) have been studied intensively in the wild at the Poço das Antas Reserve in Brazil's Atlantic coast forest (Baker *et al.*, 1993; Dietz & Baker, 1993; Dietz *et al.*, 1994; Baker & Dietz, 1996). Eighty per cent of groups there contained more than two adults (Dietz & Baker, 1993). Dietz and Baker (1993) considered golden lion tamarin groups to be polygynous when they contained two reproductive females and to remain polygynous as long as two such females remained in the group, even if they did not both continue to be reproductive in subsequent breeding seasons. This definition is consistent with other authors' definitions of social polygyny. The combined frequency of social and reproductive polygyny in this population was 10.6%. Most other studies of callitrichids have focused on reproductive polygyny, and thus Dietz and Baker's figure on the frequency of polygyny is difficult to compare with other studies of callitrichids.

Nonetheless, Dietz and Baker's (1993) study yielded interesting insights into the circumstances in which polygyny occurred. Most socio- and reproductively polygynous females were closely related, usually mothers and daughters. Indeed, in 544 group-months no adult female immigrated into a group already containing a reproductive female. One female (usually the older one) was usually dominant in relationship to the other female. In most groups with more than one adult female, the subordinate female left the group after a year or two, sometimes after repeated aggression from the dominant female. Daughters still living in their natal groups when they reached two years of age were more likely to become reproductive if their groups contained unrelated males than if they contained only their fathers (50% vs. 21%). This was similar to findings of Savage *et al.* (1996*a*) on cotton-top tamarins.

Groups of golden lion tamarins were more likely to contain multiple adult females when home ranges were larger (Dietz & Baker, 1991, cited in Rylands, 1993), suggesting that polygyny is more likely when subordinate females' opportunities to find breeding positions are reduced. Dietz and Baker (1993) modelled the costs to dominant females of allowing subordinate females to breed in their groups. The model showed that it would never benefit females to allow unrelated females to join their groups and breed. On the other hand, by allowing daughters to breed in their natal groups, dominant females would often benefit slightly through inclusive fitness if the daughters' chances of breeding were otherwise small, and the groups contained unrelated males. Dietz and Baker (1993) suggested that the frequency of reproductive and social polygyny in their study population may have been due, in part, to the isolated and saturated nature of the reserve, and thus the resulting lack of dispersal opportunities for subordinate females.

In the golden lion tamarins at Poço das Antas, about 40% of the groups contained two non-natal adult males, and in some of these groups both males participated in copulations with the groups' breeding females (Baker *et al.*, 1993). However, unlike saddle-back tamarins (Goldizen, 1989), most pairs of adult male golden lion tamarins included one that was clearly dominant over the other. In addition, males appeared to be able to detect females' periods of likely ovulation, though the mechanism for this detection is unclear. During periods of likely ovulation, subordinate males were rarely able to copulate with females, and Baker *et al.* (1993) thus believed that only the dominant males were likely to father offspring.

In order to analyse the factors affecting the formation and maintenance of golden lion tamarin groups, Baker and Dietz (1996) reported on patterns of immigration in 17 groups at Poço das Antas that were each studied from 10 to 76 months. The study groups received an average of 0.48 immigrants per year, and immigration was significantly male-biased. Most immigrations either occurred soon after the disappearance or death of the same-sex breeder in the group, or precipitated the emigration of the same-sex breeder. Of the six immigrants which joined 'intact' groups that already contained breeders of both sexes, five were males. In some of these cases, the new and old males remained together in the group for at least one breeding season. In the single case in which a female immigrant joined an intact group, the previous female had been resident for only one month and left soon after the immigration event. While females clearly had few opportunities to immigrate into groups, all four documented cases of inheritance of breeding positions in natal groups involved females.

Marmosets and pygmy marmosets

The smallest callitrichid is the pygmy marmoset (*Cebuella pygmaea*), which weighs *c.* 120 g. Soini (1988, 1993) studied about 80 groups of pygmy marmosets in northeastern Peru and found that most consisted of a socially monogamous pair of adults and offspring of different ages. Some groups contained extra adults that were not progeny of the mated pair, but only one female was ever observed to be reproductively active. In one group with two adult males, both males attempted to mate with the female but one maintained exclusive copulatory access to her during her likely oestrous period (Soini, 1987). It thus appears that multiple female breeders are rare or non-existent in pygmy marmosets, but further data are required before we can know whether multiple male breeders ever occur and, if so, how often. In an analysis of previously published data on this species, Heymann and Soini (1999) found a significant positive relationship between the numbers of six- to 12-month-old juveniles and the numbers of all adult and subadult group members across 21 groups, but no relationship between numbers of young and of adult males. The cause of the relationship between group size and reproductive success is, however, difficult to evaluate because high reproductive success could be either a cause or an effect of large group size.

The species of marmoset that has been best studied in the wild is the common marmoset (*Callithrix jacchus*). One study of this species, carried out at EFLEX-IBAMA, a forestry research station in Rio Grande do Norte in Brazil, found that all three study groups were socially and reproductively polygynous; in each group, two females gave birth during the same year, with five of the six females successfully rearing young (Digby & Barreto, 1993; Digby & Ferrari, 1994). Digby and Ferrari (1994) also reviewed observations of groups with multiple female breeders at three of four other field sites where common marmosets had been studied.

Roda and Mendes Pontes (1998) found that reproductively polygynous females in a group of common marmosets in the state of Pernambuco, Brazil, were very aggressive towards each other, and that at least one infant died from infanticide during aggressive encounters between these females. Digby (1995) also reported that dominant females were often aggressive to subordinate females in reproductively polygynous groups, and that at least one subordinate female's infant died from infanticide, most likely perpetrated by the dominant female. Among the three groups studied at EFLEX-IBAMA, the only young of subordinate females that survived were those that were sufficiently different in age from the dominant females' young so that the sets of young did not require intensive parental care at the same time (Digby, 1995).

In the three study groups of common marmosets at EFLEX-IBAMA, breeding females copulated with only one male group member (Digby, 1999). However, Digby (1999), Lazaro-Perea *et al.* (2000), and Lazaro-Perea (2001) all reported frequent intergroup sexual interactions in this population, involving both breeding and non-breeding individuals. These interactions occurred both during intergroup territorial interactions and during forays by individuals into neighbouring territories (Lazaro-Perea, 2001). Each group's dominant male appeared to guard his females during their likely conception periods, remaining closer to them than usual, and repeatedly mounting them during intergroup encounters (Digby, 1999). Indeed, none of the extra-group copulations involving breeding females occurred during times when they were likely to have conceived. The extra-group copulations involving non-breeding females suggested that, behaviourally, they were reproductively suppressed in their own groups, but not when they were with extra-group males. However, reproduction was thought to be physiologically suppressed in these non-breeding females, despite their participation in extra-group copulations, because none was known to have become pregnant (Digby, 1999).

Genetic analyses using microsatellite loci have been done on the three study groups of Digby (1999) (Nievergelt *et al.*, 2000). These analyses were not conclusive, due to the lack of samples from a few group members and a low level of polymorphism. However, the results of parentage analyses suggested that groups frequently had two breeding females whose infants were most likely to have been fathered by the group's dominant male. The groups' subordinate adult males could be excluded as fathers of most of their groups' infants. The low power of the loci for paternity exclusion did not allow any conclusions about the occurrence of extra-group parentage. Analyses of relatedness supported the conclusion from field observations that common marmoset groups in the population tended to be stable, extended family groups consisting mainly of closely related individuals.

Digby's (1999) finding that none of her three common marmoset groups had more than one breeding male appears to be typical of this species, as she found no reports of multiple breeding males in other populations of this species. Paradoxically, Koenig (1995) reviewed published data on wild common marmoset groups and found that groups with three or more adult males had significantly more juveniles under five months of age than did those with only one or two adult males. This finding appears to suggest that groups with extra males are able to rear more young and thus there might be a benefit to males from mating polyandrously. However, it is also possible that some groups have better territories than others and are able to raise more young, eventually having more adult males as a result. Sociosexual polyandry has been observed in common marmosets at EFLEX-IBAMA since Digby's study, but only in situations when groups were unstable (Lazaro-Perea *et al.*, 2000).

Other *Callithrix* species have not been studied nearly as extensively as common marmosets. One group of buffy-headed marmosets (*C. flaviceps*) was studied in Brazil and found to have multiple adults of both sexes (Ferrari, 1992). During the 1985–91 study period, this group never had more than one reproductive female at a time (Ferrari *et al.*, 1996). In a group of tassel-ear marmosets (*C. intermedia*) studied in Mato Grosso in Brazil, three of four males were seen copulating (Rylands, 1982), and in a group of the buffy-tufted-ear marmoset (*C. aurita*) studied in southern São Paolo, multiple adults of both sexes occurred in the group and two females reproduced simultaneously (Roda, 1989, cited in Ferrari *et al.*, 1996).

COSTS OF INFANT CARE IN CALLITRICHIDS

A number of researchers have carried out detailed investigations of the possible costs of lactation and infant carrying in callitrichids because of the proposed links between these costs and the variable mating patterns and cooperative breeding exhibited by this family of primates. The need for male parental care has been proposed to select for monogamous mating systems in a number of species (Orians, 1969; Emlen & Oring, 1977). For example, in a study of Djungarian hamsters (*Phodopus campbelli*), 95% of pups survived when cared for by both parents, while only 47% survived if cared for only by their mothers (Wynne-Edwards, 1987). Male care is also thought to be essential in fat-tailed dwarf lemurs (*Cheirogaleus medius*) (Fietz, chapter 14). It is, however, unclear whether high costs of parental care have been the cause of the original evolution of monogamy in any mammals (Brotherton & Komers, chapter 3).

Lactation is clearly very costly to female callitrichids. Nievergelt and Martin (1998) found that female cotton-top tamarins increased their energy intake by up to 100% and lost an average of 8% of their body weight during the 10 weeks after they gave birth. Garber and Teaford (1986) reported that wild female saddle-backed tamarins lost up to 22% of their body weight during lactation.

In contrast, the findings on the costs of infant carrying have been somewhat contradictory. Tardif and Harrison (1990) estimated that infant carrying by any given individual would increase its energetic expenditure by only 2–10% above basic maintenance costs, while Schradin and Anzenberger (2001) estimated this cost at 20%. Nievergelt and Martin (1998), who measured weight losses and food intakes in captive common marmosets, concluded that the energetic costs of carrying were low. They found that males did not lose weight or increase their energy intakes during the 10 weeks after infants were born.

Other studies, however, have provided evidence of significant costs of infant care. Sanchez *et al.* (1999) and Achenbach and Snowdon (2002) both found that fathers and male helpers in captive cotton-top tamarins lost significant amounts of body weight during the care-giving period. In one study, males lost up to 10.8% of their body weight (Achenbach & Snowdon, 2002). Moreover, a given individual's weight loss was related to its proportion of the group's carrying efforts (Sanchez *et al.*, 1999) and inversely related to the number of helpers in the group (Achenbach & Snowdon, 2002). Individuals also spent significantly less time feeding and had lower energy intakes while carrying than while not (Sanchez *et al.*, 1999). In another study of cotton-top tamarins in captivity, Price (1992) documented differences in the behaviour of individuals carrying infants compared with those not carrying: carriers spent significantly less time feeding, foraging, moving, and engaging in social activities than did non-carriers. Carriers also spent more time in concealed areas, suggesting an effort to reduce predation risk. Schradin and Anzenberger (2001) found that the distances that marmosets could leap were reduced by 17% when they were carrying newborn twins. Reduced leaping ability could affect individuals' abilities to catch prey and to evade predators themselves.

Most of the studies just described involved captive callitrichids living in cages measuring, for example, $2 \times 1 \times 2$ m or $1 \times 0.6 \times 1.4$ m (Nievergelt & Martin, 1998), or 12 square metres and 2.4 m high (Sanchez *et al.*, 1999). Such housing conditions are completely different from the experiences of wild callitrichids, which often travel 1–2 km daily, in addition to climbing many tall trees. Any measurements of the costs of infant carrying using weight loss or energy consumption of captive callitrichids must significantly underestimate the costs experienced by wild individuals. While carrying infants, most wild callitrichids do not have the option of reducing their activity to reduce the costs of carrying; in most species, carriers must stay with their group and carry infants the full distances travelled by their group. All three of the potential costs of infant carrying that have been discussed for callitrichids (direct energetic costs, reduced foraging time, and increased predation risk) must thus affect wild individuals far more than captive ones.

In a review of the relative costs of infant care in small-bodied neotropical primates, Tardif (1994) argued that infant carrying is likely to carry a particularly high cost in terms of reduced foraging time and foraging success in small primates, such as callitrichids, that rely on being cryptic to reduce risks of predation. She further argued that because of the high costs of lactation to callitrichid females, mothers would suffer the most from the reduced foraging time caused by infant transport. Indeed, many studies show that lactating mothers carry infants very little, leaving most of the carrying to adult males and younger helpers (e.g., Goldizen, 1987; Savage *et al.*, 1996*b*, for wild callitrichids). While it is not yet clear how the three suggested costs of infant carrying compare in relative importance for callitrichids, it is clear that these costs, and those of lactation, must be high.

The costs of infant care in callitrichids may vary due to ecological differences experienced by genera, species, and/or populations, and this variation in infant care costs may cause differences in the frequencies of different mating patterns. There are not yet sufficient quantitative data on mating patterns, costs of infant care, and ecological parameters from enough populations to test this idea rigorously, but we now have sufficient information to generate some specific hypotheses for further testing.

ECOLOGICAL FACTORS, INFANT CARE COSTS, AND SOCIAL MATING SYSTEMS IN CALLITRICHIDS

Pygmy marmosets

Pygmy marmosets (*Cebuella pygmaea*) usually inhabit seasonally inundated, floodplain forests, and have very small home ranges. Seven groups had home ranges varying from 0.1 to 0.5 ha in size (Soini, 1982, 1993). Groups tend to change home ranges after a period of months or up to a few years, as resources become depleted (Ramirez *et al.*, 1978; Soini, 1982, 1993). Pygmy marmosets feed on insects and exudates, which they obtain by gouging holes in exudate trees and then returning later to harvest the exudates. A group usually has a single main exudate tree, and spends much of its day in that tree (Soini, 1993). Small home ranges, with a focus on a single tree, result in short daily travel distances, averaging 290 m (Heymann & Soini, 1999).

One aspect of the reproductive strategies of pygmy marmosets that is unique among callitrichids, and is due to their small home ranges and specialization on exudate feeding, is that adults 'park' infants near their foraging sites (Soini, 1988). Infants spend most of their time 'parked' during their first six weeks of life, and thus pygmy marmosets do not bear the same high costs of infant carrying as do the other callitrichids. This may explain why Heymann and Soini (1999) found no relationship between the numbers of infants or juveniles and the numbers of adult males in groups. The reduced costs of infant care may not make reproductive polyandry sufficiently advantageous to either dominant or subordinate males for it to occur frequently, if at all. Given the reduced costs of infant care, it is less clear why multiple-breeding females have not been seen in pygmy marmoset groups. However, the tending of 'parked' young and then of independently locomoting young, and the provisioning of insects to infants, may be too demanding for a group to be able to look after two sets of young simultaneously. These aspects of its ecology appear to explain why pygmy marmosets seem to exhibit social monogamy more often than other callitrichids.

Callithrix

The marmosets in the genus *Callithrix* occupy a variety of habitats in the southern and eastern Amazon basin, but all use disturbed forest, edge habitats, and/or patches of secondary growth (Rylands & de Faria, 1993). All marmosets include a significant amount of plant exudates, mostly gums, in their diets (reviewed by Ferrari & Lopes Ferrari, 1989), but the species appear to vary in the degree to which they specialise in exudates. Common marmosets, black-tufted-ear marmosets (*C. penicillata*), and buffy-headed marmosets eat mostly exudates and little fruit, Santarem marmosets (*C. humeralifer*) eat more fruit, and Wied's marmosets (*C. kuhli*) are intermediate. Different populations within a species are also likely to vary in the relative importance of exudates and fruit in their diets, due to differences among habitats (reviewed by Rylands & de Faria, 1993).

The home range sizes of marmoset groups appear to be inversely correlated with the relative importance of exudates in their diets and are generally smaller than the home ranges of tamarins. Groups/species that primarily eat exudates rather than fruits have small home ranges, as mentioned above for pygmy marmosets (reviewed by Ferrari & Lopes Ferrari, 1989; Rylands & de Faria, 1993). Reported home range sizes are 0.5–6.5 ha for common marmoset groups and 1.2–3.5 ha for black-tufted-ear marmosets, whereas the more frugivorous Santarem marmosets have larger ranges (28 ha), and Wied's marmosets have ranges of 10–12 ha. Buffy-headed marmosets have large home ranges and thus do not fit this pattern, but Rylands and de Faria (1993) suggested that this may be because the buffy-headed marmoset groups studied relied on small, widely dispersed sources of exudates that flowed naturally.

The frequency of multiple-breeding females in common marmoset groups may be explained both by their small home range sizes and their heavy reliance on exudates. Exudates are less seasonally variable than fruits, which may make it easier for two females in the same group to have offspring a few months apart so that multiple litters do not have to be carried during the same period. Small home ranges and dependence on only a small number of exudate trees also result in shorter daily travel distances (0.5–1.0 km per day: Hubrecht, 1985) than those of other *Callithrix* and *Saguinus* species (reviewed by Ferrari & Lopes Ferrari, 1989).

Digby and Barreto (1996) suggested three reasons that adult common marmosets may experience lower costs of infant carrying than do other callitrichids. First, their small home ranges (3.9–5.2 ha in their study population) may allow carriers to rest in central areas and thus reduce their travel costs while still remaining in vocal communication with their groups. Second, common marmoset infants achieve locomotory independence sooner than do other callitrichid infants (Tardif *et al.*, 1993). Third, the large average size of common marmoset groups means that infant carrying is divided among more individuals, reducing the burden on any one. Reduced costs of infant carrying may help explain the rarity of sociosexual polyandry and the frequency of reproductive polygyny in common marmosets. However, this argument again raises the question of why multiple-breeding females occur in common marmosets but not in pygmy marmosets.

Saguinus

The most speciose genus of callitrichids, *Saguinus*, inhabits a variety of forest types from Panama through to Peru, Bolivia and northern Brazil. A number of tamarin species have been found to include a variety of habitat types within individual home ranges, including some primary forest and some secondary or edge forest. All tamarin species' diets focus on fruits and insects, with small amounts of exudates and nectar eaten, especially during seasons of fruit shortages (e.g., Terborgh, 1983; Terborgh & Stern, 1987, reviewed by Garber, 1993). The main fruiting species used by tamarins tend to be small to medium-sized trees that have only small numbers of fruits available at any one time (Terborgh, 1983, reviewed by Garber, 1993). Tamarin groups thus have to visit many fruiting trees each day. While many tamarins consume plant exudates, their dentition does not allow them to gnaw holes in trees to stimulate the flow of exudates. They are thus only able to utilize natural sources of exudate flow, or to steal from the holes gnawed by marmosets or pygmy marmosets (Coimbra-Filho & Mittermeier, 1978).

Studies of wild tamarins have generally reported home ranges of 20–40 ha (reviewed in Garber, 1993), though saddle-backed and emperor tamarins (*S. imperator*) in mixed-species groups in Manu, Peru, had home ranges of up to 120 ha (Terborgh, 1983). Thus, tamarins generally have larger home ranges than do marmosets and pygmy marmosets. Reported mean day range lengths for tamarins range from 1000 m to over 2000 m and appear to be related to the distributions of their plant resources (reviewed by Garber, 1993). These day ranges are two-dimensional travel distances, of course, and do not include the vertical travel of the tamarins up and down trees. Long daily travel distances and large home range sizes must surely increase the costs of infant care, reducing the probability that monogamous pairs of tamarins could rear offspring without assistance.

Leontopithecus

All species of *Leontopithecus* (Rosenberger & Coimbra-Filho, 1984) live in the Atlantic forests of coastal Brazil. Like the tamarins, lion tamarins rely on fruits and insects, but eat small amounts of exudates when they are available (Rylands, 1993). The insect foraging styles of lion tamarins, especially their focus on foraging in epiphytic bromeliads and their reliance on tree holes for sleeping sites, suggest that they evolved to inhabit mature forests (Rylands, 1993, 1996). The lion tamarins have the largest home ranges of the callitrichids, with reported home ranges covering 40–200 ha (reviewed by Rylands, 1993). The large sizes of their home ranges are presumably due to the seasonality of the habitats in which they live and the resulting scarcity of fruits during poor seasons, combined with low densities of sleeping and foraging sites. Daily travel distances are longer than those of marmosets and similar to those reported for tamarins.

Mating system variations and breeding opportunities

This general review of the ecological characteristics of different callitrichid species suggests that variations in mating systems may be related to variations in the costs of infant care that are caused by ecological factors. A comparative analysis using independent contrasts calculated with the Comparative Analysis by Independent Contrasts (CAIC) program (Purvis & Rambaut, 1994) found a significant relationship between the numbers of adult males in groups and daily travel path lengths in callitrichids (Heymann, 2000), providing support for the idea that, at the very least, costs of infant carrying affect group compositions.

Ecological factors may also affect the availability of breeding opportunities, which would in turn affect mating patterns. Reproductive skew theories (Vehrencamp, 1983, reviewed by Magrath & Heinsohn, 2000; Vehrencamp, 2000) predict the circumstances in which dominant males in group-living species should share mating opportunities with subordinates rather than monopolizing chances of paternity. These models predict that the degree to which a dominant male should share reproduction with a subordinate male should be related, in part, to the subordinate male's chances of obtaining a breeding position elsewhere if he left the group. Saddle-back tamarins and golden lion tamarins are the two callitrichid species in which sociosexual cooperative polyandry has been most intensively studied in the wild, and the different patterns of polyandry found in these two species may be explained by reproductive skew theories.

Socially polyandrous male golden lion tamarins tend to have clear dominance relationships, with only the dominant males suspected of fathering young (Baker *et al.*, 1993). Modelling by Baker *et al.* (1993) suggested that even if male golden lion tamarins had no chance of obtaining paternity while subordinate, such males would, on average, have higher lifetime reproductive success by waiting for a chance to obtain the dominant male breeding position in their current group or a neighbouring group, rather than leaving their group to search for a breeding position elsewhere. The reduced success of males that left groups to search for breeding opportunities was attributed to the degree of habitat saturation at the Poço das Antas Reserve, and to the lower survival rate of the males that did establish new groups, probably owing both to the small size of those groups and the poor quality of their territories. The model also suggested that subordinate males were more likely to benefit from remaining in their groups if they were related to the dominant males.

In contrast, sociosexually polyandrous male saddle-back tamarins appeared to be quite equal in rank and to copulate with their females at similar frequencies (Goldizen, 1987, 1989). Males also frequently immigrated into other groups (Goldizen *et al.*, 1996), suggesting that breeding opportunities were not scarce for males. It is possible that the habitats inhabited by saddle-back tamarins have a tendency to be more widespread, interconnected, and less saturated than those of golden-lion tamarins. If so, male saddle-back tamarins that roamed in search of breeding positions would have better chances of acquiring such positions than would male golden lion tamarins. This may have led to the evolution of a more egalitarian form of polyandry in saddle-back tamarins, with dominant males having to offer significant reproductive concessions to other males in order to keep them in the group.

MECHANISMS OF FEMALE REPRODUCTIVE SUPPRESSION AND THE FREQUENCY OF POLYGYNY

The occurrence of relatively frequent social and sometimes reproductive polygyny in *Leontopithecus* is consistent with the finding that reproductive suppression of subordinate female golden lion tamarins by dominant females occurs through behavioural rather than physiological mechanisms (French *et al.*, 1989, reviewed by French, 1997). Subordinate female lion tamarins tend to exhibit normal ovarian cycles in the presence of dominant females. In other callitrichids that have been studied, including common marmosets, cotton-top tamarins, and saddle-back tamarins, olfactory cues produced by a dominant female initiate a physiological suppression of reproduction in subordinate females. Subordinate females either do not ovulate or exhibit abnormal ovarian cycles (reviewed by Abbott *et al.*, 1993).

French (1997) argued that the different mechanisms of female reproductive suppression in *Callithrix* and *Saguinus* compared with *Leontopithecus* may be a result of differences in the costs of gestation. He reviewed data showing that the annual gestational costs of producing offspring (measured by the ratio of total offspring weights at birth to female weight), range in ascending order from *Leontopithecus*, through *Saguinus*, then *Callithrix* to *Cebuella*. The occurrence of endocrine suppression of female reproduction in genera with the higher costs of gestation, and behavioural suppression in the genus with the lowest costs of gestation, is consistent with the idea that endocrine reproductive suppression mechanisms have evolved in species in which subordinate individuals would pay the highest costs from unsuccessful reproductive attempts. Creel and Creel (1991) found a similar pattern when they compared singularly breeding and plural breeding carnivores.

Rylands (1993) proposed a different explanation for the occurrence of behavioural mechanisms of reproductive suppression in *Leontopithecus*. He proposed that the stricter habitat requirements of *Leontopithecus* compared with other callitrichids may have selected for a more flexible mechanism of reproductive suppression in this genus, allowing dominant females to assess their daughters' options and then in some circumstances permit them to reproduce in their natal groups.

The apparent high frequency of polygyny in wild common marmosets is not yet fully understood. Digby and Ferrari (1994) showed that the home ranges of this species tend to be considerably smaller than those of other marmoset species, resulting in higher population densities. They proposed that for young females in a high-density population, the best strategy for acquiring breeding positions is to wait in their natal groups to inherit breeding status there. As Dietz and Baker (1993) proposed for golden lion tamarins, Digby and Ferrari (1994) suggested that it may sometimes be advantageous for a young female common marmoset to begin breeding in her natal group, even if her chances of success are small. The costs and benefits of such premature breeding have yet to be quantified for subordinate females in different circumstances.

Further work is required to more fully understand the differences in the forms of reproductive suppression exhibited by different callitrichid genera and to test the prediction that strong physiological suppression of reproduction must occur in pygmy marmosets (French, 1997). This prediction is based on the high costs of gestation in pygmy marmosets, plus the fact that breeding by subordinate females has never been observed in the wild. One problem with our understanding of mechanisms of reproductive suppression in this family is that most of the work has been done on only one species in each of three genera; thus, there is virtually no information on whether mechanisms of reproductive suppression vary within genera. Nonetheless, this comparison of mechanisms of female reproductive suppression across the callitrichid genera suggests that female reproductive strategies, and thus mating systems, may be inherently more flexible in *Leontopithecus* than in the other callitrichid genera.

CONCLUSIONS

This review suggests, simplistically, that marmosets and pygmy marmosets have smaller home ranges, due to their heavy use of exudates, while tamarins and lion tamarins have larger home ranges and daily travel distances. As a result, the costs of infant carrying must be higher for tamarins and golden lion tamarins, explaining the apparently higher frequency of social polyandry in at least some species in these genera. The form of polyandry exhibited appears to be more egalitarian in saddle-back tamarins than in golden lion tamarins, perhaps due to a greater availability of breeding positions for male saddle-back tamarins. The different frequencies of polygyny in the four genera appear to be a result of different mechanisms of reproductive suppression, as well as differences in the availability of breeding positions for young females. Variation among species within genera, and among populations within species, would be expected to relate to ecological differences in similar ways.

While a lot has been learned about the biology of wild callitrichids during the last 15 years, there is still a great deal that we do not know about these intriguing miniature primates. This review of data on wild callitrichids suggests that the following aspects of the relationships between mating patterns and ecological factors particularly warrant further investigation.

1 ***Genetic parentage of offspring in polyandrous groups***: Genetic analysis is required to establish whether reproductive polyandry ever occurs in callitrichids and, if so, whether it is more common in some species than in others (e.g., more common in saddle-back tamarins than golden lion tamarins). In species that do exhibit reproductive polyandry, it will be interesting to see whether the provision of parental care by individual males is related to their proportion of their female's copulations and to their parentage of her young. Finally, how does the care provided by individual males relate to their total share of their female's copulations versus their share of copulations during the female's fertile period? This information would elucidate whether female callitrichids use copulations outside their fertile period to increase parental care contributions from 'helpers'.
2 ***Extra-pair copulations and extra-pair paternity***: Further genetic analyses are also needed to determine whether extra-pair paternity ever occurs in callitrichids and, if so, in what situations extra-pair copulations lead to extra-pair paternity. The sexual monomorphism and importance of male care in callitrichids predict that extra-pair paternity should be rare (Møller, chapter 2), but observations

of extra-pair copulations in some species of callitrichids suggest that this requires investigation.

3 *Mechanisms of female reproductive suppression*: These should be studied in more callitrichid species, and in particular in *Cebuella*, to test the hypothesis that physiological mechanisms are more effective in suppressing the reproductive activities of subordinate females in *Cebuella* than in *Callithrix*, perhaps because of the higher costs of gestation in *Cebuella*.
4 *Relationship between the costs of infant carrying and the frequency of reproductive polygyny*: In high-density populations of common marmosets in which groups have small home ranges, infant carriers may reduce travel distances by remaining in central parts of their home ranges much of the time. Reducing infant carrying would in turn reduce the overall costs of infant care and increase the chances of successful reproductive polygyny. In populations with larger home ranges, do carriers have to travel with their group, making the costs of infant carrying higher and reproductive polygyny rarer?
5 *Frequencies of breeding opportunities and reproductive polyandy*: Is habitat saturation greater, and are breeding vacancies rarer, for male golden lion tamarins than for male saddle-back tamarins? If so, does this reduce breeding opportunities for subordinate males and allow dominant males to exert stronger mating skew in golden lion tamarins?
6 *Relationship between the presence of older offspring and the probability of sociosexual monogamy versus polyandry in tamarins*: Do polyandrous matings occur only when groups of tamarins do not have older offspring that can serve as helpers? If so, what happens to sociosexually polyandrous groups when they eventually produce such offspring? Do new groups without helpers always start as sociosexually polyandrous trios?

Acknowledgements

I sincerely thank Ulrich Reichard and Christophe Boesch for the invitation to attend the workshop on monogamy, and I regret that I was unable to attend at the last minute. I also thank Ulrich, Christophe and two anonymous reviewers for comments on earlier drafts of the manuscript.

References

Abbott, D. H., Barrett, J. & George, L. M. (1993). Comparative aspects of the social suppression of reproduction in female marmosets and tamarins. In *Marmosets and Tamarins: Systematics, Behaviour, and Ecology*, ed. A. B. Rylands, pp. 152–63. Oxford: Oxford University Press.

Achenbach, G. G. & Snowdon, C. T. (2002). Costs of caregiving: weight loss in captive adult male cotton-top tamarins (*Saguinus oedipus*) following the birth of infants. *International Journal of Primatology*, **23**, 179–89.

Baker, A. J. & Dietz, J. M. (1996). Immigration in wild groups of golden lion tamarins. *American Journal of Primatology*, **38**, 47–56.

Baker, A. J., Dietz, J. M. & Kleiman, D. G. (1993). Behavioural evidence for the monopolization of paternity in multimale groups of golden lion tamarins. *Animal Behaviour*, **46**, 1091–103.

Burke, T., Davies, N. B., Bruford, M. W. & Hatchwell, B. J. (1989). Parental care and mating behaviour of polyandrous dunnocks *Prunella modularis* related to paternity by DNA fingerprinting. *Nature*, **338**, 249–51.

Canavez, F. C., Moreira, M. A. M., Simon, F., Parham, P. & Seuanez, H. N. (1999). Phylogenetic relationships of the Callitrichinae (Platyrrhini, Primates) based on β_2-microglobulin DNA sequences. *American Journal of Primatology*, **48**, 225–36.

Coimbra-Filho, A. F. & Mittermeier, R. A. (1978). Tree-gouging, exudates-eating, and the "short-tusked" condition in *Callithrix* and *Cebuella*. In *The Biology and Conservation of the Callitrichidae*, ed. D. G. Kleiman, pp. 105–15. Washington, DC: Smithsonian Institution Press.

Creel, S. R. & Creel, N. M. (1991). Energetics, reproductive suppression and obligate communal breeding in carnivores. *Behavioral Ecology and Sociobiology*, **28**, 263–70.

Davies, N. B. (1985). Cooperation and conflict among dunnocks, *Prunella modularis*, in a variable mating system. *Animal Behaviour*, **33**, 628–48.

(1992). *Dunnock Behaviour and Social Evolution*. Oxford: Oxford University Press.

Davies, N. B. & Lundberg, A. (1984). Food distribution and a variable mating system in the dunnock, *Prunella modularis*. *Journal of Animal Ecology*, **53**, 895–912.

Dawson, G. A. (1978). Composition and stability of social groups of the tamarin, *Saguinus oedipus goeffroyi*, in Panama: ecological and behavioral implications. In *The Biology and Conservation of the Callitrichidae*, ed. D. G. Kleiman, pp. 23–37. Washington, DC: Smithsonian Institution Press.

Dietz, J. M. & Baker, A. J. (1993). Polygyny and female reproductive success in golden lion tamarins, *Leontopithecus rosalia*. *Animal Behaiour*, **46**, 1067–78.

Dietz, J. M., Baker, A. J. & Miglioretti, D. (1994). Seasonal variation in reproduction, juvenile growth, and adult body mass in golden lion tamarins. *American Journal of Primatology*, **34**, 115–32.

Digby, L. J. (1995). Infant care, infanticide, and female reproductive strategies in polygynous groups of common marmosets (*Callithrix jacchus*). *Behavioral Ecology and Sociobiology*, **37**, 51–61.

(1999). Sexual behavior and extragroup copulations in a wild population of common marmosets (*Callithrix jacchus*). *Folia Primatologica*, **70**, 136–45.

Digby, L. J. & Barreto, C. E. (1993). Social organization in a wild population of *Callithrix jacchus*. *Folia Primatologica*, **61**, 123–34.

(1996). Activity and ranging patterns in common marmosets (*Callithrix jacchus*): implications for reproductive strategies. In *Adaptive Radiations of Neotropical Primates*, ed. M. A. Norconk, A. L. Rosenberger & P. A. Garber, pp. 173–85. New York: Plenum Press.

Digby, L. J. & Ferrari, S. F. (1994). Multiple breeding females in free-ranging groups of *Callithrix jacchus*. *International Journal of Primatology*, **15**, 389–97.

Eason, P. K. & Sherman, P. T. (1995). Dominance status, mating strategies and copulation success in cooperatively polyandrous white-winged trumpeters, *Psophia leucoptera* (Aves: Psophiidae). *Animal Behaviour*, **49**, 725–36.

Emlen, S. T. & Oring, L. W. (1977). Ecology, sexual selection and the evolution of mating systems. *Science*, **197**, 215–23.

Faaborg, J. (1986). Reproductive success and survivorship of the Galapagos hawk *Buteo galapagoensis*: potential costs and benefits of cooperative polyandry. *Ibis*, **128**, 337–47.

Faaborg, J. & Patterson, C. B. (1981). The characteristics and occurrence of cooperative polyandry. *Ibis*, **123**, 477–84.

Faaborg, J., de Vries, T., Patterson, C. B. & Griffin, C. R. (1980). Preliminary observations on the occurrence and evolution of polyandry in the Galapagos hawk (*Buteo galapagoensis*). *The Auk*, **97**, 581–90.

Faaborg, J., Parker, P. G., DeLay, L., de Vries, T., Bednarz, J. C., Maria Paz, S., Naranjo, J. & Waite, T. A. (1995). Confirmation of cooperative polyandry in the Galapagos hawk (*Buteo galapagoensis*). *Behavioral Ecology and Sociobiology*, **36**, 83–90.

Ferrari, S. F. (1992). The care of infants in a wild marmoset (*Callithrix flaviceps*) group. *American Journal of Primatology*, **26**, 109–18.

Ferrari, S. F. & Lopes Ferrari, M. A. (1989). A re-evaluation of the social organisation of the Callitrichidae, with reference to the ecological differences between genera. *Folia Primatologica*, **52**, 132–47.

Ferrari, S. F., Correa, H. K. M. & Coutinho, P. E. G. (1996). Ecology of the "southern" marmosets (*Callithrix aurita* and *Callithrix flaviceps*). In *Adaptive Radiations of Neotropical Primates*, ed. M. A. Norconk, A. L. Rosenberger & P. A. Garber, pp. 157–71. New York: Plenum Press.

French, J. A. (1997). Proximate regulation of singular breeding in callitrichid primates. In *Cooperative Breeding in Mammals*, ed. N. G. Solomon & J. A. French, pp. 34–75. Cambridge: Cambridge University Press.

French, J. A., Inglett, B. J. & Dethlefs, T. M. (1989). The reproductive status of nonbreeding group members in captive golden lion tamarin social groups. *American Journal of Primatology*, **18**, 73–86.

Garber, P. A. (1993). Feeding ecology and behaviour of the genus *Saguinus*. In *Adaptive Radiations of Neotropical Primates*, ed. M. A. Norconk, A. L. Rosenberger & P. A. Garber, pp. 273–95. New York: Plenum Press.

(1997). One for all and breeding for one: cooperation and competition as a tamarin reproductive strategy. *Evolutionary Anthropology*, **5**, 187–99.

Garber, P. A. & Teaford, M. F. (1986). Body weights in mixed species troops of *Saguinus mystax mystax* and *Saguinus fuscicollis nigrifrons* in Amazonian Peru. *American Journal of Physical Anthropology*, **71**, 331–6.

Garber, P. A., Moya, L. & Malaga, C. (1984). A preliminary field study of the moustached tamarin monkey (*Saguinus mystax*) in northeastern Peru: Questions concerned with the evolution of a communal breeding system. *Folia Primatologica*, **42**, 17–32.

Garber, P. A., Encarnacion, F., Moya, L. & Pruetz, J. D. (1993). Demographic and reproductive patterns in moustached tamarin monkeys (*Saguinus mystax*): implications for reconstructing platyrrhine mating systems. *American Journal of Primatology*, **29**, 235–54.

Goldizen, A. W. (1987). Facultative polyandry and the role of infant-carrying in wild saddle-back tamarins (*Saguinus fuscicollis*). *Behavioral Ecology and Sociobiology*, **20**, 99–109.

(1989). Social relationships in a cooperatively polyandrous group of tamarins (*Saguinus fuscicollis*). *Behavioral Ecology and Sociobiology*, **24**, 79–89.

Goldizen, A. W., Mendelson, J., van Vlaardingen, M. & Terborgh, J. (1996). Saddle-back tamarin (*Saguinus fuscicollis*) reproductive strategies: evidence from a thirteen-year study of a marked population. *American Journal of Primatology*, **38**, 57–83.

Goldizen, A. W., Putland, D. A. & Goldizen, A. R. (1998). Variable mating patterns in Tasmanian native hens (*Gallinula mortierii*): correlates of reproductive success. *Journal of Animal Ecology*, **67**, 307–17.

Goldizen, A. W., Buchan, J. C., Putland, D. A., Goldizen, A. R. & Krebs, E. A. (2000). Patterns of mate-sharing in a population of Tasmanian native hens *Gallinula mortierii*. *Ibis*, **142**, 440–7.

Hampton, J. K. Jr & Hampton, S. H. (1965). Marmosets (Hapiladae): Breeding seasons, twinning, and sex of offspring. *Science*, **150**, 915–17.

Heymann, E. W. (1996). Social behavior of wild moustached tamarins, *Saguinus mystax*, at the Estación Biológica Quebrada Blanco, Peruvian Amazonia. *American Journal of Primatology*, **38**, 101–13.

(2000). The number adult males in callitrichine groups and its implications for callitrichine social evolution. In *Primate Males: Causes and Consequences of Variation in Group Composition*, ed. P. M. Kappeler, pp. 64–71. Cambridge: Cambridge University Press.

Heymann, E. W. & Soini, P. (1999). Offspring number in pygmy marmosets, *Cebuella pygmaea*, in relation to group size and the number of adult males. *Behavioral Ecology and Sociobiology*, **46**, 400–4.

Hubrecht, R. C. (1985). Home-range and use, and territorial behavior in the common marmoset, *Callithrix jacchus jacchus* at the Tapacura field station, Recife, Brazil. *International Journal of Primatology*, **6**, 533–50.

Izawa, K. (1978). A field study of the ecology and behavior of the black-mantle tamarin (*Saguinus nigricollis*). *Primates*, **19**, 241–74.

Jamieson, I. G., Quinn, J. S., Rose, P. A. & White, B. N. (1994). Shared paternity among nonrelatives in a result of an egalitarian mating system in the communally breeding bird, the pukeko. *Proceedings of the Royal Society of London, Series B*, **257**, 271–7.

Kay, R. F. (1990). The phyletic relationships of extant and fossil Pitheciinae (Platyrrhini, Anthropoidea). *Journal of Human Evolution*, **19**, 175–208.

Kleiman, D. G. (1977). Monogamy in mammals. *Quarterly Review of Biology*, **52**, 39–69.

Koenig, A. (1995). Group size, composition, and reproductive success in wild common marmosets (*Callithrix jacchus*). *American Journal of Primatology*, **35**, 311–17.

Koenig, W. D. & Mumme, R. L. (1987). *Population Ecology of the Cooperatively Breeding Acorn Woodpecker*. Princeton, New Jersey: Princeton University Press.

Koenig, W. D. & Stacey, P. B. (1990). Acorn woodpeckers: group-living and food storage under contrasting ecological conditions. In *Cooperative Breeding in Birds: Long-term Studies of Ecology and Behaviour*, ed. P. B. Stacey & W. D. Koenig, pp. 413–53. Cambridge: Cambridge University Press.

Lazaro-Perea, C. (2001). Intergroup interactions in wild common marmosets, *Callithrix jacchus*: territorial defence and assessment of neighbours. *Animal Behaviour*, **62**, 11–21.

Lazaro-Perea, C., Castro, C. S. S., Harrison, R., Araujo, A., Arruda, M. F. & Snowdon, C. T. (2000). Behavioral and demographic changes following the loss of the breeding female in cooperatively breeding marmosets. *Behavioral Ecology and Sociobiology*, **48**, 137–46.

Magrath, R. D. & Heinsohn, R. G. (2000). Reproductive skew in birds: models, problems and prospects. *Journal of Avian Biology*, **31**, 247–58.

Maynard Smith, J. & Ridpath, M. G. (1972). Wife-sharing in the Tasmanian native hen, *Tribonyx mortierii:* a case of kin selection? *American Naturalist*, **106**, 447–52.

Neyman, P. F. (1978). Aspects of the ecology and social organization of free-ranging cotton-top tamarins (*Saguinus oedipus*) and the conservation status of the species. In *The Biology and Conservation of the Callitrichidae*, ed. D. G. Kleiman, pp. 39–71. Washington, DC: Smithsonian Institution Press.

Nievergelt, C. M. & Martin, R. D. (1998). Energy intake during reproduction in captive common marmosets (*Callithrix jacchus*). *Physiology & Behavior*, **65**, 849–64.

Nievergelt, C. M., Digby, L. J., Ramakrishnam, U. & Woodruff, D. S. (2000). Genetic analysis of group composition and breeding system in a wild common marmoset (*Callithrix jacchus*) population. *International Journal of Primatology*, **21**, 1–20.

Orians, G. H. (1969). On the evolution of mating systems in birds and mammals. *American Naturalist*, **103**, 589–603.

Price, E. C. (1992). The costs of infant carrying in captive cotton-top tamarins. *American Journal of Primatology*, **26**, 23–33.

Purvis, A. & Rambaut, A. (1994). *Comparative Analysis by Independent Contrasts (CAIC)*, Version 2. Oxford: Oxford University Press.

Ramirez, M. F., Freese, C. H. & Revilla, C. J. (1978). Feeding ecology of the pygmy marmoset, *Cebuella pygmaea*, in northeastern Peru. In *The Biology and Conservation of the Callitrichidae*, ed. D. G. Kleiman, pp. 91–104. Washington, DC: Smithsonian Institution Press.

Ridpath, M. G. (1972). The Tasmanian native hen, *Tribonyx mortierii*. II. The individual, the group, and the population. *Wildlife Research*, **17**, 53–90.

Roda, S. A. & Mendes Pontes, A. R. (1998). Polygyny and infanticide in common marmosets in a fragment of the Atlantic forest of Brazil. *Folia Primatologica*, **69**, 372–6.

Rosenberger, A. L. & Coimbra-Filho, A. F. (1984). Morphology, taxonomic status and affinities of the lion tamarins, *Leontopithecus* (Callitrichinae, Cebinadae). *Folia Primatologica*, **42**, 149–79.

Rylands, A. B. (1982). The Behaviour and Ecology of Three Species of Marmosets and Tamarins (*Callitrichidae, Primates*) in Brazil. Ph.D. thesis, University of Cambridge.

(1993). The ecology of the lion tamarins, *Leontopithecus*: some intrageneric differences and comparisons with other callitrichids. In *Marmosets and Tamarins: Systematics, Behaviour, and Ecology*, ed. A. B. Rylands, pp. 296–313. Oxford: Oxford University Press.

(1996). Habitat and the evolution of social and reproductive behavior in Callitrichidae. *American Journal of Primatology*, **38**, 5–18.

Rylands, A. B. & de Faria, D. S. (1993). Habitats, feeding ecology, and home range size in the genus *Callithrix*. In *Marmosets and Tamarins: Systematics, Behaviour, and Ecology*, ed. A. B. Rylands, pp. 262–72. Oxford: Oxford University Press.

Rylands, A. B., Coimbra-Filho, A. F. & Mittermeier, R. A. (1993). Systematics, geographic distribution, and some notes on the conservation status of the Callitrichidae. In *Marmosets and Tamarins: Systematics, Behaviour, and Ecology*, ed. A. B. Rylands, pp. 11–77. Oxford: Oxford University Press.

Sanchez, S., Pelaéz, F., Gil-Bürmann, C. &. Kaumanns, W. (1999). Costs of infant-carrying in the cotton-top tamarin. *American Journal of Primatology*, **48**, 99–111.

Savage, A., Giraldo, L. H., Soto, L. H. & Snowdon, C. T. (1996*a*). Demography, group composition, and dispersal in wild cotton-top tamarin (*Saguinus oedipus*) groups. *American Journal of Primatology*, **38**, 85–100.

Savage, A., Snowdon, C. T., Giraldo, L. H. & Soto, L. H. (1996*b*). Parental care patterns and vigilance in wild cotton-top tamarins (*Saguinus oedipus*). In *Adaptive Radiations of Neotropical Primates*, ed. M. A. Norconk, A. L. Rosenberger & P. A. Garber, pp. 187–99. New York: Plenum Press.

Schradin, C. & Anzenberger, G. (2001). Costs of infant-carrying in common marmosets, *Callithrix jacchus*. *Animal Behaviour*, **62**, 289–95.

Sherman, P. T. (1995). Social organization of cooperatively polyandrous white-winged trumpeters (*Psophia leucoptera*). *The Auk*, **112**, 296–309.

Soini, P. (1982). Ecology and population dynamics of the pygmy marmoset, *Cebuella pygmaea*. *Folia Primatologica*, **39**, 1–21.

(1987). Sociosexual behavior of a free-ranging *Cebuella pygmaea* (Callitrichidae, Platyrrhini) troop during postpartum estrus of its reproductive female. *American Journal of Primatology*, **13**, 223–30.

(1988). The pygmy marmoset, *Cebuella*. In *Ecology and Behavior of Neotropical Primates*, Volume 2, ed. R. A. Mittermeier, A. B. Rylands, A. F. Coimbra-Filho & G. A. B. Fonseca, pp. 79–129. Washington, DC: World Wildlife Fund.

(1993). The ecology of the pygmy marmoset *Cebuella pygmaea*: some comparisons with two sympatric tamarins. In *Marmosets and Tamarins: Systematics, Behaviour, and Ecology*, ed. A. B. Rylands, pp. 257–72. Oxford: Oxford University Press.

Tardif, S. D. (1994). Relative energetic cost of infant care in small-bodied neotropical primates and its relation to infant-care patterns. *American Journal of Primatology*, **34**, 133–43.

Tardif, S. D. & Harrison, M. L. (1990). Estimates of the energetic cost of infant transport in tamarins. *American Journal of Physical Anthropology*, **81**, 306.

Tardif, S. D., Harrison, M. L. & Simek, M. A. (1993). Communal infant care in marmosets and tamarins: relation to energetics, ecology, and social organization. In *Marmosets and Tamarins: Systematics, Behaviour, and Ecology*, ed. A. B. Rylands, pp. 220–34. Oxford: Oxford University Press.

Terborgh, J. (1983). *Five New World Primates: A Study in Comparative Ecology*. Princeton, New Jersey: Princeton University Press.

Terborgh, J. & Goldizen, A. W. (1985). On the mating system of the cooperatively breeding saddle-backed tamarin (*Saguinus fuscicollis*). *Behavioral Ecology and Sociobiology*, **16**, 293–9.

Terborgh, J. & Stern, M. (1987). The surreptitious life of the saddle-backed tamarin. *American Scientist*, **75**, 260–9.

Vehrencamp, S. L. (1983). A model for the evolution of despotic versus egalitarian societies. *Animal Behaviour*, **31**, 667–82.

(2000). Evolutionary routes to joint-female nesting in birds. *Behavioral Ecology and Sociobiology*, **11**, 334–44.

Wynne-Edwards, K. E. (1987). Evidence for obligate monogamy in the Djungarian hamster, *Phodopus campbelli*: pup survival under different parenting conditions. *Behavioral Ecology and Sociobiology*, **20**, 427–37.

CHAPTER 16

Monogamy in New World primates: what can patterns of olfactory communication tell us?

Eckhard W. Heymann

INTRODUCTION

Social communication and social monogamy

Communication is an integral component of social processes and social evolution (Hahn & Simmel, 1974; Philips & Austad, 1990), and variation in social systems influences patterns of communication (Marler & Mitani, 1988). Amongst other functions, communication may serve to attract, stimulate, defend, or compete for mates, and thus is intimately related to social organization and mating systems (Hahn & Simmel, 1974; Johnstone, 1997). The intensity of signals and displays can serve as a cue to the quality of mates or opponents, and if interindividual variation of signal quality exists, sexual selection can act upon signals and displays through mate choice and intrasexual competition (Johnstone, 1995). Examples are visual and vocal displays in many mammals, birds, and other vertebrates (e.g., Clutton-Brock & Albon, 1979; Ryan, 1983; Hill, 1990). Since in most animals it is usually males who compete for females, and females who are choosy about males, sexual selection results in signals and displays being either exclusive to or exaggerated in males (Andersson, 1994). Nevertheless, mutual sexual selection can lead to signals being expressed to the same degree in both sexes (Jones & Hunter, 1993). This may be the case in monogamous animals, which are often monomorphic, both physically and behaviourally, including their communication patterns (Kleiman, 1977). Deviations from the behavioural monomorphism of signals and displays in socially monogamous animals can be expected if the selective forces driving males and females into a monogamous system differ between the sexes. These deviations could then provide hints to the mechanisms and causes underlying monogamy and might point to sexual selection operating more strongly in one or the other sex.

New World primates (Platyrrhini) represent an interesting group for examining the interrelationship between social monogamy and patterns of communication. The proportion of genera in which social monogamy is found is very high in comparison to other simian primates (van Schaik & Kappeler, chapter 4). Furthermore, in several platyrrhine genera (both socially monogamous and non-monogamous) paternal care is present in the form of infant carrying, the extent of which may even exceed the amount of carrying by mothers.

Communication by vocal and olfactory signals is elaborate in New World primates (Oppenheimer, 1977; Epple, 1986; Snowdon, 1989). The focus of this chapter is on olfactory communication. The important role of this mode of communication amongst New World primates is emphasized by the diversity of signalling behaviours (mainly scent marking and ritualized urination patterns) and associated morphological structures (scent glands) (Epple & Lorenz, 1967; Perkins, 1975; Epple, 1986). After providing an overview of socially monogamous platyrrhines (including those in which pair living occurs but is not the modal pattern), I will examine patterns of olfactory communication in relation to predictions derived from sexual selection theory. I will examine whether and how patterns of olfactory communication vary between different monogamous platyrrhine taxa and what this means for the underlying structures. Finally, I will also briefly examine patterns of vocal communication.

MONOGAMY IN NEW WORLD PRIMATES

In eight genera of New World primates social monogamy is found either as the exclusive or modal pattern or as one of several patterns within a flexible system (Figure 16.1). These are the titi monkeys (*Callicebus* spp.) and the saki

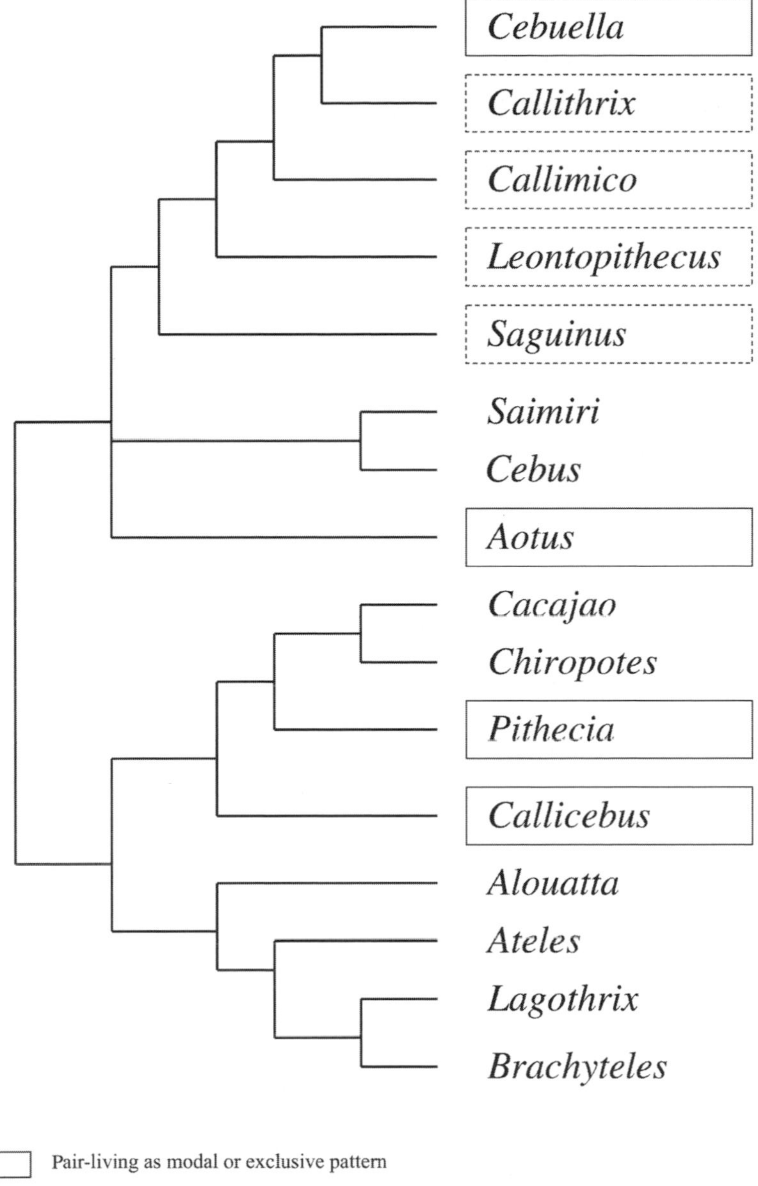

Figure 16.1. Distribution of social monogamy amongst New World primates (phylogenetic tree following von Dornum & Ruvolo, 1999).

monkeys (*Pithecia* spp.) from the family Pitheciidae (following the taxonomy of Rylands *et al.*, 2000), the night monkeys (*Aotus* spp.) from the family Aotidae, and the tamarins (*Saguinus* spp.), lion tamarins (*Leontopithecus* spp.), Goeldi's monkey (*Callimico goeldii*), marmosets (*Callithrix* spp.), and pygmy marmoset (*Cebuella pygmaea*) from the family Callitrichidae. All other New World monkeys live in socially polygynous, multi-male/multi-female or fission–fusion groups (Crockett & Eisenberg, 1986; Robinson & Janson, 1986).

When examining social monogamy in New World primates, a problem arises with regard to the database.

Table 16.1. *Body mass of wild New World primates in which social monogamy is the exclusive or modal pattern, or is one pattern within a flexible system*

	Adult body mass (kg)		Sample size		
Taxon	Males	Females	Males	Females	Reference
Aotus lemurinus	0.918	0.874	18	12	Smith & Jungers, 1997
A. nancymaae	0.795	0.78	32	24	Aquino & Encarnación, 1986
A. vociferans	0.708	0.698	20	20	Montoya *et al.*, 1990
Callicebus torquatus	1.28	1.21	15	21	Smith & Jungers, 1997
Callimico goeldii	0.366	0.355	3	5	Encarnación & Heymann, 1998
Callithrix jacchus	0.32	0.32	69	86	Araújo *et al.*, 2000
Cebuella pygmaea	0.11	0.122	36	27	Soini, 1988*b*
Leontopithecus rosalia	0.62	0.598	(471)[a]	(185)[a]	Dietz *et al.*, 1994*a*
Pithecia monachus	2.61	2.11	16	10	Smith & Jungers, 1997
P. pithecia	1.94	1.58	10	4	Smith & Jungers, 1997
Saguinus fuscicollis	0.343	0.358	69	55	Soini, 1990
S. geoffroyi	0.482	0.502	55	40	Dawson & Dukelow, 1976
S. mystax	0.505	0.535	161	104	Soini & de Soini, 1990
S. midas	0.49	0.5	87[b]		Richard-Hansen *et al.*, 1999
S. oedipus	0.417	0.404	37	29	Savage *et al.*, 1993

[a] Includes repeated measurements of the same individuals.
[b] Total number of individuals.

For several taxa, only demographic data are available from the wild, while detailed behavioural information is restricted to captive studies. Furthermore, genetic data (relatedness, paternity) have only recently become available for one species of the genus *Callithrix* (Nievergelt *et al.*, 2000).

Titi monkeys: *Callicebus*

Titi monkeys are small primates (*c.* 1 kg) with no sexual dimorphism in body mass or external appearance (Table 16.1). They live in groups of two to five individuals, the number of adults is almost always two, and the sex ratio is balanced (Table 16.2). This pattern seems to be consistent in all species for which quantitative information is available from the wild (five species from a total of 13).

Titi monkeys show strong pair bonding in captivity and this is most likely true in the wild as well (Mason, 1966; Anzenberger, 1988, 1993). Pair mates sing in coordinated duets, either spontaneously or in response to other groups, but at dawn, male titi monkeys may give loud calls, often alone (Kinzey *et al.*, 1977; Robinson, 1979). Males are the main caregivers for the single offspring, apart from the lactating mother, accounting for about 80% to almost 100% of all infant carrying, both in captivity and in the wild (Wright, 1984; Welker *et al.*, 1998; E.R. Tirado Herrera & E.W. Heymann, unpublished data). Observations of wild *Callicebus cupreus* suggest that extra-pair copulations occur (Mason, 1966). In a captive pair of *C. cupreus*, the female engaged in scent marking with the subcaudal gland three times more frequently than the male (Jantschke, 1992); information on this behaviour has not been reported from the wild. In wild *C. cupreus*, sternal scent marking occurs during intergroup encounters, but no information is available on the frequency in males and females (Mason, 1966); in captivity, this behaviour seems to be very rare (Moynihan, 1966; Jantschke, 1992). The sternal scent gland is developed in both sexes (Epple & Lorenz, 1967).

Night monkeys: *Aotus*

Night monkeys are the only nocturnal simian primates. They are small (0.6–1.2 kg) and do not show sexual dimorphism (Table 16.1). Night monkeys live in small family groups of two to six (on rare occasions seven) individuals (Table 16.2). Average group size in *Aotus*

Table 16.2. *Demographic parameters for New World primates*

Taxon	Group size Mean	Group size Range	Number of adults Mean	Number of adults Range	Adult sex ratio (males:females)	Number of groups	References
Aotus azarai, Peru	4.5	2–5			1:1	9	Wright, 1985
A. azarai, Paraguay		2–7				11	Fernandez-Duque *et al.*, 2001
A. nancymaae[a]	3.4	2–6	2.2	2–4	1:1.05	39	Aquino *et al.*, 1990
A. vociferans[a]	3.1	2–5	2.1	2–3	1:0.90	38	Aquino *et al.*, 1990
Callicebus brunneus, Peru	4.2	2–5	2.0			6	Wright, 1985
C. brunneus, Brazil	2.3					116[b]	Ferrari *et al.*, 2000
C. caligatus	2.2					5	Ferrari *et al.*, 2000
C. cupreus[a]	3.1	2–4	2.0		1:1?	9	Mason, 1966
C. moloch	2.5					15[b]	Ferrari *et al.*, 2000
C. personatus[a]		2–5		2–3	1:1?	4	Kinzey & Becker, 1983; Müller, 1995; Heiduck, 1998; Price & Piedade, 2001
C. torquatus[a]	3.0	2–4	2.1	2–3		10	Defler, 1983
Callimico goeldii, Peru[a]	8.0	6–10	4.5	4–5	1:2.1	2	Encarnación & Heymann, 1998
C. goeldii, Bolivia[a]	6.0	3–9	5.7	2–6	1:0.9	3	Porter, 2001
Callithrix jacchus[a]	7.4[c]	3–9	3.7[c]	3–5	1:0.6[c]	10	Hubrecht, 1984
	8.7[d]	5–11	7.0[d]	5–8	1:1.6	3	Digby & Barreto, 1993
Cebuella pygmaea[a]	5.1	2–9	2.8	2–5	1:1	80	Soini, 1993
Leontopithecus chrysomelas	5.2[e]	3–9	3.4	3–5	1:0.7	5	Dietz *et al.*, 1994*b*
L. chrysopygus	4.8	3–7	3.5	2 5	1.0.56	4	Valladares-Padua, 1993
L. rosalia	5.4[f]	2–11	3.5	2–7	1:0.75	206[g]	Dietz & Baker, 1993
Pithecia albicans	4.6	2–7				5	Peres, 1993
P. irrorata[a]	4.5	2–6				4	Buchanan-Smith, 1990
P. monachusa	3.6	2–8	2.7	2–5	1:0.9	21[h]	Soini, 1988*a*
P. monachus	4.5	4–5	2.0		1:1	4	Happel, 1982
P. pithecia[a]	2.7	2–3	2.1	2–3		3	Oliveira *et al.*, 1985
	5.3	5–6				1[i]	Setz & Gaspar, 1997
	6.8	5–8				1[j]	Homburg, 1998
Saguinus fuscicollis, SE Peru	5.1	3–8	3.2	2–5	1:0.68	46[k]	Goldizen *et al.*, 1996
S. fuscicollis, NE Peru[a]	5.8	2–12	3.2	2–5	1:0.83	62	Soini, 1990, 1993
S. geoffroyia	6.4				1:0.88	5	Dawson, 1977
S. labiatus	5.6[l]	2–10	4.1[l]		1:0.71[l]	38	Puertas *et al.*, 1995
S. mystax[a]	5.5	3–9	3.7	2–6	1:0.76	44	Soini & de Soini, 1990
S. nigricollis[a]	5.2	2–7	2.6	2–4	1:1	10	Izawa, 1978
	5.0	2–9	2.1	2–3	1:1	8	de la Torre *et al.*, 1995
S. oedipus	5.8	2–10	3.6[m]	3–4	1:0.8	13[n]	Savage *et al.*, 1996*a*

[a] Excluding carried infants.

[b] Number of sightings during census work.

[c] Group size calculated from Table 1 in Hubrecht, 1984; number of adults and sex ratio calculated from Table 2 in Hubrecht, 1984; $N = 3$ groups for these two parameters.

[d] Calculated from Figures 1–3 in Digby & Barreto, 1993, using group composition at beginning of study.

[e] Calculated from Table 1, in Dietz *et al.*, 1994b, solitaries excluded.

[f] Calculated from Table 2, in Dietz & Baker, 1993, groups containing adults of only one sex excluded.

[g] Number of repeated group samples.

[h] Repeatedly censused over 8 years.

[i] Repeatedly censused over 4 years.

[j] Calculated from Figure 2.4 in Homburg, 1998, using group composition at the beginning of each of 4 years.

[k] Forty-seven complete group-year compositions over a period of 13 years.

[l] Calculated from Table 3, in Puertas *et al.*, 1995 (complete groups only).

[m] Calculated from Table 2, in Savage *et al.*, 1996b.

[n] Thirteen complete group-year compositions over a period of 5 years.

[?] Adult sex ratio unknown but information provided in cited references suggests a ratio close to 1:1.

nancymaae and *A. vociferans* is slightly more than three individuals, the number of adults is slightly above two (Aquino *et al.*, 1990), and adult sex ratios are nearly 1:1.

In *A. azarai* in the extreme south of its geographic range, group size was larger in groups bordering riverine habitats compared with groups without access to the riverbank (Fernandez-Duque *et al.*, 2001). Increased habitat productivity in riverine habitats may delay dispersal in adult offspring and thus increase group size (Fernandez-Duque *et al.*, 2001). There was, however, no evidence that increased group size resulted in deviations from monogamous breeding.

The pair bond is probably strong, but detailed studies and experiments are lacking. Captive studies indicate that pair mates spend as much time in body contact as do titi monkey pair mates (Jantschke, 1992; Welker *et al.*, 1998). Males provide 80–90% of all infant carrying, both in captivity and in the wild (Dixson & Fleming, 1981; Wright, 1984; Welker *et al.*, 1998).

Night monkeys do not duet. Long-distance hoots may be given by both sexes, but only one individual night monkey, either male or female, hoots from each territory (Wright, 1985). Night monkeys possess a subcaudal and a sternal scent gland. In captive *A. lemurinus*, scent marking with the subcaudal gland seems to be female-biased (Moynihan, 1964). In captive *A. azarai*, females tend to scent-mark more often than males (Jantschke, 1992). Pair mates may also scent-mark each other (Jantschke, 1992). In wild *A. nigriceps*, subcaudal scent marking was observed during an agonistic intergroup encounter (Wright, 1978). As in titi monkeys, detailed information is lacking from the wild. A sternal scent gland is present in both sexes, but scent marking by this gland is not reported in the literature.

Saki monkeys: *Pithecia*

Saki monkeys are medium-sized (1.5–3 kg), with males generally larger than females (Table 16.1). Sexual dichromatism is present in at least two of the five recognized species and most pronounced in the Guianan saki (*Pithecia pithecia*) (Hershkovitz, 1977). In *P. pithecia*, dichromatism is either present at birth or develops shortly after birth, but appears later in *P. monachus* (Hanif, 1967; Stott, 1976; Soini, 1986).

Saki monkey groups range in size between two and eight individuals (Table 16.2). Detailed and extensive long-term data are only available for *P. monachus* (Soini, 1988*a*); information for other species stems either from studies of single or only a few groups, or from census work (which tends to underestimate real group size). Mean group size and mean number of adults in *P. monachus* are slightly larger than in *Aotus* and *Callicebus*, but the sex ratio seems to be balanced (Table 16.2). The large group size reported from two studies of *P. pithecia* are both from groups in isolated habitats, either an island or a forest fragment, where animals could not disperse to other areas (Setz & Gaspar, 1997; Homburg, 1998). In both of these groups, more than one female reproduced successfully.

Very little information is available on social relationships between adult males and females. Levels of aggression are low, and pair mates can be the preferred partner in affiliative interactions like allogrooming and huddling (Homburg, 1989). Mating patterns have not yet been studied in any detail in the wild. In the group observed by Homburg (1998), only one of the adult males copulated with each of two reproducing females. Adult males do not carry infants (Claussen, 1982; Oliveira *et al.*, 1985). An indirect form of paternal care is perhaps the behaviour of adult males against potential predators: they may display by standing bipedally and swaying from side to side, or they may lead predators away while the rest of the group hides (Oliveira *et al.*, 1985; Soini, 1986; author's personal observation).

Saki monkeys possess sternal scent glands (Epple & Lorenz, 1967; Brumloop *et al.*, 1994). Observations of both captive and wild *P. pithecia* indicate that scent-marking behaviour is strongly male-biased and increases during the breeding season (Homburg, 1998; Setz & Gaspar, 1997). A sexual communication function has been attributed to scent marking in *P. pithecia* (Setz & Gaspar, 1997). While the vocal repertoire has been described in detail (Buchanan, 1978), no information is available on calling rates, either from captive or from wild sakis.

Tamarins and marmosets: *Saguinus*, *Leontopithecus*, *Callimico*, *Callithrix*, and *Cebuella*

With body masses between 100 and 650 g, callitrichids are the smallest New World primates. In many species, females are up to 10% heavier than males (Table 16.1). Comprehensive information about social organization and mating patterns of the Callitrichidae is provided by Garber (1997), Goldizen (1986, chapter 15), Heymann (2000), and in contributions to the symposium on

'Callitrichid social structure and mating system: evidence from field studies' (*American Journal of Primatology*, **38**, 1996). Therefore, only a short overview is provided here.

Mean group size varies between 4.8 and 8.7 individuals (Table 16.2), and single groups may comprise up to 15 individuals. The adult sex ratio is mainly male-biased, but can also be female-biased or balanced, and varies within and between populations and species. Observed mating patterns include polyandry, monogamy, and polygyny. Extra-group copulations have been reported in *Callithrix jacchus*, *Saguinus fuscicollis*, and *S. mystax* during long-term behavioural studies (Digby, 1999; E. R. Tirado Herrera, unpublished data; P. Löttker & M. Huck, unpublished data). In *C. jacchus* extra-group copulations do not result in extra-group paternity (Nievergelt *et al.*, 2000), but no pertinent information is available for other callitrichids.

Callitrichid reproductive biology is characterized by habitual dizygotic twinning, except in the monotypic *Callimico*. The litter mass at birth may account for up to 20% of maternal body mass (Tardif, 1994; Garber & Leigh, 1997). All group members contribute to infant carrying, but depending on the genus, either adult males/fathers are the principal infant carriers, or adult male and mother contributions to infant carrying are balanced (e.g., Goldizen, 1988; Stevenson & Rylands, 1988; Savage *et al.*, 1996*a*; Garber, 1997; for overview see Heymann, 2003).

Callitrichids possess an extensive vocal repertoire, including loud or long calls (Snowdon, 1989, 1993). These calls are employed in within-group and between-group communication and may play a role in sexual competition and mate choice (Moynihan, 1970; Snowdon, 1989). In several groups of captive *Saguinus oedipus* and one pair of captive *S. mystax*, adult females have been observed uttering almost three times more loud calls than males (Heymann, 1985; McConnell & Snowdon, 1986). In contrast, captive male and female common marmosets (*Callithrix jacchus*) and captive golden lion tamarins (*Leontopithecus rosalia*) emit loud calls at identical rates (McLanahan & Green, 1977; Norcross & Newman, 1993). No data are available from wild animals.

Of all New World primates, the Callitrichidae possess perhaps the richest repertoire of olfactory communication, with scent glands located in the anogenital, suprapubic, and sternal regions, and specific marking behaviour associated with each of these glands (Epple *et al.*, 1986, 1993). Extensive captive data and recently emerging field data indicate that depending on the species or genus, scent-marking behaviour is either female-biased or balanced between the sexes (Heymann, 2003).

SEXUAL SELECTION AND PARENTAL INVESTMENT

Sexual selection acts through competition for mates and through mate choice. The sex that invests more heavily in offspring is expected to be the choosier sex, while the sex with less or no investment in offspring should be the competitive sex (Trivers, 1972). Since in most organisms females invest more heavily in offspring than males, sexual selection usually operates more strongly upon males through male–male competition and female choice, resulting in secondary sex traits in males (Andersson, 1994). However, when infant care is provided by males, females may compete for males, and males may therefore become choosy (Burley, 1977; Petrie, 1983; Parker & Simmons, 1996), resulting in sexual selection acting upon females and leading to the evolution of secondary sex traits in females (Amundsen, 2000; Amundsen & Forsgren, 2001). It can therefore be anticipated that where males invest in offspring, the usual patterning of sexual selection is modified in New World primates.

SEXUAL SELECTION AND SCENT MARKING

Communication by means of olfactory signals, particularly by scent marking, is a major mode of information transfer in mammals (for review see Brown & MacDonald, 1985; Eisenberg & Kleiman, 1972). It may function in territory or resource advertisement and defence, advertisement of social status and reproductive condition, social regulation, and mate attraction (e.g., Gosling, 1982; Gosling & Wright, 1994; Heise & Rosenfeld, 1999; Roberts & Dunbar, 2000; Rosell, 2002). Therefore, in more general terms, olfactory signals play a role in reproductive competition and mate choice and thus can be expected to be subject to sexual selection (e.g., Johnstone *et al.*, 1997; Penn, 2002). However, while vocal and visual signals and displays are often examined in the light of sexual selection theory, much less consideration has been given to olfactory signals. Darwin (1871) long ago suggested that scent glands evolved through sexual selection, and Blaustein (1981) argued that sexual selection should act upon scents just as it acts upon visual signals, and that scents may represent secondary sex traits. This reasoning can be extended to the corresponding signalling behaviour, i.e., scent marking.

SCENT MARKING AND PATERNAL CARE IN NEW WORLD PRIMATES

Based on these considerations, one can examine whether a relationship exists between the occurrence and degree of paternal care and the direction of sexual dimorphism in scent marking in New World monkeys. This relationship is, however, not considered to be a direct one; rather, the degree of reproductive competition within the sexes and the relative importance of mate choice are seen as the intervening variables. Specifically, the expectation is that lack of paternal care is associated with male-biased scent marking and occurrence of paternal care with female-biased or unbiased scent marking, depending on whether male care exceeds or equals female care. Given the scarcity of information on scent marking in most New World monkeys, a test of this expectation can be made only tentatively at the moment. I extracted both quantitative and qualitative information on paternal care and scent marking from the literature. New World primate genera were then classified as having male-biased, female-biased, or unbiased rates of scent marking, and having no male care, equal contributions of males and females to infant care, or male care exceeding female care (see Heymann, 2003, for a detailed description of quantitative methods). I included New World primate genera that are not monogamous but for which at least some information on scent marking and paternal care is available.

All genera in which paternal care exceeds maternal care (*Saguinus*, *Aotus*, and *Callicebus*) show female-biased scent marking, and all genera that lack paternal care (*Pithecia*, *Cebus*, *Alouatta*, and *Lagothrix*) show male-biased scent marking. In genera with balanced maternal and paternal care, scent marking is not biased towards either one of the sexes (*Callithrix*, *Leontopithecus rosalia*) or is male-biased (*L. chrysomelas*) (Figure 16.2). Taxa that are almost exclusively or predominantly socially monogamous are found in the female-biased and in the male-biased category. Therefore, neither can the social organization and mating system predict the degree and direction of sexual dimorphism in scent marking, nor can patterns of communication predict

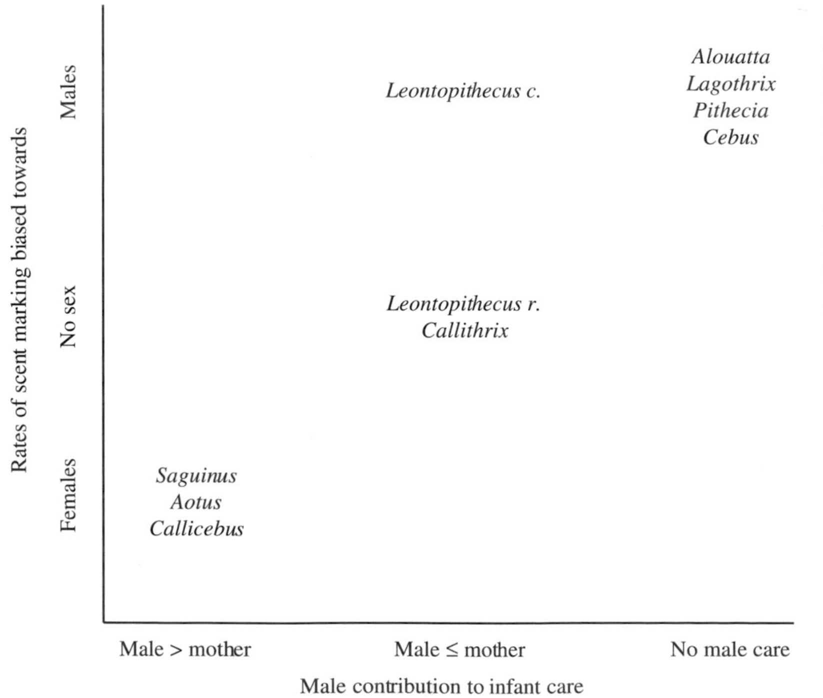

Figure 16.2. Direction of bias in rates of scent marking in relation to male contribution to infant care. Sources: *Alouatta*: Braza *et al.*, 1981, Sekulic & Eisenberg, 1983; *Cebus*: Dobroruka, 1972; *Lagothrix*: White *et al.*, 2000; *Leontopithecus*: Jan Verpooten, personal communication; see text and Heymann (2003) for references on all other taxa.

the type of social organization. However, if we accept that sexual dimorphism in displays like scent marking provides evidence for intrasexual competition and/or intersexual mate choice, my analyses suggest that the structures underlying social monogamy in New World monkeys differ between taxa.

A NOTE OF CAUTION

There are compelling reasons to view the results of this analysis with caution. First, for most New World monkeys, except for several callitrichids, the database is quite limited and in several cases qualitative rather than quantitative. Second, I have used overall rates of scent marking and did not differentiate between the different types of marking. Different types of marking may convey different messages (Epple *et al.*, 1993), and the relative frequency of use of different types of scent marking may vary even between closely related species (Heymann, 2001). Third, my analysis did not control for potential phylogenetic biases. However, even if taxa from the same clade showing no difference in the direction of sex bias in scent marking and in male contribution to infant care were lumped into a single data point, respectively (e.g., *Leontopithecus* and *Callithrix*; *Ateles* and *Alouatta*; see Figure 16.2), the same trend would emerge. Thus, phylogeny probably does not bias the result. Phylogeny also does not seem to constrain the results: *Callicebus* and *Pithecia* from the same family are completely different with regard to the sex bias in scent marking and in male contribution to infant care.

Another problem is that only one part of the communication process – signal emission through scent marking – has been examined. However, a complete picture can only emerge if the whole communication process is examined (Kappeler, 1998). For example, it is necessary to examine responses to scent marks (e.g., through sniffing or overmarking/countermarking), and whether this is mainly by same-sex competitors within and outside their own groups or by opposite-sex mates and potential mates. Currently, there is even less information available on this aspect of the communication process than on signal emission. In *Saguinus mystax*, the only species for which field information is available, most responses to scent marks are by males to female scent marks, while the comparatively fewer male scent marks are mainly visited by females (Heymann, 1998). In captive *S. labiatus*, males visited the scent marks of their female partners, but females did not respond to male scent marks (Smith & Gordon, 2002). These observations suggest an intersexual function of scent marking in these tamarins.

A relationship similar to the one found here between scent marking and parental care could emerge if the predators' cueing in on scent marks represents a major threat. In this case, infant carriers should scent-mark less frequently to avoid being attacked by an olfactory-oriented predator. However, this explanation is unlikely to account for the observed relationship. The principal predators of New World primates are raptors (Terborgh, 1983; Heymann, 1990; Hart, 2000), which are visually/auditorily-oriented hunters. Furthermore, scent marks are stationary signals and do not predict a group's direction of travel. Finally, even if predators were cued in by scent marks, they would be led to the group as a whole rather than to a single animal, since groups of all species examined here are usually rather cohesive.

VOCAL COMMUNICATION

As I have already mentioned (see sections above on Night monkeys; Tamarins and Marmosets), New World primates possess a rich vocal repertoire that includes 'long calls' or 'loud calls' in many but not all species. Since these calls may play a role in sexual competition and mate choice (Moynihan, 1970; Snowdon, 1989), analyses of sex biases in loud call emissions may provide additional or complementary information to the analyses of sex biases in scent marking. The analyses of sex bias in loud calling reveals a trend similar to that found for scent marking (Figure 16.3). While *Callicebus* stands out by having a male bias in loud calling, among other genera for which information is available on both olfactory and vocal communication, the two occupy the same relative position. However, as with olfactory communication, the database is very scanty and does not allow for further quantitative analyses.

CONCLUSIONS

No consistent relationship has been found between patterns of olfactory communication and the type of social organization/mating system in New World primates. However, there is a relationship between the direction of the sex bias in rates of scent marking and patterns of male

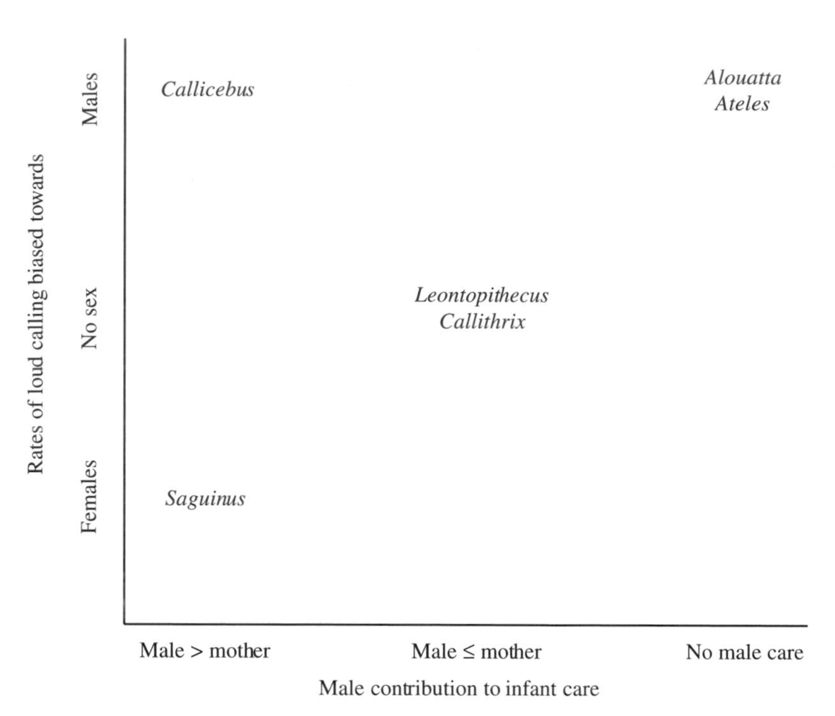

Figure 16.3. Direction of bias in rates of loud call (long call) emission in relation to male contribution to infant care. Sources: *Alouatta*: Sekulic, 1983; *Ateles*: van Roosmalen, 1985; see text for references on all other taxa.

care for infants. Female-biased or unbiased rates of scent marking in species with extensive male care suggest that female–female competition for males (as helpers in infant care) and/or male choice of females are selective forces equal to or exceeding the strength of male–male competition and female choice of males in these taxa. In both monogamous and non-monogamous taxa without paternal care, male-biased rates of scent marking conform to the more 'classical' pattern of male–male competition and/or female choice of males. Different monogamous taxa show either male-biased or female-biased rates of scent marking, indicating that there is no uniform pattern of sexual competition and mate choice in monogamous New World primates. Analyses of patterns of vocal communication largely coincide with the findings on scent marking, but also show one contrast, namely female-biased rates of scent marking but male-biased rates of loud calling in *Callicebus*. This finding adds to the conclusion that monogamy in New World monkeys is based on different underlying factors. These factors may be related both to the ecology and to the life history of the species under consideration. The lack of field data on the ecology, mating patterns and their genetic consequences, and the patterns of social behaviour and communication of most monogamous New World primates currently inhibits the clear identification of these underlying structures, and thus a deeper understanding of the evolution of monogamy in this taxon.

Acknowledgements

I thank Ulrich Reichard and Christophe Boesch for inviting me to the monogamy workshop. Discussions with workshop participants and comments by four anonymous reviewers greatly helped to improve the manuscript. I also thank Jan Verpooten for personal communications on rates of scent marking in *Leontopithecus*.

References

Amundsen, T. (2000). Why are female birds ornamented? *Trends in Ecology and Evolution*, **15**, 149–55.

Amundsen, T. & Forsgren, E. (2001). Male mate choice selects for female coloration in a fish. *Proceedings of the National Academy of Sciences of the USA*, **98**, 13155–60.

Andersson, M. (1994). *Sexual Selection*. Princeton, New Jersey: Princeton University Press.

Anzenberger, G. (1988). The pairbond in the titi monkey (*Callicebus moloch*): intrinsic versus extrinsic contributions of the pairmates. *Folia Primatologica*, **50**, 188–203.

(1993). Social conflict in two monogamous New World primates: pairs and rivals. In *Primate Social Conflict*, ed. W.A. Mason & S. P. Mendoza, pp. 291–329. New York: State University of New York Press.

Aquino, R. & Encarnación, F. (1986). Population structure of *Aotus nancymai* (Cebidae: Primates) in Peruvian Amazon lowland forest. *American Journal of Primatology*, **11**, 1–7.

Aquino, R., Puertas, P. & Encarnación, F. (1990). Supplemental notes on population parameters of northeastern Peruvian night monkeys, genus *Aotus* (Cebidae). *American Journal of Primatology*, **21**, 215–21.

Araújo, A., Arruda, M. F., Alencar, A. I., Albuquerque, F., Nascimento, M. C. & Yamamoto, M. E. (2000). Body weight of wild and captive common marmosets (*Callithrix jacchus*). *International Journal of Primatology*, **21**, 317–24.

Blaustein, A.R. (1981). Sexual selection and mammalian olfaction. *American Naturalist*, **117**, 1006–10.

Braza, F. Alvarez, F. & Azcarate, T. (1981). Behaviour of the red howler monkey (*Alouatta seniculus*) in the llanos of Venezuela. *Primates*, **22**, 459–73.

Brown, R. E. & MacDonald, D. W. (1985). *Social Odours in Mammals*. Oxford: Clarendon Press.

Brumloop, A., Homburg, I., Peetz, A. & Riehl, R. (1994). Gular scent glands in adult white-faced saki, *Pithecia pithecia pithecia*, and field observations on scent-marking behaviour. *Folia Primatologica*, **63**, 212–15.

Buchanan, D. B. (1978). Communication and Ecology of Pithecine Monkeys with Special Reference to *Pithecia pithecia*. Ph.D. thesis, Wayne State University, Detroit.

Buchanan-Smith, H. M. (1990). Observations of Gray's bald face saki, *Pithecia i. irrorata*, in Bolivia. In *Regional (U.K.) Studbook for the White-Faced Saki Pithecia p. pithecia*, Volume 2, ed. S. S. Waters, pp. 2–8.

Burley, N. (1977). Parental investment, mate choice, and mate quality. *Proceedings of the National Academy of Sciences of the USA*, **74**, 3476–9.

Claussen, R. (1982). Beobachtungen zur Aufzucht eines männlichen Weißgesichtsakis (*Pithecia pithecia*) im Familienverband. *Zoologischer Garten*, **52**, 188–94.

Clutton-Brock, T. H. & Albon, S. D. (1979). The roaring of red deer and the evolution of honest advertisement. *Behaviour*, **69**, 145–70.

Crockett, C. M. & Eisenberg, J. F. (1986). Howlers: variations in group size and demography. In *Primate Societies*, ed. B. B. Smuts, D. L. Cheney, R. M. Seyfarth, R. W. Wrangham & T. T. Struhsaker, pp. 54–68. Chicago: University of Chicago Press.

Darwin, C. (1871). *The Descent of Man and Selection in Relation to Sex*. London: John Murray.

Dawson, G. A. (1977). Composition and stability of social groups of the tamarin, *Saguinus oedipus geoffroyi*, in Panama: ecological and behavioral implications. In *The Biology and Conservation of the Callitrichidae*, ed. D. G. Kleiman, pp. 23–37. Washington, DC: Smithsonian Institution Press.

Dawson, G. A. & Dukelow, W. R. (1976). Reproductive characteristics of free-ranging Panamanian tamarins (*Saguinus oedipus geoffroyi*). *Journal of Medical Primatology*, **5**, 266–75.

Defler, T. R. (1983). Some population characteristics of *Callicebus torquatus lugens* (Humboldt, 1812) (Primates: Cebidae) in eastern Colombia. *Lozania: Acta Zoológica Colombiana*, **38**, 1–9.

de la Torre, S., Campos, F. & de Vries, T. (1995). Home range and birth seasonality of *Saguinus nigricollis graellsi* in Ecuadorian Amazonia. *American Journal of Primatology*, **37**, 39–56.

Dietz, J. M. & Baker, A. J. (1993). Polygyny and female reproductive success in golden lion tamarins, *Leontopithecus rosalia*. *Animal Behaviour*, **46**, 1067–78.

Dietz, J. M., Baker, A. J. & Miglioretti, D. (1994*a*). Seasonal variation in reproduction, juvenile growth, and adult body mass in golden lion tamarins (*Leontopithecus rosalia*). *American Journal of Primatology*, **34**, 115–32.

Dietz, J. M., de Sousa, S. N. F. & da Silva, J. R. O. (1994*b*). Population structure and territory size in golden-headed lion tamarins, *Leontopithecus chrysomelas*. *Neotropical Primates*, **2**, 21–3.

Digby, L. J. (1999). Sexual behavior and extragroup copulations in a wild population of common marmosets (*Callithrix jacchus*). *Folia Primatologica*, **70**, 136–45.

Digby, L. J. & Barreto, C. E. (1993). Social organization in a wild population of *Callithrix jacchus*. I. Group composition and dynamics. *Folia Primatologica*, **61**, 123–34.

Dixson, A. F. & Fleming, D. (1981). Parental behaviour and infant development in owl monkeys (*Aotus trivirgatus lemurinus*). *Journal of Zoology, London*, **194**, 25–39.

Dobroruka, L. J. (1972). Social communication in the brown capuchin. *International Zoo Yearbook*, **12**, 43–5.

Eisenberg, J. F. & Kleiman, D. G. (1972). Olfactory communication in mammals. *Annual Review of Ecology and Systematics*, **3**, 1–32.

Encarnación, F. & Heymann, E. W. (1998). Body mass of wild *Callimico goeldii*. *Folia Primatologica*, **69**, 368–71.

Epple, G. (1986). Communication by chemical signals. In *Comparative Primate Biology: Behavior, Conservation and Ecology*, Volume 2A, ed. J. Erwin, pp. 531–80. New York: Alan R. Liss.

Epple, G. & Lorenz, R. (1967). Vorkommen, Morphologie und Funktion der Sternaldrüse bei den Platyrrhini. *Folia Primatologica*, **7**, 98–126.

Epple, G., Belcher, A. M., Küderling, I., Zeller, U., Scolnick, L., Greenfield, K. L. & Smith, A. B. III (1993). Making sense out of scents: species differences in scent glands, scent-marking behaviour, and scent-mark composition in the Callitrichidae. In *Marmosets and Tamarins: Systematics, Behaviour, and Ecology*, ed. A. B. Rylands, pp. 123–51. Oxford: Oxford University Press.

Epple, G., Belcher, A. M. & Smith, A. B. III (1986). Chemical signals in callitrichid monkeys – a comparative review. In *Chemical Signals in Vertebrates: Ecology, Evolution and Comparative Biology*, Volume 4, ed. D. Duvall, D. Müller-Schwarze & R. M. Silverstein, pp. 653-72. New York: Plenum Press.

Fernandez-Duque, E., Rotundo, M. & Sloan, C. (2001). Density and population structure of owl monkeys (*Aotus azarai*) in the Argentinian Chaco. *American Journal of Primatology*, **53**, 99–108.

Ferrari, S. F., Iwanaga, S., Messias, M. R., Ramos, E. M., Ramos, P. C. S., da Cruz Neto, E. H. & Coutinho, P. E. G. (2000). Titi monkeys (*Callicebus* spp., Atelidae: Platyrrhini) in the Brazilian state of Rondônia. *Primates*, **41**, 229–34.

Garber, P. A. (1997). One for all and breeding for one: cooperation and competition as a tamarin reproductive strategy. *Evolutionary Anthropology*, **6**, 187–99.

Garber, P. A. & Leigh, S. R. (1997). Ontogenetic variation in small-bodied New World primates: implications for patterns of reproduction and infant care. *Folia Primatologica*, **68**, 1–22.

Goldizen, A. W. (1986). Tamarins and marmosets: communal care of offspring. In *Primate Societies*, ed. B. B. Smuts, D. L. Cheney, R. M. Seyfarth, R. W. Wrangham & T. T. Struhsaker, pp. 34–43. Chicago: University of Chicago Press.

(1988). Tamarin and marmoset mating systems: unusual flexibility. *Trends in Ecology and Evolution*, **3**, 36–40.

Goldizen, A. W., Mendelson, J., van Vlaardingen, M. & Terborgh, J. (1996). Saddle-back tamarin (*Saguinus fuscicollis*) reproductive strategies: evidence from a thirteen-year study of a marked population. *American Journal of Primatology*, **38**, 57–83.

Gosling, L. M. (1982). A reassessment of the function of scent marking in territories. *Zeitschrift für Tierpsychologie*, **60**, 89–118.

Gosling, L. M. & Wright, K. H. M. (1994). Scent marking and resource defence by male coypus (*Myocastor coypus*). *Journal of Zoology, London*, **234**, 423–36.

Hahn, M. E. & Simmel, E. C. (1974). *Communicative Behavior and Evolution*. New York: Academic Press.

Hanif, M. (1967). Notes on breeding the white-headed saki monkey *Pithecia pithecia* at Georgetown Zoo. *International Zoo Yearbook*, **7**, 81–2.

Happel, R. E. (1982). Ecology of *Pithecia hirsuta* in Peru. *Journal of Human Evolution*, **11**, 581–90.

Hart, D. L. (2000). Primates as Prey. Ecological, Morphological and Behavioural Relationships between Primate Species and their Predators. Ph.D. thesis, Washington University, Saint Luis.

Heiduck, S. (1998). *Nahrungsstrategien Schwarzköpfiger Springaffen (*Callicebus personatus melanochir*)*. Göttingen: Cuvillier Verlag.

Heise, S. R. & Rozenfeld, F. M. (1999). Reproduction and urine marking in laboratory groups of female common voles. *Journal of Chemical Ecology*, **25**, 1671–85.

Hershkovitz, P. (1977). *Living New World Monkeys* (Platyrrhini), Volume 1. Chicago: University of Chicago Press.

Heymann, E. W. (1985). Untersuchungen zur Vokalen und Olfaktorischen Kommunikation des Schnurrbarttamarins *Saguinus mystax mystax* (Spix, 1823) (Primates: *Callitrichidae*). Ph.D. thesis, University of Giessen, Giessen.

(1990). Reactions of wild tamarins, *Saguinus mystax* and *Saguinus fuscicollis*, to avian predators. *International Journal of Primatology*, **11**, 327–37.

(1998). Sex differences in olfactory communication in a primate, the moustached tamarin, *Saguinus mystax* (Callitrichinae). *Behavioral Ecology and Sociobiology*, **43**, 37–45.

(2000). The number of adult males in callitrichine groups and its implication for callitrichine social evolution. In *Primate Males*, ed. P. M. Kappeler, pp. 64–71. Cambridge: Cambridge University Press.

(2001). Interspecific variation of scent-marking behaviour in wild tamarins, *Saguinus mystax* and *Saguinus fuscicollis*. *Folia Primatologica*, **72**, 253–67.

(2003). Scent marking, paternal care, and sexual selection in callitrichines. In *Sexual Selection and Reproductive Competition in Primates: New Perspectives and Directions. Special Topics in Primatology*, Volume 4, ed. C. B. Jones. New York: Alan R. Liss (in press).

Hill, G. E. (1990). Female house finches prefer colourful males: sexual selection for a condition-dependent trait. *Animal Behaviour*, **37**, 665–73.

Homburg, I. (1989). Soziale Strukturmerkmale von Weißgesichts-Sakis (*Pithecia pithecia*). Diplom thesis, University of Bielefeld, Bielefeld.

(1998). *Ökologie und Sozialverhalten von Weißgesicht-Sakis. Eine Freilandstudie in Venezuela*. Göttingen: Cuvillier Verlag.

Hubrecht, R. C. (1984). Field observations on group size and composition of the common marmoset (*Callithrix jacchus jacchus*), at Tapacura, Brazil. *Primates*, **25**, 13–21.

Izawa, K. (1978). A field study of the ecology and behavior of the black-mantled tamarin (*Saguinus nigricollis*). *Primates*, **19**, 241–74.

Jantschke, B. (1992). Vergleichende Untersuchungen zum Sozialverhalten des Springaffen *Callicebus cupreus* und des Nachtaffen *Aotus azarae*. Ph.D. thesis, University of Kassel, Kassel.

Johnstone, R. A. (1995). Sexual selection, honest advertisement and the handicap principle: reviewing the evidence. *Biological Reviews*, **70**, 1–65.

(1997). The evolution of animal signals. In *Behavioural Ecology*, 4th Edition, ed. J. R. Krebs & N. B. Davies, pp. 155–78. Oxford: Blackwell Science.

Johnstone, R. E., Sorokin, E. S. & Ferkin, M. H. (1997). Scent counter-marking by male meadow voles: females prefer the top-scent male. *Ethology*, **103**, 443–53.

Jones, I. L. & Hunter, F. M. (1993). Mutual sexual selection in a monogamous bird. *Nature*, **362**, 238–9.

Kappeler, P. M. (1998). To whom it may concern: the transmission and function of chemical signals in *Lemur catta*. *Behavioral Ecology and Sociobiology*, **42**, 411–21.

Kinzey, W. G. & Becker, M. (1983). Activity patterns of the masked titi monkey, *Callicebus personatus*. *Primates*, **24**, 337–43.

Kinzey, W. G., Rosenberger, A. L., Heisler, P. S., Prowse, D. L. & Trilling, J. S. (1977). A preliminary field investigation of the yellow-handed titi monkey, *Callicebus torquatus torquatus*, in northern Peru. *Primates*, **18**, 159–81.

Kleiman, D. G. (1977). Monogamy in mammals. *Quarterly Review in Biology*. **52**, 39–69.

Marler, P. & Mitani, J. (1988). Vocal communication in primates and birds: parallels and contrasts. In *Primate Vocal Communication*, ed. D. Todt, P. Goedeking & D. Symmes, pp. 3–14. Heidelberg: Springer-Verlag.

Mason, W. A. (1966). Social organization of the South American monkey, *Callicebus moloch*: a preliminary report. *Tulane Studies in Zoology*, **13**, 23–8.

McConnell, P. B. & Snowdon, C. T. (1986). Vocal interactions between unfamiliar groups of captive cotton-top tamarins. *Behaviour*, **97**, 273–96.

McLanahan, E. B. & Green, K. M. (1977). The vocal repertoire and an analysis of the contexts of vocalizations in *Leontopithecus rosalia*. In *The Biology and Conservation of the Callitrichidae*, ed. D. G. Kleiman, pp. 251–69. Washington, DC: Smithsonian Institution Press.

Montoya, E., Málaga, C. & Villavicencio, E. (1990). Evaluación de una colonia de reproducción de *Aotus vociferans*. In *La Primatología en el Perú: Investigaciones Primatológicas (1973–1985)*, ed. DGFF/IVITA/OPS, pp. 616–24. Lima: Imprenta Propaceb.

Moynihan, M. (1964). Some behavior patterns of platyrrhine monkeys. I. The night monkey (*Aotus trivirgatus*). *Smithsonian Miscellaneous Collections*, **146**, 1–84.

(1966). Communication in the titi monkey, *Callicebus*. *Journal of Zoology, London*, **150**, 77–127.

(1970). Some behavior patterns of platyrrhine monkeys. II. *Saguinus geoffroyi* and some other tamarins. *Smithsonian Contributions to Zoology*, **28**, 1–77.

Müller, K.-H. (1995). *Langzeitstudie zur Ökologie von Schwarzköpfigen Springaffen (*Callicebus personatus melanochir, *Cebidae, Primates) im Atlantischen Küstenregenwald Ostbrasiliens*. Aachen: Verlag Shaker.

Nievergelt, C. M., Digby, L. J., Ramakrishnan, U. & Woodruff, D. S. (2000). Genetic analysis of group composition and breeding system in a wild common marmoset (*Callithrix jacchus*) population. *International Journal of Primatology*, **21**, 1–20.

Norcross, J. L. & Newman, J. D. (1993). Context and gender-specific differences in the acoustic structure of common marmoset (*Callithrix jacchus*) phee cells. *American Journal of Primatology*, **30**, 37–54.

Oliveira, J. M. S., Lima, M. G., Bonvincino, C., Ayres, J. M. & Fleagle, J. G. (1985). Preliminary notes on the ecology and behavior of the Guianan saki (*Pithecia pithecia*, Linnaeus 1766; Cebidae, Primates). *Acta Amazonica*, **15**, 249–63.

Oppenheimer, J. R. (1977). Communication in New World monkeys. In *How Animals Communicate*, ed. T. A. Sebeok, pp. 851–89. Bloomington: Indiana University Press.

Parker, G. A. & Simmons, L. W. (1996). Parental investment and the control of sexual selection: predicting the direction of sexual competition. *Proceedings of the Royal Society of London, Series B*, **263**, 315–21.

Penn, D. J. (2002). The scent of genetic compatibility: sexual selection and the major histocompatibility complex. *Ethology*, **108**, 1–21.

Peres, C. A. (1993). Notes on the ecology of buffy saki monkeys (*Pithecia albicans*, Gray 1860): a canopy seed-predator. *American Journal of Primatology*, **31**, 129–40.

Perkins, E. M. (1975). Phylogenetic significance of the skin of New World monkeys (order Primates, infraorder Platyrrhini). *American Journal of Physical Anthropology*, **42**, 395–424.

Petrie, M. (1983). Mate-choice in role-reversed species. In *Mate Choice*, ed. P. Bateson, pp. 167–79. Cambridge: Cambridge University Press.

Philips, M. & Austad, S. N. (1990). Animal communication and social evolution. In *Interpretation and Explanation in the Study of Animal Behavior: Interpretation, Intentionality and Communication*, Volume 1, ed. M. Bekoff & D. Jamieson, pp. 254–68. Boulder, Colorado: Westview Press.

Porter, L. A. (2001). Social organization, reproduction and rearing strategies of *Callimico goeldii*: new clues from the wild. *Folia Primatologica*, **72**, 69–79.

Price, E. C. & Piedade, H. M. (2001). Ranging behavior and intraspecific relationships of masked titi monkeys

(*Callicebus personatus personatus*). *American Journal of Primatology*, **53**, 87–92.

Puertas, P., Encarnación, F., Aquino, R. & García, J. E. (1995). Análisis poblacional del pichico pecho anaranjado, *Saguinus labiatus*, en el sur oriente peruano. *Neotropical Primates*, **3**, 4–7.

Richard-Hansen, C., Vie, J. C., Vidal, N. & Keravec, J. (1999). Body measurements on 40 species of mammals from French Guiana. *Journal of Zoology, London*, **247**, 419–28.

Roberts, S. C. & Dunbar, R. I. M. (2000). Female territoriality and the function of scent-marking in a monogamous antelope (*Oreotragus oreotragus*). *Behavioral Ecology and Sociobiology*, **47**, 417–23.

Robinson, J. G. (1979). An analysis of the organization of vocal communication in the titi monkey *Callicebus moloch*. *Zeitschrift für Tierpsychologie*, **49**, 381–405.

Robinson, J. G. & Janson, C. H. (1986). Capuchins, squirrel monkeys, and atelines: socioecological convergence with Old World primates. In *Primate Societies*, ed. B. B. Smuts, D. L. Cheney, R. M. Seyfarth, R. W. Wrangham & T. T. Struhsaker, pp. 69–82. Chicago: University of Chicago Press.

Rosell, F. (2002). The Function of Scent Marking in Beaver (*Castor fiber*) Territorial Defence. Ph.D. thesis, Norwegian University of Science and Technology, Trondheim.

Ryan, M. J. (1983). Sexual selection and communication in a neotropical frog, *Physalaemus pustulosus*. *Evolution*, **37**, 261–72.

Rylands, A. B., Schneider, H., Langguth, A., Mittermeier, R. A., Groves, C. P. & Rodríguez-Luna, E. (2000). An assessment of the diversity of New World primates. *Neotropical Primates*, **8**, 61–93.

Savage, A., Giraldo, H., Blumer, E. S., Soto, L. H., Burger, W. & Snowdon, C. T. (1993). Field techniques for monitoring cotton-top tamarins (*Saguinus oedipus oedipus*) in Colombia. *American Journal of Primatology*, **31**, 189–96.

Savage, A., Snowdon, C. T., Giraldo, L. H. & Soto, L. H. (1996*a*). Parental care patterns and vigilance in wild cotton-top tamarins (*Saguinus oedipus*). In *Adaptive Radiation of Neotropical Primates*, ed. M. A. Norconk, A. L. Rosenberger & P. A. Garber, pp. 187–99. New York: Plenum Press.

Savage, A., Shideler, S. E., Soto, L. H., Causado, J., Giraldo, L. H., Lasley, B. L. & Snowdon, C. T. (1996*b*). Reproductive events of wild cotton-top tamarins (*Saguinus oedipus*) in Colombia. *American Journal of Primatology*, **43**, 329–37.

Sekulic, R. (1983). The effect of female call on male howling in red howler monkeys (*Alouatta seniculus*). *International Journal of Primatology*, **4**, 291–305.

Sekulic, R. & Eisenberg, J. F. (1983). Throat-rubbing in red howler monkeys (*Alouatta seniculus*). In *Chemical Signals in Vertebrates*, Volume 3, ed. D. Müller-Schwarze & R. M. Silverstein, pp. 347–50. New York: Plenum Press.

Setz, E. Z. F. & Gaspar, D. A. (1997). Scent-marking behaviour in free-ranging golden-faced saki monkeys, *Pithecia pithecia chrysocephala*: sex differences and context. *Journal of Zoology, London*, **241**, 603–11.

Smith, R. J. & Jungers, W. L. (1997). Body mass in comparative primatology. *Journal of Human Evolution*, **32**, 523–59.

Smith, T. E. & Gordon, S. J. (2002). Sex differences in olfactory communication in *Saguinus labiatus*. *International Journal of Primatology*, **23**, 429–41.

Snowdon, C. T. (1989). Vocal communication in New World monkeys. *Journal of Human Evolution*, **18**, 611–33.

(1993). A vocal taxonomy of the callitrichids. In *Marmosets and Tamarins: Systematics, Behaviour, and Ecology*, ed. A. B. Rylands, pp. 78–94. Oxford: Oxford University Press.

Soini, P. (1986). A synecological study of a primate community in the Pacaya-Samiria National Reserve, Peru. *Primate Conservation*, **7**, 63–71.

(1988*a*). El huapo (*Pithecia monachus*): dinámica poblacional y organización social. *Informe de Pacaya*, **27**, 1–24.

(1988*b*). The pygmy marmoset, *Cebuella*. In *Ecology and Behavior of Neotropical Primates*, ed. R. A. Mittermeier, A. B. Rylands, A. F. Coimbra-Filho & G. A. B. Fonseca, pp. 79–129. Washington, DC: World Wildlife Fund.

(1990). Ecología y dinámica poblacional de pichico *Saguinus fuscicollis* (Primates, Callitrichidae). In *La Primatología en el Perú. Investigaciones Primatológicas (1973–1985)*, ed. DGFF/IVITA/OPS, pp. 202–53. Lima: Imprenta Propaceb.

(1993). The ecology of the pygmy marmoset, *Cebuella pygmaea*: some comparisons with two sympatric tamarins. In *Marmosets and Tamarins: Systematics, Behaviour, and Ecology*, ed. A. B. Rylands, pp. 257–61. Oxford: Oxford University Press.

Soini, P. & de Soini, M. (1990). Distribución geográfica y ecología poblacional de *Saguinus mystax*. In *La Primatología en el Perú. Investigaciones Primatológicas (1973–1985)*, ed. DGFF/IVITA/OPS, pp. 272–313. Lima: Imprenta Propaceb.

Stevenson, M. F. & Rylands, A. B. (1988). The marmosets, genus *Callithrix*. In *Ecology and Behavior of Neotropical Primates*, ed. R. A. Mittermeier, A. B. Rylands, A. F. Coimbra-Filho & G. A. B. Fonseca, pp. 131–222. Washington, DC: World Wildlife Fund.

Stott, K. Jr (1976). Sakis – imps of the rain forest. *Zoonooz*, **49**, 4–8.

Tardif, S. D. (1994). Relative energetic cost of infant care in small-bodied neotropical primates and its relation to infant-care patterns. *American Journal of Primatology*, **34**, 133–43.

Terborgh, J. (1983). *Five New World Primates.* Princeton, New Jersey: Princeton University Press.

Trivers, R. D. (1972). Parental investment and sexual selection. In *Sexual Selection and the Descent of Man, 1871–1971*, ed. B. Campbell, pp. 136–79. London: Heinemann.

Valladares-Padua, C. B. (1993). The Ecology, Behavior and Conservation of the Black Lion Tamarins (*Leontopithecus chrysopygus* Mikan, 1823). Ph.D. thesis, University of Florida, Gainesville.

van Roosmalen, M. G. M. (1985). Habitat preferences, diet, feeding strategy and social organization of the black spider monkey (*Ateles paniscus paniscus* Linnaeus 1758) in Surinam. *Acta Amazonica*, **15**, 1–238.

von Dornum, M. & Ruvolo, M. (1999). Phylogenetic relationships of the New World monkeys (Primates, Platyrrhini) based on nuclear G6PD DNA sequences. *Molecular Phylogenetics and Evolution*, **11**, 459–76.

Welker, C., Jantschke, B. & Klaiber-Schuh, A. (1998). Behavioural data on the titi monkey *Callicebus cupreus* and the owl monkey *Aotus azarae boliviensis*. A contribution to the discussion on the correct systematic classification of these species. *Primate Report*, **51**, 1–71.

White, B. C., Dew, S. E., Prather, J. R., Stearns, M., Schneider, E. & Taylor, S. (2000). Chest-rubbing in captive woolly monkeys (*Lagothrix lagotricha*). *Primates*, **41**, 185–8.

Wright, P. C. (1978). Home range, activity pattern, and agonistic encounters of a group of night monkeys (*Aotus trivirgatus*) in Peru. *Folia Primatologica*, **29**, 43–55.

Wright, P. C. (1984). Biparental care in *Aotus trivirgatus* and *Callicebus moloch*. In *Female Primates: Studies by Women Primatologists*, ed. M. F. Small, pp. 59–75. New York: Alan R. Liss.

Wright, P. C. (1985). The Costs and Benefits of Nocturnality for *Aotus trivirgatus* (the Night Monkey). Ph.D. thesis, City University of New York.

Index